Chemical Demonstrations

Volume 2

Volume 2 Collaborators and Contributors

JERRY A. BELL, PH.D.
Professor of Chemistry, Simmons College; Fellow, Institute for Chemical Education, University of Wisconsin–Madison, 1983–1984

HENRY A. BENT, PH.D.
Professor of Chemistry, North Carolina State University

GEORGE M. BODNER, PH.D.
Associate Professor of Chemistry, Purdue University

GLEN E. DIRREEN, PH.D.
Coordinator of the General Chemistry Program; Associate Director, Institute for Chemical Education, University of Wisconsin–Madison

THOMAS J. GREENBOWE, PH.D.
Assistant Professor of Chemistry and Director of Freshman Program, Southeastern Massachusetts University; Lecture Demonstrator, Purdue University, 1979–1983

FREDERICK H. JUERGENS, M.A.T.
Lecture Demonstrator, University of Wisconsin–Madison

LENARD J. MAGGINNIS, B.S.
Lecture Demonstrator, Purdue University, 1983–1984

RICHARD M. NOYES, PH.D.
Professor of Chemistry, University of Oregon

RONALD I. PERKINS, M.S.T.
Senior Teacher, Greenwich High School, Connecticut; Visiting Lecturer, University of Wisconsin–Madison, Summer 1983; Resident Fellow and Assistant Director, Institute for Chemical Education, University of Wisconsin–Madison, June 1984—July 1985.

RODNEY SCHREINER, PH.D.
Project Associate; Resident Fellow, Institute for Chemical Education, University of Wisconsin–Madison

A. TRUMAN SCHWARTZ, PH.D.
DeWitt Wallace Professor of Chemistry, Macalester College; Visiting Professor, University of Wisconsin–Madison, 1979–1980, Summer 1980, Summer 1984

EARLE S. SCOTT, PH.D.
Professor of Chemistry, Ripon College; Visiting Professor, University of Wisconsin–Madison, June–December 1980, Summer 1981, Summer 1982, Summer 1984

LEE R. SHARPE, B.S.
Undergraduate Project Assistant, Ripon College, 1982–1983; Project Assistant, University of Wisconsin–Madison, June 1983—May 1984

LLOYD G. WILLIAMS, PH.D.
Associate Professor of Chemistry, Hampshire College; Project Assistant and Lecturer, University of Wisconsin, 1973–1978

Chemical Demonstrations

A Handbook for
Teachers of Chemistry

Bassam Z. Shakhashiri

VOLUME **2**

THE UNIVERSITY OF WISCONSIN PRESS

The University of Wisconsin Press
1930 Monroe Street, 3rd floor
Madison, Wisconsin 53711-2059

3 Henrietta Street
London WC2E 8LU, England

www.wisc.edu\wisconsinpress

13   12   11   10

Printed in the United States of America

Library of Congress Cataloging-in-Publication Data
(Revised for vol. 2)
Shakhashiri, Bassam Z.
   Chemical demonstrations.
   Includes bibliographical references.
   1. Chemistry—Experiments.
QD43.S5   1983   540′.7′8   81-70016
ISBN 0-299-08890-1 (v. 1)
ISBN 0-299-10130-4 (v. 2)

ISBN-13: 978-0-299-10130-5 (v. 2: alk. paper)

To
Henry E. Bent,
a master teacher,
whose lecture experiments and demonstrations
inspired me to learn more chemistry.

B.Z.S.

Contents

6 CHEMICAL BEHAVIOR OF GASES   101

Rodney Schreiner, Bassam Z. Shakhashiri, Glen E. Dirreen,
and Lenard J. Magginnis

Preface

This is the second in a series of volumes aimed at providing teachers of chemistry at all educational levels with detailed instructions and background information for using chemical demonstrations in the classroom and in public lectures. Volume 1 included demonstrations in the areas of thermochemistry, chemiluminescence, polymers, and metal ion precipitates and complexes. The demonstrations in this volume deal with the physical behavior of gases, the chemical behavior of gases, and chemical oscillating reactions. Additional volumes, now in preparation, will include demonstrations on acids, bases, colligative and other properties of solutions, corridor exhibits, colloids, clock reactions, cryogenics, electrochemistry, and other topics.

The reception of Volume 1, which was published in 1983, has been gratifying. Teachers of chemistry in different parts of the world have used the book and have found it to be a valuable resource. The book has been favorably received by reviewers for scientific and educational journals published in the United States, Canada, Europe, and elsewhere. The most gratifying response has come from secondary school teachers and from college and university professors who comment about the usefulness of Volume 1 and who impatiently ask about Volumes 2, 3, and so on. In the spring of 1984 I was appointed Assistant Director of the National Science Foundation and I took a leave of absence from my position as professor of chemistry and Director of the Institute for Chemical Education at the University of Wisconsin–Madison. Although my responsibilities at the National Science Foundation have made it necessary for me to curtail my other activities and to rely even more than ever on my collaborators, we are determined to publish a volume every other year until the entire series is completed.

Several individuals have contributed to the development of this volume. Rod Schreiner deserves major credit for drafting several versions of Chapters 5, 6, and 7. Earle Scott's tackling of the conceptually difficult area of oscillating reactions has made Chapter 7 a reality; preliminary work on the Briggs-Rauscher reaction procedures was done by Lloyd Williams, now a faculty member at Hampshire College, and by Robert Olsen, now a graduate student with Irving Epstein at Brandeis University. Guy Laidig tested the procedures in Chapter 7. Fred Juergens and Vince Genna tested every procedure in this volume and provided helpful suggestions for modifications. Ron Perkins, an outstanding teacher from Greenwich High School, in Connecticut, reviewed the entire manuscript and provided advice about the overall balance of the text and level of presentation. Chapters 5 and 6 were first drafted by George Bodner with the assistance of Lenard J. Magginnis and Thomas J. Greenbowe of Purdue University and on the basis of demonstration files from Wisconsin and Purdue. After Chapter 6 had been redrafted at Wisconsin, Bodner decided to withdraw as a coauthor of that chapter.

I am grateful for the suggestions and comments made by colleagues from across the country about the contents of this volume. In particular, I thank Aaron Ihde of the University of Wisconsin–Madison, Art Campbell of Harvey Mudd College, Henry Bent of North Carolina State University, Truman Schwartz of Macalester College, Jerry Bell of Simmons College (now at NSF), Art Breyer of Beaver College, Tom Lippincott of the University of Arizona (now at the University of Wisconsin–Madison as Director of the Institute for Chemical Education), and Derek Davenport of Purdue University for reviewing Chapters 5 and 6 at various stages of completion. Also, I thank Richard Noyes of the University of Oregon, Irving Epstein of Brandeis University, and Peter Bowers of Simmons College for reviewing Chapter 7.

The staff of the University of Wisconsin Press were helpful in their dealings with us. In particular, I wish to express special gratitude to Elizabeth A. Steinberg, the chief editor, and to Robin Whitaker.

I express thanks to Patti Puccio of the University of Wisconsin–Madison General Chemistry Office for her assistance in proofreading and in other matters which helped to complete this volume.

Last, but not least, I thank Glen Dirreen for his continuing assistance in all aspects of this project and for his collaboration in so many other related projects. His dedication and support are invaluable.

Washington, D.C.   Bassam Z. Shakhashiri
April 1985   Assistant Director for Science &
                                                         Engineering Education
                                                     National Science Foundation

What Do I Remember?

The Role of Lecture Demonstrations in Teaching Chemistry†

Henry A. Bent

My father told a story, a few years ago when we were sharing a speakers' platform, about a lecture demonstration he called one of his worst failures in teaching.

"Usually we do not talk about our failures," he began, "perhaps for fear we will lose our jobs. But after retirement such fear is gone. It may be that a description of a big failure will encourage some beginning lecturer.

"One year I came to the subject of flames just before Christmas vacation. So I told the class on Friday morning at 7:30 on the last day before vacation that as a Christmas present I was going to spend the whole hour on flames and explosions; that I felt these were the most interesting experiments I would be doing in the whole course; and that I wanted them to enjoy them. For that purpose I had prepared and was handing out a mimeographed sheet with notes on it so that they could watch and not spend time writing. Finally, I announced that to eliminate worry with regard to questions on future examinations, I would assure the class that there would be no questions on today's lecture experiments.

"As I paused to start the first experiment, about one third of the class arose, walked down toward me and out the door at the side of the front of the room. It probably took a minute for them to escape, during which time I was barely able to recover from the shock sufficiently to continue with the day's program. I felt that I had never been such a failure, and had great difficulty in continuing with the experiments.

"But the sequel is even more interesting. At 11:30 of the same day, the last class before vacation which started at noon, I had another section of similar students and similar size, about 160 individuals. With three hours to think about my experience, I decided to proceed in precisely the same manner as at 7:30. I said exactly the same things and then made one change.

"After my announcement that there would be no examination questions on the day's reactions, I began immediately with the first experiment without a pause. Everything went beautifully, as in previous years. When I came to the last experiment and my

†Adapted with additions and deletions from an article that appeared in the *Journal of Chemical Education* 57:609–18 (September 1980). The original article was based on a joint presentation made by Henry A. Bent and his father, Henry E. Bent, on the occasion of Henry A. Bent's receiving the 1980 American Chemical Society Award in Chemical Education. Henry E. Bent served on the faculty of the University of Missouri, where he was Dean of the Graduate School for 28 years, and is now Emeritus Professor of Chemistry.

best wishes for the Christmas, I received a big hand, perhaps the most enthusiastic I had ever received. And the students just sat there as though they were waiting for more.

"And so ended my most successful lecture and my most dismal failure, both on the same day and both with the same lecture. I tell this to emphasize that there is much art in a good chemistry lecture [beyond manual manipulations], and also that we in the teaching profession [who do lecture experiments] are treading a dangerous [and, therefore, exhilarating] path with unknown pitfalls [and unexpected peaks]."

1.25†   I myself remember, for example, the time that I planned to show a class the striking luminosity of sodium when it burns in chlorine. In haste to finish before the end of the period, as demonstrator I failed to remove all of the kerosine-impregnated porous oxide coating on the sodium sample. Produced instead of a brilliant yellow glow was a copious evolution of *black soot*! Happily, as instructor, I had followed Joel Hildebrand's advice and prefaced the experiment, not with the remark, "Now I will show you . . ." but, rather, with "Let's see what happens when . . ." What happened was that, unexpectedly, we learned something about the chlorination of hydrocarbons.

6.16   I remember, also, the time I was preparing to use Priestley's NO-test for the "goodness of air" on a sample of air exhaled by a student, when the bell rang on a Friday afternoon. (With lecture experiments, class periods always seem to end too soon.) The following Monday we found a higher percentage of oxygen in the "bad air" than expected. "Of course!" I—the lecturer—thought, later—like the student who sees the answer to an exam question immediately on walking out of the room. Over the weekend, the bad air in the graduated cylinder inverted over water had been equilibrating with good air in the room. Unexpectedly, we learned something about the solubility and diffusion of air in water; something about Henry's Law, equilibrium, and partial pressures; something about Henry E.'s Law of Lecture Experiments: Doing experiments in lecture frequently yields unexpected, interesting, and memorable observations; and something about Whitehead's First Law of Higher Education: "The chief aim of a university professor [should be] to exhibit himself in his own true character—that is, as an ignorant man thinking."

6.24   I remember, too, the times the HCl-fountain "hasn't worked" (as expected). Once, filling the large round-bottomed flask with hydrogen chloride in preparation for a lecture, I noticed some condensation within the flask, but gave it little thought, until the flask—which hadn't been thoroughly dried—failed to fill with water, at all. Unexpectedly, we learned something about fog formation. On another occasion, to hasten evolution of hydrogen chloride from salt and sulfuric acid, the reaction vessel was overheated. A sudden evolution of gas blew out the stopper and spilled concentrated sulfuric acid over the lecture bench. Unexpectedly the lecturer learned something about haste and waste, and the class something about safety equipment, the dessicating property of concentrated sulfuric acid, and the fizz reaction of acidified bicarbonates.

4.10   And I remember the first time I dropped dry ice into a cylinder of lime water. The solution, as expected, soon turned milky white. Pleased, as instructor I turned to the blackboard to write a chemical equation for formation of insoluble calcium carbonate. The timing was exquisite. The moment I finished, the solution behind my back became perfectly clear. "Oh," remarked a bemused student, "another experiment that didn't work." The experiment did work, of course. Nature always does her thing. Nature's an

†References in the margins are to demonstrations in this series.

ideal lecture assistant. *She* never fails. (Excess CO_2 plus water yields carbonic acid. Carbonic acid plus calcium carbonate yields soluble calcium bicarbonate.) What may fail, however, is the lecturer's imagination. There are no failed experiments, only unimaginative responses to unexpected occurrences.

Chemistry at the lecture bench is, in fact, at least for the students, one unexpected occurrence after another. A sweet-tasting, water-soluble, low-melting, white solid (sugar) heated yields a colorless, tasteless liquid (water) and a *highly refractory, insoluble, black mass* (carbon). A colorless, not easily condensed, fuming, highly toxic gas (hydrogen chloride) dissolves with enormous ease in water, yielding *electrically conducting stomach acid* (hydrochloric acid). A rocklike, high-melting solid (calcium carbide) fizzes in a nonflammable, colorless liquid (water), yielding a *highly combustible gas* (acetylene).

It's the unexpected occurrences in life that we remember. Several years ago on a visit to Newfoundland I had a conversation with the soon-to-retire president of Memorial University at St. Johns. He was, as I recall, a political geographer, raised in one of the island's small, isolated outports, educated in a one-teacher, one-room school house. But he said he'd had a course in chemistry. I asked him my usual question, "What do you remember?" Immediately he replied, some 50 years after the fact, "I remember three things. I remember sodium reacting with water. I remember water being decomposed with electricity: hydrogen here," he gestured, "oxygen there. That was incredible! And," he added, "I remember magnesium burning in air. That was amazing!"

Even more amazing is magnesium burning in fire-extinguishing steam and carbon dioxide! Despite much liberated heat and light, the white ash remaining actually weighs more than the starting magnesium did! And the other combustion products, hydrogen and carbon, are themselves highly combustible! The carbon appears as solid, black specks that one can *see*. And that is important.

1.37

Chemistry has to be seen to be believed. "It is difficult to conceive," says Berthelot, "how bodies [sodium and chlorine] endowed with properties so little like those of sea-salt, are yet the only true elements of it." It is difficult to conceive that an ordinary condiment, table salt, plus a liquid, sulfuric acid, yields a highly toxic, invisible gas. Chemical transformations are amazing: "The water a cow drinks turns to milk; the water a snake drinks turns to poison" (Zen saying).

6.22

"Ohne Phosphor, kein Gedanke" ("without phosphorus, no thought"). The soul never thinks without an image (Aristotle). What means "nitric acid acts on copper" wondered young Ira Remsen. *Nihil intellectu quod non prius in sensu:* Nothing's in the mind that's not previously been in the senses (Latin proverb from Aristotle). All knowledge has its origin in perception (Leonardo). First-hand knowledge is the ultimate basis of intellectual life (Whitehead).

Chemical literacy requires literal chemistry. When we use the intellect [and lectures] alone, we arrive nowhere [in chemical education]; thought deepens intuition, but can in no way substitute for it (Blyth). All knowledge begins with observation (Roger Bacon). Observations precede analysis, examples precede rules (Comenius). Good judgment proceeds from clear understanding, clear understanding comes from reason derived from sound rules, and sound rules are the daughters of experience (Leonardo). [Chemical] philosophy is not supposed to work *out* of concepts, but *into* them (Schopenhauer).

Placing principles before properties is "jobbing backwards" (Ziman). In commencing the study [or presentation] of a physical science, we ought to form [or advance] no idea but what is a necessary consequence, and immediate effect, of an experiment or observation (Lavoisier).

> Faith [and Theory] is a fine invention
> For gentlemen [and scientists] who see;
> But microscopes [and experiments] are prudent
> In [a classroom] emergency.
>   Emily Dickinson [augmented]

When all else fails, try a demonstration. Drop copper into nitric acid, dry ice into phenolphthalein-spiked aqueous sodium hydroxide, flaming brimstone into water in a darkened (and well-ventilated) room. No amount of writing [or lecturing] can produce the equal of such image[s], either in form or in power (Leonardo). One case from experience [one demonstration] teaches more than many a pupil is taught by a thousand lectures that he *knows* [for exams] but does not really *understand* [or remember] (Schopenhauer).

The basic content of any science consists not in proofs but in unproved apprehensions of the perceptions (Schopenhauer). That there exist substances in more or less stationary states of being (such as copper and nitric acid) that have distinctive properties (solid redness, colorless liquidity) and that can become other substances (nitrogen dioxide, aquated cupric ions) that have different properties (brown gaseousness, blue liquidity) is a fact to be apprehended, not a theorem to be proved.

He who hears only lectures and does not see experiments [and demonstrations] will never attain to the least degree of mastery of chemistry, said Gerber. To make a fact his own, said Faraday, he had to see it. A fact seen is a fact possessed.

The eye is the window of the soul (Leonardo). An understanding of chemical terminology has to be acquired by an inspection of the objects of chemistry (Whewell).

Display a sample of sodium. A vast world of chemical thought is latent in that exhibition of sodium's properties. The solid sodium is stored beneath a colorless liquid. Water? No. (We may see in a moment what happens between sodium and water.) The liquid is a hydrocarbon, perhaps kerosine, one of the paraffins. Par-affin means "without affinity." Paraffins are relatively unreactive. They have a long residence time in nature. Natural gas and liquid fossil fuels are paraffins. The liquid hydrocarbon helps to protect the surface of the sodium from action of the atmosphere (just as a film of oil on iron parts of guns and tools helps to keep the iron from rusting). For sodium oxide is a stable compound—so stable that sodium could not be released from it until the era of modern chemistry. Indeed, a plot of heats of formation of metal oxides against the dates of discovery of the metals is approximately a straight line. Metals with high heats of oxide formation—all Group I and II metals, for example—were discovered relatively recently. Their names end, systematically, with *ium*. (*Aluminum* is spelled in other English-speaking countries *aluminium*). On the other hand, metals with low heats of oxide formation, particularly the coinage metals copper, silver, and gold, were known to the ancients. All such metals, found native, have low oxygen-affinities—and incomplete shells of d-electrons. Such metals were not named systematically. No metal known in antiquity has a name ending in *ium*. The whole chemical history—and modern electronic theory—of metals is latent in the suffix *-ium*.

1.34   Another object of chemistry terminologically enlightening to display is brilliant orange-red ammonium dichromate, $(NH_4)_2Cr_2O_7$. Now heat it (Caution!). Produced (spectacularly) is, among other things, chrome oxide green, Cr_2O_3, one of the most enduring of all paint and textile pigments. Memorably illustrated are two reasons for the name of an element common to the two compounds, chromium (from *chroma* "color" + *ium*). All chromium compounds are colored. Clearly, however, the color of chromium compounds depends on the chromium's chemical environment. Rubies, for example, are corundum (aluminum oxide, Al_2O_3) that contains a small amount of chromium. Sapphires, too, are corundum that contains a small amount of chromium, but in a slightly different location in the crystal lattice. Chromium, it may be noted, is one of the transition elements. Their existence and the color of their compounds arise from the presence of incomplete shells of d-electrons. The whole unfolding of the long periods of the Periodic Table and much of modern inorganic chemistry is latent in the colors exhibited in the thermal decomposition of ammonium dichromate.

Yet another terminologically enlightening, simple, striking display is an exhibition of one of the most beautiful colors in chemistry: the color of iodine vapor—from which the element's name is derived. Iodine's congener chlorine is named, also, for its color (*chlor* "yellow-green"). And hence, for example, we have the name *chlorophyll*, from *chlor* + *phylle*: green leaf. Names mean something. An understanding of the meanings encoded in chemical names is part of the charm of a modern liberal education.

Probably no introduction to chemistry is complete without an explanation of the meaning of *oxygen*. The required demonstrations are simple, short, safe, striking, inexpensive, and fundamental to chemical thought. Neither phosphorus, sulfur, nor carbon, to pick three common examples, is acidic. All three elements, however, are combustible. They burn nicely in oxygen. The "phosphorus moon" experiment is one of my favorites. Produced are acidic oxides. Hence the name *oxy-gen*: acid-former.

And probably every introductory chemistry course should demonstrate the meaning of *hydrogen*. Below is my father's description of one of his favorite demonstrations of the *water-generating* property of hydro-gen.

"In the experiment to show the reaction of hydrogen and oxygen, we will fill this bell jar (or equivalent jar [or plastic bottle] with its bottom removed) with hydrogen. At   6.7
the top of the bell jar is a 5 mm i.d. glass tube. Since hydrogen is much lighter than air, we can fill the bell jar with hydrogen by downward displacement of air. With someone (a student) to help, I will ask you to hold your hand at the bottom of the bell jar. When hydrogen is escaping, you can tell from the feeling that the bell jar is full. Hydrogen being a better conductor of heat than air makes your hand feel cooler.

"Now we will turn off the hydrogen and light the gas at the top of the tube, hoping that we have fairly pure hydrogen and not a mixture of hydrogen and air for reasons you will all be aware of [this demonstration follows previous demonstrations of flames]. We have about 45 seconds before the bang. Yes we do: There is our hydrogen-air diffusion flame.

"Next, we will try to imagine what is happening in the bell jar. As the hydrogen escapes, air takes its place in the bell jar. This means first that the gas in the bell jar is (on the average) getting denser and the buoyancy effect is diminishing, and so the flame gets smaller. In addition, air is mixing with the hydrogen to give eventually an explosive composition. This issuing premixed gas burns like Bunsen's flame, with inner and outer cones. Finally, as the flame gets smaller and unstable, you can often hear, if the room is quiet, a vibration of the flame . . . BANG!

"And there are now two observations for my assistant. First, what do you see on the inside of the bell jar? What do you think it is? How much would you guess there is? And, second, feel the bell jar. What would you guess its temperature might be? Thank you.

"Usually students want the experiment repeated. We will do it again using a longer tube, adding an organic halide, introducing safety-lamp gauze, trying methane . . . Many things can be illustrated with $H_2 + O_2$."

An important point illustrated by Henry E.'s hydrogen-in-a-bell-jar lecture experiment is that it is precisely that: *an experiment with a lecture*. The two aspects of the experience, the sensations and the thoughts, are both important. Especially important is their close conjunction in time. For learning chemistry is like learning a first language: it's learning new terms for new sensations. And an important part of learning a first language, writes Wittgenstein, consists in the teacher's pointing to objects [and events], directing the pupil's attention to them, and *at the same time* [emphasis added] uttering a word [or words] of explanation.

Sense data [flames, explosions] without concepts [hydrogen, diffusion] are blind [Chemysteries], while concepts [sea of air, oxidation] without sensory data [flame height, dew formation] are empty [Chemiseries]. By their union only can knowledge— [chemistry with a capital C, Capital Chemistry]—be acquired (Kant).

"Simultaneity!" Humpty Dumpty might say; else "Impenetrability!" Lectures Mondays, Wednesdays, and Fridays, labs Tuesdays or Thursdays, don't do the trick. We've all heard the student remark about labs, "I didn't know what I was doing," and about lectures, "I didn't understand what I heard." The unassisted eye [or merely textbook-assisted eye] and unassisted mind [the merely lectured-to mind] have little power (Francis Bacon).

There's much more to seeing than meets the eye. Seeing as a scientist sees is a theory-laden activity. It is theory that makes an observation relevant (Hanson). Believing [chemical theory] is not simply seeing [chemical phenomena], but *seeing and translating* (Langer).

In the Tao it is written that the wheelwright, the carpenter, the butcher, the bowman, the swimmer [and, we might add, the student and the scientist] achieve their skill [and insights] not by accumulating facts concerning their art, nor by the energetic use either of muscles or outward senses [in the laboratory], but through utilizing the *fundamental kinship* that unites their own Primal Stuff [their minds' thoughts] to the Primal Stuff of the medium in which they work [the body's sensations produced by laboratory experiments].

The kinetics of the synthesis of thoughts and sensations are simple: synthesis rate is proportional to collision frequency. Reactants (thoughts and sensations) must be in the reaction vessel (the mind) *at the same time*: d(Knowledge)/dt = k(Thoughts)(Sensations). The product is greater than its factors. Seeing in science is first order in visual experiences and first order in thoughts.

6.17   Needed are demonstrations *with* lectures and lectures *with* demonstrations. Needed, for example, are NO_2/N_2O_4 sealed-tube experiments with lectures on chemical equilibrium and Le Chatelier's Principle, and for lectures on, for example, the discovery of oxygen, air pollution, the oxides of nitrogen, third-order reactions, and mechanisms of nitration, the Brown Bottle demonstration (Box 1).

What is said about a demonstration is as important as what is shown. Science is *organized* knowledge (H. Spencer). An observation is not a *systematic* observation until it is embedded in an intellectual *system*. From showing—in a few minutes, with a light bulb in series with two electrodes—that the electrical conductivity of 0.1 M $HC_2H_3O_2$ is about the same as the conductivity of 0.001 M HCl, and much greater than the conductivity of 0.001 M NaCl, one says—after some minutes of discussion—that evidently the concentration of H_3O^+ in 0.1 M $HC_2H_3O_2$ is about the same as the concentration of H_3O^+ in 0.001 M HCl; whence $K_{HC_2H_3O_2} \simeq 0.001 \times 0.001/0.1 = 10^{-5}$.

Box 1. The Brown Bottle Experiment

NO and 30–100 mL of H_2O in a 1–3-liter stopped flask

Unstopper Colorless Flask

$2\,NO + O_2 \longrightarrow 2\,NO_2$ (brown)

Stopper and Shake Colored Flask until Colorless

$2\,NO_2 + H_2O \longrightarrow H^+ + NO_3^- + HNO_2$

Wait

$3\,HNO_2 \longrightarrow 2\,NO(\uparrow) + H^+ + NO_3^- + H_2O$

Repeat

T h e   B o t t o m   L i n e

$4\,NO + 3\,O_2 + 2\,H_2O \longrightarrow 4\,H^+ + 4\,NO_3^-$
(See Demonstration 6.14 in this volume.)

Box 2. Uses of the Brown-Bottle Experiment

$NO/O_2/H_2O$ has everything (van Vleck, slightly paraphrased)

Thermodynamics and Descriptive Chemistry
Colors of NO, NO_2
Solubilities of NO, NO_2
Relative Stabilities
 NO, O_2/NO_2
 NO_2, H_2O/HNO_3, HNO_2
 HNO_2/NO, HNO_3, H_2O
 NO, O_2, H_2O/HNO_3

General Chemistry
Oxidation-Reduction
Disproportionation
Acid Anhydrides
Acid Strengths
pH; Nitrate Test
Conductivity, Ionization

Industrial Chemistry
Scrubbing
SO_2 Oxidation; H_2SO_4, etc.
N_2 Fixation
 Atmospheric
 Catalytic Oxidation of NH_3
 Haber Process
  N-Economy
  World War I
  NH_4NO_3 Decomposition

Environmental Chemistry
Photo-Chemical Smog
Catalytic Converters
Acid Rain
Ozone Layer/UV/Cancer/SST

History of Chemistry
Priestley's Discovery of Oxygen
Dalton's Atomic Theory
Linnett's Double-Quartets

Kinetics and Mechanisms
Negative E_a for NO Oxidation
NO^+ or NO_2^+ in $NO_2 + H_2O$?
HNO_2 Decomposition: R_f and R_b
$R_t = k_f(HNO_2)^4/P_{NO}^2 \Rightarrow$
 (Complex)$^{\ddagger} \sim N_2O_4$

Structure and Bonding
N–O Bonds: Orders and Lengths

H_2NOH NO_2^- NO_3^-

RNO NO_2 N_2O

NO^+ NO_2^+ R_3NO

RNO_2, $RONO$, and N_2O_4

O_2NNO_2 $ONOONO$

$ONONO_2$
N_2O_2 and the Third Law

"Wonderful are the capacities of experiments to lead us into various departments of knowledge." (Faraday)

To interpret [chemical experiments] is to think; seeing [as a chemist sees] is a state [of mind] (Wittgenstein).

Such is a lecture demonstrator's opportunity to make sense of sensations, to (in Cassirer's words) "saturate elements of experience with diverse functions of meaning." Only through "ideation," only through a "mode of vision," does a sensory experience take on a kind of "spiritual articulation," a "certain nonintuitive signification." Brownian motion didn't move physicists mentally until Einstein articulated its symbolic meaning.

Nature doesn't express itself in words (Tao). Experience, dumb in itself, needs a mouthpiece (Suzuki). There is nothing good or bad, trivial or momentous, transient or memorable, but thinking makes it so. An observed phenomenon (to paraphrase John Wheeler) is not a *significant* phenomenon until it is a *thought-linked* phenomenon. Remarked Faraday in his Christmas Discourses on the Chemical History of the Candle, "We come here [to the Royal Institute's Lecture Demonstration Facilities] to be philosophers." The role of a chemical demonstrator is to be philosophical: to show what chemists can show, to say what chemists can say, and to describe what chemists imagine.

An experiment is not simply the observation of a phenomenon, remarks Duhem. It is in addition a theoretical interpretation of the phenomenon. The often-asked question, "What are you doing?" (Sterilizing a thermometer in that boiling water?) is asking often, "What are you *thinking*?" (That boiling points depend on molar concentrations; and that molar concentrations depend on molar masses.)

Thoughts are spectacles *behind* the eyes (Hanson). We want to see *What's so:* e.g., 5.1 a cooled steam-filled can collapsing. And we want to know *So what?:* Heated, the collapsed, liquid-containing can expands, raising the atmosphere [again]; which falling [as before, with can-cooling] can be used to raise, seesaw fashion, another ponderable body, e.g., water in a mine (Newcomen); thereby harnessing the 'motive power of fire,' more efficiently with a separate condensor (Watt); whence, on wheels, fossil-fuel–guzzling substitutes for grass-grazing horses; and, thence, industrial revolutions, population explosions, environmental pollution, resource depletion, and international tensions.

We want new sights and new insights. The important thing in science [and in lecture demonstrations] is not so much to obtain new facts [and new demonstrations] as to discover new ways of thinking about them (W. L. Bragg). Discoveries [and improvements in teaching] consist in seeing [and doing] what everybody has seen [and done] and thinking [and saying] what nobody has thought [and said] (A. Szent-Gyorgyi).

The context is the thing. The truth is whole. Every sensation strikes a thousand connections.

Extinction of a candle in a falling flask recalls Galileo's law, gas viscosities, diffu- 6.13 sion rates, reaction velocities, and Faraday's observations regarding flames and the volatilities of the oxides of the elements.

Neutralization of sulfuric acid spilt preparing hydrogen chloride offers opportunities to discuss antacids, rising dough, and sparkling water; fire extinguishers, the carbonate test, and limestone caves; hard water, water softening, and polyelectrolytes; buffers, blood bicarbonate, and CO_2-transport; dry ice, triple points, and phase diagrams; the Solvay Process, the Solvay Conferences, and Atomism Extended; proton transfers, acid-base strength, and molecular structure; rate and equilibrium constants, carbonic anhydrase, and catalysis.

Concrete cases are containers (Robert Frost). As individual experiences are woven into a whole, they gain power to represent that whole, and transitions to the whole become possible from any single factor (Cassirer)—albeit not directly; nor with the compulsion of pure logic. For—

Chemistry is a fact-rich, theory-lean science. Evidence for its theories is voluminous, but indirect. Thus, no experiment can be used anywhere definitively (Duhem). But almost any experiment can be used almost anywhere illustratively (Larry Strong).

When one thing is taken up [in depth], all things are taken with it; one flower is spring; a falling leaf has the whole of autumn (Suzuki). From the Brown Bottle Experiment (Box 1), a hundred thoughts blossom (Box 2).

Every case represents innumerable cases (Schopenhauer). The One embraces All, and All is merged in the One. The One pervades All, and All is in the One. This is so with every object, and with every existence (Zen Buddhist Principle). An expert in the *I Ching* [and Inorganic Chemistry] can "see" a hexagram [and an octahedron] in almost anything (Watts). An artist can see in a wall spotted with stains or with a mixture of stones various landscapes, battles, lively postures of strange figures, and an infinite number of things (Leonardo). Anything can trigger off and start something (Henry Moore). One can reach the center from any point of the compass (Jung). In the formula for sulfuric acid, it's been said, is summarized the history of mankind.

The lecture demonstrator thinks, "Here's what I can show. What can I say?" Almost anything. Almost any chemical experiment can be connected to almost any chemical conception. "I use $H_2 + O_2$ to illustrate all the principles of chemistry," says Gil   1.42
Haight with characteristic enthusiasm, to which one might add, in the same spirit: Lecture demonstrators can use demonstrations to illustrate almost any principle of *Life*.

"How many precautions and observations, what resource and invention, what delicacy and vigilance," Whewell remarks, "are requisite in Chemical Manipulations." Would we want it otherwise? Fail-safe demonstrations, like data-proof lectures, would be, in Robert Frost's phrase, like playing tennis with the net down.

Neatness in experimental work establishes the Principle of Determinism, Determinism provides the philosophical grounds for Rationality. And Rationality diminishes the domain of mankind's fear of the supernatural.

Lecture experiments make chemical lectures demanding for lecturers, meaningful for philosophers, and interesting for students. They are highly motivational. They have immense heuristic value, tremendous rhetorical power, overwhelming persuasive force (Ziman). For the effort of knowing is guided by a sense of obligation toward the truth (Polanyi). If you don't see it [water freezing by boiling under reduced pressure by cool-   5.21
ing], you won't believe it (anon.). And if you don't believe it, you won't understand it (Saint Augustine). And if you don't understand it, you won't long remember it (Ushinsky). The senses are important, not only for first discovering, but for receiving knowledge (Harvey).

But to be useful, knowledge received must be knowledge retained. A mind less memory is mindlessness. Who remembers nothing thinks about nothing. A mind is like a computer: no memory, no output. Who remembers not that air contains oxygen that reacts with colorless, water-insoluble nitric oxide to produce brown, water-soluble ni-   6.14
trogen dioxide thinks about little of chemical interest in the Brown Bottle Experiment (Box 1).

An experience is not a useful experience, in science, unless it is somehow a recorded experience, a part of the public record. Be published or perish. Science as an institution is that which has been institutionalized.

In the same way, a demonstration is not a useful demonstration for students, on tests, unless it is a memorable demonstration, a part of their notes. Unless there is a linguistic component to seeing, nothing we see, observe, witness has the slightest relevance to our knowledge (Hanson). To paraphrase Faraday's injunction: (Work. Finish. Publish.)—Watch what is shown. Listen to what is said. Record what can be recorded.

6.2   Translation of personal sensations into written records is not trivial, in practice or in principle (Ziman). It is not a trivial exercise to summarize experiments with dry ice, for example, with such remarks as—

> White vapors issuing forth from the bottom of a column of liquid water over solid carbon dioxide sink in the room and eventually disappear.

> Strong-acid acidification of a CO_2-neutralized basic solution produces a vigorous evolution of a flame-extinguishing gas.

Certain things about seeing [as chemists see] are puzzling [to students] because we [chemists] do not find the whole business of seeing [chemically] puzzling enough (Wittgenstein). A profound and mysterious linguistic facility is required to collapse a phenomenon's wavefunction onto the printed page in such fashion as this—

> Tweezer-held solid sodium dropped into phenolphthalein-spiked, colorless, nonconducting liquid water forms a bead, produces a colorless, odorless, less-dense-than-air, highly flammable gas and a pink, electrically conducting solution, and eventually disappears.

Transformation of chemical demonstrations into informative student notes could transform inarticulate students into observant, noteworthy, expressive human beings.

When asked why he kept a volume of Goethe on his desk, Werner replied, "To improve my style."

Rutherford used to say he felt sorry for the humanists. They had no lab to go to. A lecture demonstration course fully utilized could serve as a core course in a core curriculum in general education for all students. We could hardly ask for more: full utilization of a chemist's skills through chemistry in the service of the humanities.

Needed, however, is a new sense of values among chemists. The product of mental labor—a balanced equation, for example, for the reaction of sodium with water—always stands with students and teachers far below its true value because, as Marx remarks, the labor-time necessary to reproduce it has no relation at all to the labor-time required for its original production. After the fact, it is merely a fact.

In fact, the highest value, Goethe felt, would be to recognize that "Everything factual is itself theoretical." A fact is merely a familiar theory (Whewell). All observation involves interpretation in the light of our theoretical knowledge. Pure observational knowledge, unadulterated by theory, would, if at all possible, be utterly futile and barren (Popper).

Chemistry is chemistry. "Pure descriptive chemistry" is merely an effortless application of familiar theoretical conceptions. Today's descriptive chemistry was yesterday's theoretical chemistry. Today's theoretical chemistry will be tomorrow's descriptive chemistry.

A study of science, particularly its more descriptive [that is to say, its more familiar, well-understood] parts, gives precision to the senses (Whewell). The mind works critically only on what the mind understands thoroughly; i.e., on what the soul believes; i.e., on what the body sees.

What monstrosities of 'scholarship' philosophers, in other respects of the subtlest genius, may produce by neglecting experiments. Experience alone is the Sole Mistress of Truth; the Reconciler of Difficulties; the Dissolver of all Doubts, Paradoxes, and Antitheses in Education in Chemistry (Bacon, augmented).

Lecture experiments are a universal solvent for dissolving the interface, the boundary, the gap, the cut between thoughts and sensations, telling and showing, the blackboard and the lab bench; between the pure and the applied, science and technology, academia and industry; between principles and properties, theoretical chemistry and descriptive chemistry, the quantitative and the qualitative; between left-brain/classical/ Apollonian/analytical/reductionist/formal/abstract/exact/numerical/content-oriented/ training-mode/testable knowledge and right-brain/romantic/Dionysian/synthetical/ holistic/concrete/real/relational/topological/form-oriented/education-mode/ineffable knowledge; between the logical and the creative, the mathematical and the linguistic; between theories and concepts, decoding and encoding, deduction and induction; between disciplined studies and transdisciplinary studies, the scientific and the artistic; between examinations and criticism, right/wrong and good/bad, grades and judgments, GPAs and values, scientists and humanists.

Lecture experiments are for everyone. There is no such thing as hydrogen-plus-oxygen-for-premedical-students. CHEMISTRY is CHEMISTRY.

But when all has been said and shown, not all that can be done is done. To learn chemistry, students need to see chemistry, hear chemistry, *and think chemistry*. A lecture experiment is not a memorable, thought-registered phenomenon *in students' minds* until it is exam-tested phenomenon. To make chemistry memorable, a teacher must show, tell, and *test*. I spend more time preparing tests for students to test their understanding of Nature than I do preparing demonstrations of Nature's nature.

Testing corresponds in Zen Buddhist terminology to the final step of the Noble Eightfold Path to "a brightening up of the mind-works" and attainment of Insight and Enlightenment. Tests are, so to speak, classroom *koan* (questions and paradoxes) that turn the mind inward upon its own resources to "pierce the veils of sensory perception [demonstrations] and conceptual thought [lectures] in order to arrive at an intuitive perception of reality," which we call understanding.

Below are illustrative *koan* for a demonstration lecture in which tank hydrogen was rolled into the classroom to inflate balloons (later ignited) and to fill a spark-activated hydrogen cannon. Even with extensive remarks and hints in class, most beginning students are not entirely certain of the answers for most of the following questions.

True or False?

Tank hydrogen is under high pressure. (True)
    CAUTION: Tanks containing gases at high pressure are potential projectiles! (True)
      Gas tanks are so shaped that it is difficult to tip them over. (False)
        Tank hydrogen is very cold. (False)
          Tank hydrogen is largely in the liquid state. (False)
            Hydrogen has a high boiling point. (False)
              Intermolecular forces in hydrogen are relatively weak. (True)
                Gases at 25°C and 1 atm are mostly empty space. (True)

Hydrogen at high pressures is highly explosive. (False. Would it have been brought into the room if it were?)
    Hydrogen is a colorless, odorless, highly toxic gas. (False. The lecturer squirted it in his face.)

Hydrogen gas at room temperature rapidly attacks iron, brass, glass, rubber, and air. (False. It was in contact with all of those substances during inflation of the balloons, without noticeable effect.)

The rate of the hydrogen-oxygen reaction increases with rising temperature. (True. It was initiated with a flame.)

The product of the reaction ends up at room temperature. (True)

Hydrogen-oxygen mixtures are kinetically but not thermodynamically stable at room temperature. (True)

The reaction is exothermic. (True)

All flame-reactions are exothermic. (True)

One should not vent hydrogen around electric motors. (True)

One should not vent large volumes of hydrogen up fume hoods. (True)

It is dangerous to pass pure hydrogen over electrical sparks. (False)

Hydrogen has an allotropic form much more stable than gaseous hydrogen, to which the latter, if activated, reverts in a highly exothermic, explosively rapid reaction. (False)

Explosion limits are regions of space where, for reasons of national security, hydrogen-oxygen mixtures may not be exploded. (False)

Hydrogen can be burned anywhere on the surface of the earth. (True)

Canned hydrogen is a potential source of heat-energy at high temperatures. (True)

Heat-energy at high temperatures can be used to produce steam. (True)

Steam-driven turbines can be used to generate electricity. (True)

Electrical energy is highly transformable energy. (True)

Electrical energy is converted to mechanical energy by motors, to heat-energy by resistors, to chemical energy by electrochemical cells. (True)

Combustion of hydrogen in air produces air pollution and acid rain. (False)

The product(s) of combustion of hydrogen in air is (or are) colorless, odorless, and nontoxic. (True)

Water is the sole product of combustion of hydrogen in air. (True)

Hydro-gen means "generator (with oxygen) of water." (True)

Hydrogen can be obtained from water by electrolysis. (True)

Electro-lysis means "breaking apart with electricity." (True)

Electrolysis is a transformation of electrical energy into chemical energy. (True)

Production of hydrogen from water and electricity, gas-pipe transportation of hydrogen, on-site combustion of hydrogen back to water, and production of steam to produce electricity is, in effect, wireless transporation of electrical energy. (True)

It is possible to produce more electrical energy in the combustion of hydrogen than is consumed in the hydrogen's production by electrolysis. (False)

The "Hydrogen Economy" is a way of producing and transporting energy economically, thereby making a nation independent of external sources of transformable energy. (False)

In summary, and to repeat, wonderful are the capacities of experiments to lead us into various departments of knowledge. Every experiment strikes a thousand connections. When one thing is taken up in depth, all things are taken with it. The center of chemical philosophy can be reached from any of the lecture experiments described in this volume.

ANNOTATED BIBLIOGRAPHY

Arnheim, Rudolph. *Visual Thinking*, University of California Press: Berkeley, California (1969).

> A persuasive brief for the "intelligence of perception" and the falseness of the ancient dichotomy between thinking and seeing, lectures and laboratory, by a noted scholar of the psychology of art.

Bernard, Claude. *An Introduction to the Study of Experimental Medicine*, translated by Henry Copley Green, Dover Publications: New York (1957).

> A forceful statement of the case for determinism in the laboratory, the inseparability in science of head and hand, and the importance of ideas in making facts great. A classic. As splendid a statement of the basic features of scientific research as has ever been written (I. B. Cohen). Recommended reading for gremlin-plagued demonstrators.

Blyth, R. H. *Haiku*, Vol. 1, Hokuseido Press: Tokyo (1981).

> A sensitive, scholarly discussion of the oriental "return to things," to the "felt indubitable certainty of experience." *The history of the human spirit* [and development of an individual intelligence] *consists of two elements: an escape from this world* [of things] *to another* [of thoughts]; *and a return to it.* Lecture experiments are like Chinese haiga: *small sketches that endeavor to express with simple apparatus what haiku* [and lectures] *do in words. Our poetry*—our science—*of things is not something superimposed on experiences, but is brought out of them as the sun and rain bring the tender leaf out of the hard bud.* A good examination question, like a felicitous haiku, *shows us what we knew all the time, but did not know we knew; it shows that we are poets*—and philosopher-scientists—*in so far as we live*—and think—*at all.*

Carroll, Lewis. Complete Works, Vintage Books: New York (1976).

> Philosophical dry wit for any occasion.

Cassirer, Ernst. *The Philosophy of Symbolic Forms*, Vol. 3: *The Phenomenology of Knowledge*, Yale University Press: New Haven, Connecticut (1957).

> A scholarly, original, provocative, profound, and articulate, if not always instantly lucid, study of mankind's creative symbolizing activities, by an outstanding philosopher of the physical and cultural sciences.

Duhem, Pierre. *The Aim and Structure of Physical Theory*, Princeton University Press: Princeton, New Jersey (1954).

> It was the merit of Duhem that, with extraordinary sharpness, he was the first to show the intellectual mediations through which we must pass if we are to gain physical theorems and judgments from a mere observation of individual phenomena (Cassirer).

Faraday, Michael. *The Chemical History of a Candle: A Course of Lectures Delivered Before a Juvenile Audience at the Royal Institution*, The Viking Press: New York (1960; out of print).

> A superb example of exploring chemistry with lecture demonstrations. Outstanding chemically, philosophically, and pedagogically. A loved labor of love. A classic. Tops. A$^+$. Faraday illustrates nearly all of the principles—and many of the reactions—of chemistry with "$(CH_2)_n + O_2$". Inspirational—and humbling. Chemical education has fallen far below Faraday's standard of excellence. Few modern chemistry courses can hold a candle to Faraday's course on the candle.

Hanson, Norwood. *Perception and Discovery: An Introduction to Scientific Inquiry*, Freeman, Cooper and Company: San Francisco, California (1969).

> An engaging discussion by a noted philosopher-physicist of this paper's principle thesis: that *seeing and knowing are interdependent concepts.*

Hume, David. *An Inquiry Concerning Human Understanding*, edited by C. W. Hendel, Bobbs-Merrill Educational Publishers, Indianapolis, Indiana (1955).

A lecture demonstrator's philosopher. Anticipator of many of the present paper's theses. *The most lively thought is still inferior to the dullest sensation. Where we find no impressions, we find no ideas.* [Reasoning] *amounts to no more than the faculty of compounding, transposing, augmenting, or diminishing the materials afforded by the senses* [and memory]. *From similar causes, similar effects follow.* [Chemical] *effects are totally different than their causes: No man imagines that the explosion of gunpowder could ever be discovered by arguments a priori; nor that the color, consistency, and other qualities of bread have any connection with its secret power of nourishment.* In sum: DETERMINISM plus EXPERIENCES of Determinism plus MEMORY of Experiences of Determinism yield SURVIVAL.

Jung, Carl. *Man and His Symbols*, Doubleday and Company, Inc.: New York (1964).

Perceptive insights into the impact of a mathematically advanced discipline (physics) upon a mathematically less-articulated discipline (chemistry).

Anthropologists have often described what happens to a primitive society [such as classical chemistry] *when its spiritual values are exposed to the impact of modern civilization* [$H\Psi = E\Psi$]. *Its people* [and phenomena] *lose the meaning of their lives* [descriptive chemistry is used passively, if at all, to *illustrate* physical principles, rather than aggressively, to *capture* ideas], *their social organization* [the Periodic Table] *disintegrates* [into an exercise in the aufbau principle], *and they themselves morally decay* [to telling rather than showing].

Kauffman, George B. *Alfred Werner: Founder of Coordination Chemistry*, Springer-Verlag: New York (1966).

A concentrated lode of interesting information about a dedicated lecture demonstrator. Cites Werner's frightening of examination candidates with, *"Es gibt keine Chemie für Mediziner! Chemie ist Chemie!"* (p. 60).

Leonardo da Vinci. *Notebooks* edited with commentaries by I. Richter, Oxford University Press: New York (1977).

Leonardo combines an artist's sensitivity for the UNITY OF VISUAL EXPERIENCES with a scientist's desire for unity of knowledge. *The poet* [and lecturer] *cannot attain with the pen* [and tongue] *what the painter* [and demonstrator] *attains with the brush* [and apparatus]. *The poet* [and lecturer] *in describing any figure* [or event] *can only reveal it to you CONSECUTIVELY, bit by bit, while the painter* [and demonstrator] *will display it ALL AT ONCE. It is beyond the poet's* [and lecturer's] *power to say DIFFERENT THINGS SIMULTANEOUSLY, as the painter* [and demonstrator] *does (EMPHASIS added).*

Meyerson, Emile. *Identity and Reality*, translated by K. Loewenberg, Dover Publications, Inc.: New York (1962).

Brilliant studies in the theory of knowledge (Einstein). One of the most interesting philosophical studies of the development of physical theories (Popper)—by one of the few philosopher–historians of science with a profound knowledge and love of chemistry.

Polanyi, Michael. *Personal Knowledge*, The University of Chicago Press: Chicago, Illinois (1958).

Many sage observations by a physical chemist, philosopher, and professor of social studies. "Connoisseurship can be communicated only by example, not by precept."

Popper, Karl. *Conjectures and Refutations: The Growth of Scientific Knowledge*, Basic Books, Publishers: New York (1962).

One of the most important books in philosophy of science (Hanson). "How do we know?" asks Popper. *Neither by observation* [alone], *nor by reason* [alone], but rather, *by guessing* [hypotheses] *and criticizing* [with experiments].

Schopenhauer, Arthur. *The World as Will and Representation*, translated by E. F. J. Payne, Dover Publications, Inc.: New York (1969).

The mature fruit of a lifetime's reflection on the nature of *thought as a re-presentation of* the data of senses. A two-volume ode to lecture demonstrations. *The whole world of thought rests on the world of perceptions. Perceptions furnish us with the real content of our thinking. Perceptions are the ready money, concepts the notes. All great minds always think in the presence of perceptions. All original thinking is done in pictures and images. Observation and contemplation of everything ACTUAL is more instructive than all reading and hearing about it. From a book* [or lecture], *we obtain the truth only second-hand at best, and often not at all. We have a thorough understanding of things and their relations only in so far as we are capable of representing them to ourselves in purely distinct perceptions without the aid of words.*

Schwartz, Richard. *Samuel Johnson and the New Science*, The University of Wisconsin Press: Madison, Wisconsin (1971).

Johnson, author of "Life of Boerhaave," and connoisseur of science generally *(nothing spreads more fast than science, when rightly . . . cultivated)* and of chemistry particularly (A friend once found him "all covered with soot . . . in a little room, with an intolerable heat and strange smell . . . making *aether*"), was another anticipator of many of this paper's theses: *Knowledge* [is] *nothing but as it is communicated. Triflers may find or make any Thing a Trifle; but . . . it is the great Character of a wise Man to see Events in their Causes. A chymist is locally at rest; but his mind is hard at work. Useless . . . are experiments, unless theory brings her light to direct their application.* Of an especially striking chemical transformation, Johnson wrote, "Who, when he saw the first sand or ashes, by a casual intenseness of heat melted into a metalline form, rugged with excrescences, and clouded with impurities, would have imagined, that in this shapeless lump lay concealed so many conveniences of life, as would in time constitute a great part of the happiness of the world?"

Suzuki, Daisetz. *The Essentials of Zen Buddhism* edited with an introduction by Bernard Phillips, Greenwood Press: Westport, Connecticut (1973).

A selection from the writings of the chief interpreter for the West of the spirit of Far Eastern philosophical thought. *We must have language* [and lectures], *but it is this language that stands between reality and ourselves and misleads us by making us think the pointing finger* [and theory] *more real than the moon* [and phenomena] *to which it points. Reality must be taken with the naked hands, not with the gloves of language. Our love of Nature* [and our students' love of chemistry] *unfolds itself as* [together] *we come in contact with its objects* [and its chemical transformations].

Ushinsky, K. D. *Man As the Object of Education: An Essay in Pedagogical Anthropology*, Progress Publishers: Moscow (English translation 1978).

The chief literary work of Russia's leading educational reformer of the last century. A little gem. Contains excellent chapters on Habits, Attention, Memory, Imagination, Thinking, Feelings, and Will.

Watts, Alan. *The Way of Zen*, Vintage Books: New York (1957).

A lucid and concise introduction to the philosophy of *seeing reality directly*, via passage through four Dharmadhatus, namely: *(1) SHIH, the unique, individual "thing-events"* [experiments; sensations]; *(2) LI, the "principle"* [or thoughts] *underlying the multiplicity of things; (3) LI SHIH WU AI, "between principle and things* [lecture *and* demonstration] *no obstruction*; and, thence, ultimately, to *(4) SHIH SHIH WU AI, "between thing and thing no obstruction"* [connections intuitively seen]. Our deepest insights are direct perceptions of connections between things in their natural *"suchness."* Theories are merely means of connecting concrete cases to concrete cases.

Whewell, William. *The Philosophy of the Inductive Sciences, Founded Upon Their History*, Part 2, Frank Cass and Co.: London (1967; first published in 1847).

Deserves wider study, more frequent citation. Contains particularly insightful commentaries on Frances Bacon and the Fundamental Antithesis between thoughts and things.

Whitehead, Alfred. *The Aims of Education and Other Essays*, Macmillan Publishing Company: New York (1957).

>Fresh, provocative, timeless. *In your* [experimental] *results you cannot be too concrete. In your* [lecture] *methods you cannot be too general*. Keep it simple. And milk everything you can from a demonstration. *The best education is to be found in gaining the utmost information from the simplest apparatus.*

Wittgenstein, Ludwig. *Philosophical Investigations*, translated by G. E. M. Anscombe, Macmillan Publishing Company: New York (1958).

>The "precipitate" of 16 years of philosophical investigations into the essence of human language. Catalyst for Hanson's study of perception and discovery. A philosophical "lecture demonstration." Philosophy presented as a process, not a product.

Yost, D. M., and H. Russell. *Systematic Inorganic Chemistry of the Fifth-and-Sixth-Group Nonmetallic Elements*, Prentice-Hall: New York (1946).

>An exemplary treatment of theoretical descriptive chemistry, particularly $NO/O_2/H_2O$ systems. Many years ahead of its time. Highly recommended.

Ziman, John. *Public Knowledge: An Essay on the Social Dimension of Science*, Cambridge University Press: New York (1968).

>A sprightly stated thesis by a theoretical physicist that the need to communicate experimental observations determines the essential character of science—and, by inference, education.

Introduction [†]

Bassam Z. Shakhashiri

Lecture demonstrations help to focus students' attention on chemical behavior and chemical properties, and to increase students' knowledge and awareness of chemistry. To approach them simply as a chance to show off dramatic chemical changes or to impress students with the "magic" of chemistry is to fail to appreciate the opportunity they provide to teach scientific concepts and descriptive properties of chemical systems. The lecture demonstration should be a process, not a single event.

In lecture demonstrations, the teacher's knowledge of the behavior and properties of the chemical system is the key to successful instruction, and the way in which the teacher manipulates chemical systems serves as a model not only of technique but also of attitude. The instructional purposes of the lecture dictate whether a phenomenon is demonstrated or whether a concept is developed and built by a series of experiments. Lecture experiments, which some teachers prefer to lecture demonstrations, generally involve more student participation and greater reliance on questions and suggestions, such as "What will happen if you add more of . . .?" Even in a lecture demonstration, however, where the teacher is in full control of directing the flow of events, the teacher can ask the same sort of "what if" questions and can proceed with further manipulation of the chemical system. In principle and in practice, every lecture demonstration is a situation in which teachers can convey their attitudes about the experimental basis of chemistry, and can thus motivate their students to conduct further experimentation, and lead them to understand the interplay between theory and experiment.

Lecture demonstrations should not, of course, be considered a substitute for laboratory experiments. In the laboratory, students can work with the chemicals and equipment at their own pace and make their own discoveries. In the lecture hall, students witness chemical changes and chemical systems as manipulated by the teacher. The teacher controls the pace and explains the purposes of each step. Both kinds of instruction are integral parts of the education we offer students.

In teaching and in learning chemistry, teachers and students engage in a complex series of intellectual activities. These activities can be arranged in a hierarchy which indicates their increasing complexity [1]:

(1) observing phenomena and learning facts
(2) understanding models and theories
(3) developing reasoning skills
(4) examining chemical epistemology

This hierarchy provides a framework for the purposes of including lecture demonstrations in teaching chemistry.

At the first level, we observe chemical phenomena and learn chemical facts. For example, we can observe that, at room temperature, sodium chloride is a white crys-

†Reprinted with minor modifications from Volume 1.

talline solid and that it dissolves in water to form a solution with characteristic properties of its own. One such property, electrical conductivity, can be readily observed when two wire electrodes connected to a light bulb and a source of current are dipped into the solution. There are additional phenomena and facts that can be introduced: the white solid has a very high melting point; the substance is insoluble in ether; its chemical formula is NaCl; etc.

At the second level, we explain observations and facts in terms of models and theories. For example, we teach that NaCl is an ionic solid compound and that its aqueous solution contains hydrated ions: sodium cations, $Na^+(aq)$, and chloride anions, $Cl^-(aq)$. The solid, which consists of Na^+ and Cl^- particles, is said to have ionic bonds, that is, there are electrostatic forces between the oppositely charged particles. The ions are arranged throughout the solid in a regular three-dimensional array called a face-centered cube. Here, the teacher can introduce a discussion of the ionic bond model, bond energy, and bond distances. Similarly, a discussion of water as a molecular covalent substance can be presented. The ionic and covalent bonding models can be compared and used to explain the observed properties of a variety of compounds.

At the third level, we develop skills which involve both mathematical tools and logic. For example, we use equilibrium calculations in devising the steps of an inorganic qualitative analysis scheme. We combine solubility product, weak acid dissociation, and complex ion formation constants for competing equilibria which are exploited in analyzing a mixture of ions. The logical sequence of steps is based on understanding the equilibrium aspects of solubility phenomena.

At the fourth level, we are concerned with chemical epistemology. We examine the basis of our chemical knowledge by asking questions such as, "How do we know that the cation of sodium is monovalent rather than divalent?" and "How do we know that the crystal structure of sodium chloride can be determined from x-ray data?" At this level we deal with the limits and validity of our fundamental chemical knowledge.

Across all four levels, the attitudes and motivations of both teacher and student are crucial. The attitude of the teacher is central to the success of interactions with students. Our motivation to teach is reflected in what we do and, as well, in what we do not do, both in and out of the classroom. Our modes of communicating with students affect their motivation to learn. All aspects of our behavior influence students' confidence and their trust in what we say. Our own attitudes toward chemicals and toward chemistry itself are reflected in such matters as how we handle chemicals, adhere to safety regulations, approach chemical problems, and explain and illustrate chemical principles. In my opinion, the single most important purpose that lectures serve is to give teachers the opportunity to convey an attitude toward chemistry—to communicate to students an appreciation of chemistry's diversity and usefulness, its cohesiveness and value as a central science, its intellectual excitement and challenge.

PRESENTING EFFECTIVE DEMONSTRATIONS

In planning a lecture demonstration, I always begin by analyzing the reasons for presenting it. Whether a demonstration is spectacular or quite ordinary I undertake to use the chemical system to achieve specific teaching goals. I determine what I am going to say about the demonstration and at what stage I should say it. Prior to the lecture, I practice doing the demonstration. By doing the demonstration in advance, I often see aspects of the chemical change which help me formulate both statements and questions that I then use in class.

Because one of the purposes of demonstrations is to increase the students' ability to make observations, I try to avoid saying, "Now I will demonstrate the insolubility of barium sulfate by mixing equal volumes of 0.1M barium chloride and 0.1M sodium sulfate solutions." Instead, I say, "Let us mix equal volumes of 0.1M barium chloride and 0.1M sodium sulfate solutions and observe what happens." Rather than announcing what should happen, I emphasize the importance of observing all changes. Often, I ask two or three students to state their observations to the entire class before I proceed with further manipulations. In addition, I help students to sort out observations so that relevant ones can be used in formulating conclusions about the chemical system. Some valid observations may not be relevant to the main purpose of the demonstration. For example, when the above-mentioned solutions are mixed, students may observe that the volumes are additive. However, this observation is not germane to the main purpose of the demonstration, which is to show the insolubility of barium sulfate. However, this observation is relevant if the purposes include teaching about the additive properties of liquids.

Every demonstration that I present in lectures is aimed at enhancing the understanding of chemical behavior. In all cases, the chemistry speaks for itself more eloquently than anything I can describe in words, write on a chalk board, or show on a slide.

Wesley Smith of Ricks College, who was a visiting faculty member at the University of Wisconsin–Madison from 1974 to 1977, has outlined six characteristics of effective demonstrations which best promote student understanding [2]:

1. *Demonstrations must be timely and appropriate.* Demonstrations should be done to meet a specific educational objective. For best results, plan demonstrations that are immediately germane to the material in the lesson. Demonstrations for their own sake have limited effectiveness.

2. *Demonstrations must be well prepared and rehearsed.* To ensure success, you need to be thoroughly prepared. *All* necessary material and equipment should be collected well in advance so that they are ready at class time. You should rehearse the entire demonstration from start to finish. Do not just go through the motions or make a dry run. Actually mix the solutions, throw the switches, turn on the heat, and see if the demonstration really works. Only then will you know that all the equipment is present and that all the solutions have been made up correctly. Always practice your presentation.†

3. *Demonstrations must be visible and large-scale.* A demonstration can help only those students who experience it. Hence, you need to set up the effect for the whole class to see. If necessary, rig a platform above desktop level to ensure visibility.

Perhaps the most important factor to consider is the size of what you are presenting. Only in the very tiniest of classes can the students see phenomena on the milligram and milliliter scale. Many situations require the use of oversized glassware and specialized equipment. Solutions and liquids should be shown in full-liter volumes, and solids should be displayed in molar or multi-molar amounts.

Contrasting backgrounds help emphasize chemical changes. A collection of large white and black cards to place behind beakers and other equipment is a valuable addition to your demonstration equipment. These are inexpensive, easy to use, and can provide an extra bit of polish to your demonstration.

4. *Demonstrations must be simple and uncluttered.* A common source of distraction is clutter on the lecture bench. Make sure that the demonstration area is neat and

---

† As Fred Juergens likes to say, "Prior practice prevents poor presentation."

free of extraneous glassware, scattered papers, and other disorder. All attention must be focused on the demonstration itself.

5. *Demonstrations must be direct and lively.* Action is an important part of a good demonstration. It is the very ingredient that makes demonstrations such efficient attention-grabbers. Students are eager to see something happen, but if nothing perceptible occurs within a few seconds you may lose their attention. The longer they have to wait for results, the less likely it is that the demonstration will have maximum educational value.

6. *Demonstrations must be dramatic and striking.* Usually, a demonstration can be improved by its mode of presentation. A lecture demonstration, according to Alfred T. Collette, is like a stage play. "A demonstration is 'produced' much as a play is produced. Attention must be given to many of the same factors as stage directors consider: visibility, audibility, single centers of attention, audience participation, contrasts, climaxes" [3]. The presentation of effective demonstrations is such an important part of good education that "no instructor is doing his best unless he can use this method of teaching to its fullest potential" [4].

USING THIS BOOK

The demonstrations in this volume are grouped in topical chapters dealing with the physical behavior of gases, the chemical behavior of gases, and chemical oscillating reactions. Each chapter has an introduction which covers the chemical background for the demonstrations that follow. We confine the discussion of relevant terminology and concepts to the introduction rather than repeating it in the discussion section of each demonstration. Accordingly, when teachers read the discussion section of any particular demonstration, they may find it necessary to refer to the chapter introduction for background information. For additional information teachers may wish to consult the sources listed at the end of each chapter's introduction.

Each demonstration has seven sections: a brief summary, a materials list, a step-by-step account of the procedure to be used, an explanation of the hazards involved, information on how to store or dispose of the chemicals used, a discussion of the phenomena displayed and principles illustrated by the demonstration, and a list of references. The brief summary provides a succinct description of the demonstration. The materials list for each procedure specifies the equipment and chemicals needed. Where solutions are to be used, we give directions for preparing stock amounts larger than those required for the procedure. The teacher should decide how much of each solution to prepare for practicing the demonstration and for doing the actual presentation. The availability and cost of chemicals may also affect decisions about the volumes to be prepared.

The procedure section often contains more than one method for presenting a demonstration. In all cases, the first procedure is the one the authors prefer to use. However, the alternative procedures are also effective and valid pedagogically.

The hazards and disposal sections include information compiled from sources believed to be reliable. We have enumerated many potentially adverse health effects and have called attention to the fact that many of the chemicals should be used only in well-ventilated areas. In all instances teachers should inquire about and follow local disposal practices and should act responsibly in handling potentially hazardous material. We recognize that several chemicals such as silver and mercury can be recovered and re-used and have given references to recovery and purification procedures.

The purpose of the discussion section is to provide the teacher with information for explaining each demonstration. We include discussion of chemical equations, relevant data, properties of the materials involved, as well as a theoretical framework through which the chemical processes can be understood. Again, we remind teachers that they should refer to the introduction of each chapter for background information not included in the discussion section of each demonstration. Finally, each demonstration contains a list of references used in developing procedures and providing information for the demonstration.

A WORD ABOUT SAFETY

Jearl Walker, professor of physics at Cleveland State University and editor of the Amateur Scientist section in *Scientific American*, has been quoted in newspaper stories as saying, "The way to capture a student's attention is with a demonstration where there is a possibility the teacher may die." Walker is said to get the attention of his students by dipping his hand in molten lead or liquid nitrogen, or by gulping a mouthful of liquid nitrogen, or by lying between two beds of nails and having an assistant with a sledge hammer break a cinder block on top of him. Walker reportedly has been injured twice, once when he used a small brick instead of a cinder block in the bed-of-nails demonstration, and once when he walked on hot coals and was severely burned.

We disagree strongly with this kind of approach. Demonstrations that result in injury are likely to confirm beliefs that chemicals are dangerous and that their effects are bad. In fact, every chemical is potentially harmful if not handled properly. That is why every person who does lecture demonstrations should be thoroughly knowledgeable about the safe handling of all chemicals used in a demonstration and should be prepared to handle any emergency. A first-aid kit, a fire extinguisher, a safety shower, and a telephone must be accessible in the immediate vicinity of the demonstration area. Demonstrations involving volatile material, fumes, noxious gases, or smoke should be rehearsed and presented only in well-ventilated areas.

We recognize that any of the demonstrations in this book can be hazardous. Our procedures are written for experienced chemists who fully understand the properties of the chemicals and the nature of their behavior. We take no responsibility or liability for the use of any chemical or procedure specified in this book. We urge care and caution in handling chemicals and equipment.

REFERENCES

1. I have adapted many ideas from Paul Saltman's address at the Third Biennial Conference on Chemical Education which was sponsored by the American Chemical Society, Division of Chemical Education, and held at Pennsylvania State University, State College, Pennsylvania (1974); see *J. Chem. Educ.* 52:25 (1975).
2. *Chemical Demonstrations Proceedings*, Western Illinois University and Quincy-Keokuk Section of the American Chemical Society: Macomb, Illinois (1978).
3. A. T. Collette, *Science Teaching in the Secondary School*, Allyn and Bacon: Boston (1973).
4. R. Miller, F. W. Culpepper, Jr., *Ind. Arts Voc. Educ.* 60:24 (1971).

Sources Containing Descriptions of Lecture Demonstrations

We call attention to the following sources of information about lecture demonstrations. These lists, updated from Volume 1, are not intended to be comprehensive. Some of the books are out of print but may be available in libraries.

BOOKS

Alyea, H. N. *TOPS in General Chemistry*, 3d ed., Journal of Chemical Education: Easton, Pennsylvania (1967).

Alyea, H. N., and F. B. Dutton, Eds. *Tested Demonstrations in Chemistry*, 6th ed., Journal of Chemical Education: Easton, Pennsylvania (1965).

Arthur, Paul. *Lecture Demonstrations in General Chemistry*, McGraw-Hill: New York (1939).

Blecha, M. T. Ph.D. Dissertation, "The Development of Instructional Aids for Teaching Organic Chemistry," Kansas State University, Manhattan, Kansas (1981).

Chemical Demonstrations Proceedings, Western Illinois University and Quincy-Keokuk Section of the American Chemical Society, Macomb, Illinois, May 5–6, 1978.

Chemical Demonstrations Proceedings, Western Illinois University and Quincy-Keokuk Section of the American Chemical Society, Macomb, Illinois, May 4–5, 1979.

Chemical Demonstrations Proceedings, Western Illinois University and Quincy-Keokuk Section of the American Chemical Society, Macomb, Illinois, May 1–2, 1981.

Chemical Demonstrations Proceedings, Fifth Annual Symposium at 16th Great Lake Regional Meeting of the American Chemical Society, Normal, Illinois, June 8, 1982.

Chen, Philip S. *Entertaining and Educational Chemical Demonstrations*, Chemical Elements Publishing Co.: Camarillo, California (1974).

Faraday, M. *The Chemical History of a Candle: A Course of Lectures Delivered Before a Juvenile Audience at the Royal Institution*, The Viking Press: New York (1960).

Ford, L. A. *Chemical Magic*, T. S. Denison & Co.: Minneapolis, Minnesota (1959).

Fowles, G. *Lecture Experiments in Chemistry*, 5th ed., Basic Books, Inc.: New York (1959).

Frank, J. O., assisted by G. J. Barlow. *Mystery Experiments and Problems for Science Classes and Science Clubs*, 2d ed., J. O. Frank: Oshkosh, Wisconsin (1936).

Freier, G. D., and F. J. Anderson. *A Demonstration Handbook for Physics*, 2d ed., American Association of Physics Teachers: Stony Brook, New York (1981).

Gardner, R. *Magic Through Science*, Doubleday & Co., Inc.: Garden City, New York (1978).

Hartung, E. J. *The Screen Projection of Chemical Experiments*, Melbourne University Press: Carlton, Victoria (1953).

Herbert D. *Mr. Wizard's Supermarket Science*, Random House: New York (1980).

Herbert, D., and H. Ruchlis. *Mr. Wizard's 400 Experiments in Science*, Revised Edition, Book-Lab: North Bergen, New Jersey (1983).

Joseph, A., P. F. Brandwein, E. Morholt, H. Pollack, and J. Castka. *A Sourcebook for the Physical Sciences*, Harcourt, Brace, and World, Inc.: New York (1961).

Lippy, J. D., Jr., and E. L. Palder. *Modern Chemical Magic*, The Stackpole Co.: Harrisburg, Pennsylvania (1959).

Meiners, H. F., Ed. *Physics Demonstration Experiments*, Vols. 1 and 2, The Ronald Press Company: New York (1970).

My Favorite Lecture Demonstrations, A Symposium at the Science Teachers Short Course, W. Hutton, Chairman; Iowa State University, Ames, Iowa, March 6–7, 1977.

Newth, G. S. *Chemical Lecture Experiments*, Longmans, Green and Co.: New York (1928).

Sharpe, S., Ed. *The Alchemist's Cookbook: 80 Demonstrations*, Shell Canada Centre for Science Teachers, McMaster University: Hamilton, Ontario, undated.

Siggins, B. A. M.S. Thesis, "A Survey of Lecture Demonstrations/Experiments in Organic Chemistry," University of Wisconsin–Madison, Wisconsin (1978).

Walker, J. *The Flying Circus of Physics—With Answers*, Interscience Publishers, John Wiley and Sons: New York (1977).

Wilson, J. W., J. W. Wilson, Jr., and T. F. Gardner. *Chemical Magic*, J. W. Wilson: Los Alamitos, California (1977).

ARTICLES

Bailey, P. S., C. A. Bailey, J. Anderson, P. G. Koski, and C. Rechsteiner. Producing a chemistry magic show. *J. Chem. Educ.* 52:524–25 (1975).

Castka, J. F. Demonstrations for high school chemistry. *J. Chem. Educ.* 52:394–95 (1975).

"Chem 13 News" 81. The November issue contained a collection of chemical demonstrations. (1976).

Gilbert, G. L., Ed. Tested demonstrations. Regular column in *J. Chem. Educ.* since 1976.

Hanson, R. H. Chemistry is fun, not magic. *J. Chem. Educ.* 53:577–78 (1976).

Hughes, K. C. Some more intriguing demonstrations. *Chem. in Australia* 47:458–59 (1980).

McNaught, I. J., and C. M. McNaught. Stimulating students with colourful chemistry. *School Sci. Review* 62:655–66 (1981).

Rada Kovitz, R. The SSP syndrome. *J. Chem. Educ.* 52:426 (1975).

Schibeci, R. A., J. Webb, and F. Farrel. Some intriguing demonstrations. *Chem. in Australia* 47:246–47 (1980).

Schwartz, A. T., and G. B. Kauffman. Experiments in alchemy, Part I: Ancient arts. *J. Chem. Educ.* 53:136–38 (1976).

Schwartz, A. T., and G. B. Kauffman. Experiments in alchemy, Part II: Medieval discoveries and "transmutations". *J. Chem. Educ.* 53:235–39 (1976).

Shakhashiri, B. Z., G. E. Dirreen, and W. R. Cary. Lecture Demonstrations, in *Sourcebook for Chemistry Teachers*, pp. 3–16, W. T. Lippincott, Ed., American Chemical Society, Division of Chemical Education: Washington, D.C. (1981).

Sources of Information on Hazards and Disposal

In preparing the Hazards and Disposal sections of Volume 2, we used the following references. The order of listing reflects our degree of utilization.

Bretherick, L., Ed. *Hazards in the Chemical Laboratory*, 3d ed.; The Royal Society of Chemistry: London (1981).

Windholz, M., Ed. *The Merck Index*, 10th ed., Merck & Co., Inc.: Rahway, New Jersey (1983).

Prudent Practices for Handling Hazardous Chemicals in Laboratories, Committee on Hazardous Substances in the Laboratory, National Research Council (1981).

Laboratory Waste Disposal Manual, Manufacturing Chemists Association (1975). This book is out of print, but some of the information was available in the 1981 Reagent Catalog of MCB Manufacturing Chemists, Inc., 2909 Highland Avenue, Cincinnati, Ohio 45212.

Registry of Toxic Effects of Chemical Substances, Dept. of Health, Education and Welfare (NIOSH): Washington, D.C., revised annually. Available from Superintendent of Documents, U.S. Government Printing Office, Washington, D.C. 20402.

Fire Protection Guide on Hazardous Materials, 6th ed., National Fire Protection Association: 470 Atlantic Ave., Boston, Massachusetts 02210 (1975). New editions are published at intervals.

Safety in Academic Chemistry Laboratories, 3d ed., American Chemical Society Committee on Chemical Safety: Washington, D.C. (1979). A fourth edition is scheduled for publication. The bibliography lists many journal articles and books.

Health and Safety Guidelines for Chemistry Teachers, American Chemical Society Dept. of Educational Activities: Washington, D.C. (1979). The bibliography lists journal articles and books.

Steere, N. V. *Handbook of Laboratory Safety*, 2d ed., CRC Press: Cleveland, Ohio (1971).

Guide for Safety in the Chemical Laboratory, 2d ed., Van Nostrand Reinhold Co., Litton Educational Publishing, Inc.: New York (1972).

Steere, N. V., Ed. *Safety in Chemical Laboratory*, Journal of Chemical Education: Easton, Pennsylvania, Vol. 1 (1967), Vol. 2 (1971), Vol. 3 (1974).

Renfrew, M. M., Ed. *Safety in the Chemical Laboratory*, Journal of Chemical Education: Easton, Pennsylvania, Vol. 4 (1981).

Sax, N. I. *Dangerous Properties of Industrial Materials*, 3d ed., Van Nostrand Reinhold Co., Litton Educational Publishing, Inc.: New York (1968).

Chemical Demonstrations

Volume 2

5

Physical Behavior of Gases

George M. Bodner, Rodney Schreiner, Thomas J. Greenbowe, Glen E. Dirreen, and Bassam Z. Shakhashiri

This chapter includes demonstrations which display the relationships between pressure, volume, and temperature for material in the gaseous phase. Various laws which describe and govern the behavior of gases are discussed and illustrated. We assume that teachers know the principles of gas behavior and can refresh such knowledge by referring to chemistry textbooks or other sources. We urge fellow teachers to demonstrate the physical reality of the variables used to describe gas behavior: pressure, volume, temperature, and number of moles. The ideal gas law, its applications, and its limitations should be emphasized. In addition, we urge that the kinetic molecular theory—its postulates, consequences, and applications—be fully discussed.

The word *gas*, which was first used in the middle of the 17th century by the Belgian physician Johann van Helmont, is derived from the Greek word *chaos*. Apparently he chose *chaos* because so many of the vessels he used exploded when he tried to prepare gases by chemical reactions [1]. In the 18th century the British biologist Stephen Hales introduced the pneumatic trough apparatus, which made it possible to generate and manipulate gases experimentally [2]. Observations of gas behavior led to quantitative experiments, and gas laws were formulated to summarize the results of the experiments. Eventually, a theory—the kinetic molecular theory—was developed to explain the related observations, facts, and laws. The study of gases in the latter part of the 18th century and the early part of the 19th century resulted in significant developments in chemistry. Although gases differ as widely in their chemical behavior as do liquids and solids, their physical behavior is remarkably uniform. Regardless of their chemical properties, gases fill whatever space is available, expand and contract with changes in temperature, exert greater pressure with increasing temperature, flow rapidly, are compressible, and generally have low density. This similarity in physical behavior suggests that these properties are not independent of one another and that they can be related by a theoretical model which assumes that gases consist of particles which are in constant random motion and which collide with one another and with the walls of vessels that contain them.

Handling Samples of Gas

Demonstrations of the properties of gases can be time consuming without a convenient way to store and handle these gases. Even the ready availability of compressed gas cylinders does not always solve this problem, because many demonstrations require

3

samples of gas at atmospheric pressure. A simple method of handling gas samples at atmospheric pressure involves an apparatus constructed of common materials (see figure). Bore a large hole in a #6 1-holed rubber stopper, extract the core from the cork borer, and save both the core and the stopper. Insert the mouth of a plastic sandwich bag through the large hole in the stopper, and insert the core into the stopper to hold the bag in place. Insert a dropper through the hole in the core. Squeeze the air out of the bag and seal the bag by attaching a rubber septum to the dropper. The bag can be filled by removing the rubber septum and connecting the dropper to a gas cylinder or gas-generating apparatus with a piece of rubber tubing. When the bag is full, reseal it with the septum. The gas can be sampled by using a syringe with an 18- to 20-gauge needle, which should be inserted through the septum.

Caution Regarding Compressed Gas Cylinders

Compressed gas cylinders can be hazardous if not handled properly. Always keep the protective cap over the valve when moving or storing the cylinder. Close the valve tightly when the cylinder is not in use. Secure gas cylinders with a belt or chain at all times. Always move large cylinders with a hand truck, never by rolling or dragging them. For more information, see reference 3.

REFERENCES

1. A. J. Ihde, *The Development of Modern Chemistry*, Harper and Row: New York (1964); reprint ed., Dover Publications: New York (1984).
2. J. Parascandola and A. J. Ihde, *Isis* 60: Part 3, 351 (1969).
3. G. Pinney, *J. Chem. Educ.* 42: A976 (1965).

Physical Properties of Selected Gases

	Formula	MW (g/mol)	Freezing point (°C)	Boiling point (°C)	Gas density (g/liter) (1 atm, 25°C)
acetylene	C_2H_2	26.038		-84.0[a]	1.064
ammonia	NH_3	17.031	-77.7	-33.4	0.696
argon	Ar	39.948	-189.2	-185.88	1.633
boron trichloride	BCl_3	117.17	-107	12.4	4.789
boron trifluoride	BF_3	67.80	-127.1	-100.3	2.772
butane	C_4H_{10}	58.124	-138.3	-0.5	2.376
carbon dioxide	CO_2	44.010		-78.5[a]	1.799
carbon monoxide	CO	28.010	-199	-191.5	1.145
chlorine	Cl_2	70.906	-101.0	-34.1	2.898
dimethyl ether	$(CH_3)_2O$	46.069	-141.5	-24.7	1.883
dinitrogen oxide	N_2O	44.013	-90.8	-89.5	1.799
ethane	C_2H_6	30.070	-183.3	-88.6	1.229
ethene	C_2H_4	28.054	-169.5	-103.7	1.147
fluorine	F_2	37.997	-219.8	-188.3	1.553
freons					
freon-11	CCl_3F	137.368	-111.0	23.8	5.615
freon-12	CCl_2F_2	120.914	-158	-29.8	4.942
freon-13	$CClF_3$	104.459	-181	-81.4	4.270
genetron-23	CHF_3	70.014	-155.2	-82.2	2.862
freon-22	$CHClF_2$	86.469	-160	-40.8	3.534
genetron-21	$CHCl_2F$	102.923	-135	8.9	4.207
helium	He	4.0026	-272.0	-268.9	0.164
hydrogen	H_2	2.016		-252.9	0.082
hydrogen bromide	HBr	80.912	-86.9	-66.8	3.307
hydrogen chloride	HCl	36.461	-114.2	-85.0	1.490
hydrogen cyanide	HCN	27.026	-13.2	25.7	1.105
hydrogen fluoride	HF	20.006	-83.4	19.5	0.818
hydrogen iodide	HI	127.913	-50.8	-35.5	5.228
hydrogen sulfide	H_2S	34.08	-85.5	-60.3	1.393
krypton	Kr	83.80		-153.6	3.425
methane	CH_4	16.043	-182.6	-161.5	0.656
neon	Ne	20.179		-246.1	0.825
nitrogen	N_2	28.013		-195.8	1.145
nitrogen dioxide	NO_2	46.006	-9.3	21.2	1.880
nitrogen oxide	NO	30.006	-163.6	-151.7	1.226
oxygen	O_2	31.999		-183.0	1.308
ozone	O_3	47.998	-192.5	-111.9	1.962
phosphine	PH_3	33.998	-133.8	-87.7	1.390
propane	C_3H_8	44.097	-187.7	-42.1	1.802
sulfur dioxide	SO_2	64.06	-75.5	-10.0	2.618
sulfur hexafluoride	SF_6	146.05		-63.8[a]	5.970
xenon	Xe	131.30		-108.1	5.367

[a] Sublimation point.

5.1

Collapsing Can

Water in the bottom of a can is boiled until virtually all of the air has been driven out. The can is then sealed. It collapses as it cools to room temperature [1]. The can may be returned to near its original shape by reheating it. Alternatively, when the can is connected to a vacuum pump, it collapses as the air is pumped from it.

MATERIALS FOR PROCEDURE A

200 mL tap water

collapsible metal can (e.g., a duplicating-fluid can or Sargent-Welch #1513)

iron tripod with wire gauze

burner, Bunsen or Meker (or hot plate)

insulated gloves

solid rubber stopper to fit mouth of can

pan or tray filled with ice (optional)

MATERIALS FOR PROCEDURE B

15 mL tap water

aluminum soft-drink can, 12-ounce size, empty

iron tripod, with wire gauze

burner, Bunsen or Meker (or hot plate)

tongs

beaker, 600 mL or larger, filled with cold tap water

MATERIALS FOR PROCEDURE C

10-cm length glass tubing, with outside diameter of 8 mm

1-holed rubber stopper to fit collapsible metal can

120-cm length heavy-walled rubber vacuum tubing

vacuum pump, capable of producing pressure of 10 torr

collapsible metal can (e.g., a duplicating-fluid can or Sargent-Welch #1513)

PROCEDURE A

Preparation and Presentation

Pour 100 mL of water into the can. Place the wire gauze on the tripod, and set the can on top of the gauze. Light the burner, and place it beneath the can. Heat the can until the water inside of it boils vigorously and until the cloud of condensed water vapor escapes from the mouth of the can for at least 30 seconds. Turn off the burner, and, wearing insulated gloves, seal the can firmly with the rubber stopper. Set the can on the table top or in the pan of ice, and watch it collapse as it cools and the water vapor within it condenses.

Remove the stopper from the collapsed can, and pour another 100 mL of water into the can. Securely seat the stopper in the mouth of the can. Place the can on the tripod and reheat it. As the water inside begins to boil, the can will expand again to near its original shape. Alternatively, if the can is small (1 gallon or less), it can be restored by blowing it up as you would a balloon.

PROCEDURE B [2]

Preparation and Presentation

Pour 15 mL of water into the can. Place the wire gauze on the tripod or ring stand, and set the can on top of the gauze. Light the burner, and place it beneath the can. Heat the can until the water inside of it boils vigorously and until the cloud of condensed water vapor escapes from the mouth of the can for about 20 seconds. Using tongs, quickly lift the can from the burner and invert it in the beaker of cold water. The can will collapse instantaneously as the water vapor within it condenses.

PROCEDURE C

Preparation

Insert the glass tube through the rubber stopper. Using the heavy-walled rubber tubing, connect the glass tube to the vacuum pump.

Presentation

Seat the rubber stopper in the mouth of the can. Turn on the vacuum pump. As the pump removes the air from the can, the can will collapse.

HAZARDS

The can containing boiling water has a temperature of 100°C and can cause burns to the skin if it is touched with unprotected hands.

DISPOSAL

The can should be discarded in a solid-waste receptacle.

DISCUSSION

The difference between the atmospheric pressure acting on the outside surface of the can and the low pressure within the can leads to the collapse of the can. Atmospheric pressure is about 760 torr, which is $1.0 \times 10^5 \, N/m^2$. The surface area of the soft-drink can used in Procedure B is the sum of the area of its side plus the areas of its top and bottom. The side of the can is a rectangle, 0.12 m high and πd, 3.14×0.065 m, wide. The top and bottom are circles of area πr^2, $3.14 \times (0.033 \, m)^2$. Therefore, the area of the surface is $0.031 \, m^2$, and the pressure of the atmosphere on this surface is $3.1 \times 10^3 \, N$, which is 680 pounds. If the pressure within the can is reduced by as little as 75%, the difference between the forces on the outer surface and the inner surface is 500 pounds. Soft-drink cans are not designed to withstand such a force.

When the soft-drink can filled with water vapor is inverted in the water bath, the water seals the opening and cools the can. As the can cools, the vapor condenses, reducing the pressure inside the can. Conceivably, this reduced pressure could draw the water up into the can through its opening. However, it appears that the can is sufficiently weak and the water sufficiently viscous that the can collapses before it fills with water. It is necessary to use an all-aluminum can, because a steel or partly steel can fills with water rather than collapses.

The collapse of the can in Procedure B is virtually instantaneous, and the effect can be startling. If a more gradual collapse is desired, then Procedure A or C should be followed.

REFERENCES

1. H. N. Alyea and F. B. Dutton, Eds., *Tested Demonstrations in Chemistry*, 6th ed., Journal of Chemical Education: Easton, Pennsylvania (1965).
2. G. Kauffman, *J. College Sci. Teach.* 14:364 (1985).

5.2

Mercury Barometers

Though barometers exist in various shapes and sizes, the heights of their mercury columns are the same.

MATERIALS

225 mL (3.15 kg) mercury

85-cm length glass tubing, with outside diameter of 10 mm

120-cm length glass tubing, with outside diameter of 10 mm

2 40-cm lengths glass tubing, with outside diameter of 10 mm

10-cm length glass tubing with outside diameter of 20–30 mm

glass-working torch

3 ring stands

3 clamps with holders

plastic tray, ca. 30 cm × 40 cm

3 crystallizing dishes, 70 mm × 50 mm

3 crystallizing dishes, 125 mm × 65 mm

100-mL beaker

gloves, plastic or rubber

PROCEDURE

Preparation

Seal one end of the 85-cm length of glass tubing, and fire-polish the open end.

Bend a C into the 120-cm piece of glass tubing, as illustrated on the left in the figure. Seal one end of the tube, and fire-polish the open end.

Seal one of the 40-cm pieces of glass tubing onto each end of the 10-cm piece of tubing, as illustrated on the right in the figure. Seal one end of the resulting tube, and fire-polish the open end.

Adjust the clamps on the ring stands so that each can hold one of the glass tubes vertically with its open end 1–2 cm above the base of the stand. Over a plastic tray (to contain any spills), half fill each of the smaller crystallizing dishes with mercury. Set one of the dishes of mercury in each of the larger dishes. Fill each of the glass tubes with mercury from the 100-mL beaker to within 2 cm of the opening, and clamp it to one of the stands. Half fill the beaker with mercury.

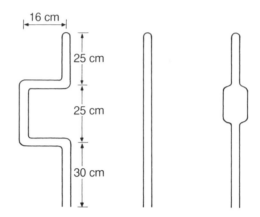

Presentation

Wearing gloves, pour mercury from the beaker into one of the tubes until the mercury is 1–2 mm from the top of the tube. Unclamp the tube, tightly cover the open end of the tube with a finger, and invert it. Immerse the covered end of the tube in the dish of mercury on the base of the stand, and uncover the opening. Some of the mercury will drain out of the tube. Clamp the tube to the stand.

Repeat this procedure with the other two tubes.

Display the three barometers side by side to show that the heights of the mercury columns in the three tubes are the same.

HAZARDS

Mercury is extremely toxic and should be handled with care to avoid prolonged or repeated exposure to the liquid or vapor. Continued exposure to the vapor may result in severe nervous disturbance, insomnia, and depression. Continued skin contact also can cause these effects as well as dermatitis and kidney damage. Mercury should be handled only in well-ventilated areas. Mercury spills should be cleaned up immediately by using a capillary attached to a trap and an aspirator. Small amounts of mercury in inaccessible places should be treated with zinc dust to form a nonvolatile amalgam.

DISPOSAL

The barometers should be disassembled by a process the reverse of that in the Procedure. Wearing gloves, place a finger over the submerged opening of the glass tube, lift it out of the dish, and invert it. Pour the mercury back into the storage container. Return the mercury from the dishes to the container.

DISCUSSION

Evangelista Torricelli (1608–1647) was the first to suggest that air has weight and that we live in a sea of air which constantly exerts a pressure on us and all objects around us. He was also the first to construct a mercury barometer to measure this pressure, and as a consequence, he was the first to prepare a vacuum (or near vacuum) at the top of one of his Torricellian barometers.

When a mercury-filled tube over 76 cm in length is inverted in a pool of mercury, the level of mercury in the tube falls, leaving a near vacuum at the top. The only pressure in the top of the tube is the vapor pressure of mercury, which is 0.0012 mm Hg at 20°C [1]. The level of mercury in the tube falls until the pressure at the surface of the pool produced by the column of mercury equals the pressure produced by the surrounding atmosphere.

The pressure produced by the column of mercury is the force that it exerts on the surface of the pool divided by the cross-sectional area at the bottom of the column. The force that the mercury exerts is equal to the weight of the mercury, which is the product of the volume of mercury multiplied by its density multiplied by the acceleration of gravity (F = ma). In a tube of constant diameter, such as the one illustrated at the center of the figure, the volume of mercury is proportional to the height of the column. The pressure is proportional to the weight, which is proportional to the volume, which is proportional to the height of the mercury. Therefore, pressure is expressed in terms of the height of a column of mercury.

As this demonstration illustrates, it is not necessary to use a straight, vertical tube with a constant cross-sectional area to construct a barometer. Tubes of various shapes can be used, and in each one the level of the mercury will be the same as in the others. This observation can be explained as follows: In each barometer, regardless of its shape, the mercury that is supported by the pressure of the atmosphere can be considered as a vertical column of constant cross-sectional area. The additional mercury contained in barometers of variant shapes is supported by the tube itself.

There is quite a variety of units used to express pressure. The torr, named in honor of Torricelli, is equivalent to a pressure sufficient to support 1 mm of mercury. The United States Weather Service reports atmospheric pressure in terms of inches of mercury. Normal atmospheric pressure at sea level supports a column of mercury about 76 cm high. A 76-cm column of mercury in a tube with a cross-sectional area of 1.0 cm^2 would contain 76 cm^3 of mercury. Because the density of mercury is 13.6 g/cm^3 [2], this column would contain 1030 g of mercury and would exert a pressure of 1030 g/cm^2. In English units, this is 14.7 pounds/$inch^2$. The SI unit of pressure, the pascal, is equivalent to a force of 1 Newton (1 kg m/s^2) acting on an area of 1 m^2. The force exerted by 1.03 kg (1030 g) of mercury is obtained by multiplying this mass by the acceleration of gravity, 9.8 m/s^2, which yields 10.1 N. This force is exerted over an area of 1×10^{-4} m^2 (1.0 cm^2), producing a pressure of 1.01×10^5 N/m^2, or 1.01×10^5 Pa.

REFERENCES

1. J. A. Dean, Ed., *Lange's Handbook of Chemistry*, 12th ed., McGraw-Hill: New York (1979).
2. R. C. Weast, Ed., *Handbook of Chemistry and Physics*, 63d ed., CRC Press: Boca Raton, Florida (1982).

5.3

Effect of Pressure
on the Size of a Balloon

A partially inflated balloon is sealed inside a large filter flask, and the flask is attached to an aspirator. The balloon expands as the flask is evacuated [1].

MATERIALS

4-liter filter flask

ring stand

iron ring, 10 cm diameter or larger

round balloon, with inflated diameter of ca. 25 cm

solid rubber stopper to fit filter flask

2-way stopcock

15-cm length of vacuum tubing

90-cm length of vacuum tubing

water aspirator or vacuum pump

PROCEDURE

Preparation

Clamp the filter flask to the ring stand with an iron ring. Partially inflate the balloon just until it becomes rigid yet will still fit through the neck of the flask. Tie the balloon closed. Insert the balloon into the flask and seal the flask with the rubber stopper. Connect the tubulation on the filter flask to a small stopcock with the short piece of vacuum tubing and connect the longer piece of tubing to the free end of the stopcock.

Presentation

Connect the vacuum tubing to the water aspirator or pump. Turn on the aspirator or pump and slowly open the stopcock. As the air in the flask is withdrawn, the balloon expands in the flask. Close the stopcock and detach the tubing from the aspirator or pump. Open the stopcock slowly and watch as the balloon again shrinks to its former size. This process may be repeated indefinitely.

HAZARDS

The filter flask, like any evacuated container, must be regarded as an implosion hazard. A grid-like wrapping of transparent tape, which will leave the contents visible, is recommended for the flask. The pressure difference between the outside and inside of the flask approaches 760 torr when a vacuum pump is used and about 740 torr with an aspirator.

DISCUSSION

A balloon will expand until the pressure of the gas inside the balloon is equal to the sum of the pressure exerted by the elasticity of the balloon plus the external pressure. As the external pressure is reduced, the gas within the balloon expands. If a good vacuum is achieved within the flask, the balloon in this demonstration may expand to as much as 20 times its original volume.

REFERENCE

1. H. N. Alyea, *J. Chem. Educ.* 32:A13 (1955).

5.4

Boyle's Law

The relationship between the pressure and volume of a fixed amount of gas at constant temperature is studied by monitoring the volume of the gas while varying the pressure, or vice versa.

MATERIALS FOR PROCEDURE A

100 mL mercury

130-cm length glass tubing, with outside diameter of 10 mm

glass-working torch

meter stick

ring stand, with clamp for glass tube

plastic tray, ca. 30 cm × 40 cm

white backdrop (e.g., piece of white poster board, 20 cm × 100 cm)

short-stemmed funnel, ca. 50 mm in diameter, nonmetallic

5-cm length rubber tubing to fit both glass tubing and stem of funnel

MATERIALS FOR PROCEDURE B

4 right-angle glass bends, with outside diameter of 7 mm, one arm 5 cm long, the other 20 cm

4 2-holed rubber stoppers to fit round-bottomed flasks

4 right-angle glass bends, with outside diameter of 7 mm and length of each arm ca. 5 cm

4 1-liter round-bottomed flasks

4 30-cm lengths copper wire, 16 gauge

3 15-cm lengths rubber tubing to fit 7-mm glass bends

4 cork rings to support flasks

2 40-cm lengths rubber tubing to fit 7-mm glass bends

mercury-filled U-tube manometer, with arms 2 m long

10 5-cm lengths copper wire, 16 gauge

PROCEDURE A

Preparation

Construct the apparatus as illustrated in Figure 1. Bend the glass tubing into a J having the dimensions in the figure. The distance between the two arms should be large enough to fit a meter stick easily between them. Seal the short end of the tube. Clamp the J-tube to the ring stand, and place the stand on a plastic tray to contain any spills. Place the white backdrop behind the tube. Attach the funnel to the open end of the tube with the rubber tubing.

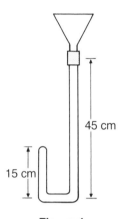

45 cm

15 cm

Figure 1.

Presentation

Pour enough mercury through the funnel to fill the curved portion at the bottom of the J-tube. As the mercury fills the bottom of the tube, it will trap the air in the sealed arm of the tube. Measure and record both the length of the column of air trapped in the short arm of the J-tube and the difference between the heights of the mercury in the two arms of the tube.

Pour a few more milliliters of mercury into the tube and repeat the two measurements. Continue to add mercury and record data until at least five sets of data have been recorded.

The data can be used to illustrate Boyle's law, as described in the Discussion section.

PROCEDURE B

Preparation

Assemble the apparatus shown in Figure 2 [1]. Insert the 20-cm end of the bends through each of the 2-holed rubber stoppers. Insert one of the short bends into each of the stoppers. Seat one of the stoppers in the mouth of each of the round-bottomed flasks. Fasten the stoppers to the flasks by wrapping wire over the top of each stopper

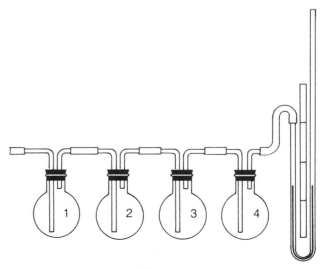

Figure 2.

and around the neck of the flask, in the manner a champagne cork is fastened. Using the short pieces of rubber tubing, connect the flasks together, the short glass bend of one to the long glass bend of the next. Support each flask on a cork ring. With longer pieces of rubber tubing, connect the free, *short* glass bend of an end flask to one arm of the manometer, and connect the free, *long* glass bend of the other end flask to a water tap. Secure all connections by tightening short pieces of wire around the rubber tubing wherever it joins the glass bends.

Presentation

Note that initially the gas in the apparatus has a volume of approximately 4 liters (the total volume of the flasks, neglecting the small volume of the tubing) at atmospheric pressure. Open the water tap and fill the first of the flasks with water. The volume of the gas within the apparatus is now 3 liters. Measure and record the difference in the levels of the mercury in the two arms of the manometer. Open the tap again, and fill the second flask. Now the volume of the gas is 2 liters. Record the difference in the mercury levels.

The third flask should not be filled with water. If the third flask were to be filled with water, the pressure of the gas would become quite high, too high for the manometer to read, causing the mercury to overflow.

The data collected in this procedure can be used to illustrate Boyle's law, as described in the Discussion section.

HAZARDS

Mercury is extremely toxic and should be handled with care to avoid prolonged or repeated exposure to the liquid or vapor. Continued exposure to the vapor may result in severe nervous disturbance, insomnia, and depression. Continued skin contact also can cause these effects as well as dermatitis and kidney damage. Mercury should be handled only in well-ventilated areas. Mercury spills should be cleaned up immediately by using a capillary attached to a trap and an aspirator. Small amounts of mercury in inaccessible places should be treated with zinc dust to form a nonvolatile amalgam.

DISPOSAL

Pour the mercury from the J-tube back into its container for reuse.

DISCUSSION

Robert Boyle (1627–1691) was fascinated by the behavior of gases, and devoted much of his life to the study of their properties. Among the properties he studied was the relationship between the pressure and the volume of a gas. He studied this relationship at pressures both above and below atmospheric pressure. To study gases at pressures higher than atmospheric, he used a J-shaped tube as used in Procedure A of this demonstration [2]. For studies at low pressures, he used a straight glass tube, sealed at one end and partially filled with mercury, with its open end immersed in a vessel of mercury. By raising the tube in the vessel of mercury, he could reduce the pressure of the gas in the tube and measure its volume. (This method is not used in this demonstration, because it requires a rather large amount of mercury.) Experiments with this equipment led him in 1662 to the law bearing his name, which states that the volume of a gas is inversely proportional to its pressure.

Expressed algebraically, Boyle's law can be represented by the equation

$$V = \frac{k}{P}$$

where V is the volume of the gas,
    P is its pressure, and
    k is a proportionality constant.

An equivalent form of the equation is

$$PV = k$$

Therefore, Boyle's law can be demonstrated by showing that the product of the volume of a sample of gas multiplied by its pressure remains constant as the pressure and volume change.

In the apparatus of Procedure A, the sample of gas is contained in the closed end of the J-tube, which is a cylindrical container. The volume of this gas is given by the equation for the volume of a cylinder,

$$V = \pi r^2 h$$

where r is the cross-sectional inside radius of the tube and
    h is the height of the column of gas in the tube.

This equation indicates that the volume of the gas is proportional to the height of the column of gas. While the actual volume of the gas in the tube can be calculated, this is unnecessary to illustrate Boyle's law. Because the height of the gas column is proportional to the volume of the gas, Boyle's law can be demonstrated by showing that the product of this height multiplied by the pressure of the gas is constant.

The pressure of the gas in the J-tube can be determined by measuring the levels of the mercury in the two arms of the tube. The gas exerts pressure on the mercury, pushing some of it into the open arm of the tube. This pressure is exerted against the pressure of the atmosphere outside the apparatus and against the weight of the mercury that is elevated above the level of the mercury in the closed arm of the tube. Therefore, the pressure of the enclosed gas is the sum of atmospheric pressure plus the pressure produced by the weight of the raised mercury:

$$P_g = P_a + P_{Hg}$$

The mass of this mercury can be found by multiplying the density of mercury (13.6 g/mL [3]) by the volume of the mercury. The volume of the column of mercury, like the volume of the enclosed gas, is given by the equation for the volume of a cylinder. Therefore, it is proportional to the height of the mercury column, and the pressure of the gas produced by the mercury column is proportional to the difference between the heights of the mercury columns in the two arms of the J-tube.

The height of the column of gas and the difference in the levels of the mercury in the two arms of the J-tube can be used to illustrate Boyle's law. The pressure of the gas in the tube is the sum of the atmospheric pressure and the pressure produced by the weight of the column of mercury, which is proportional to the difference in the levels of the mercury in the two tubes. The pressure of the gas in the tube may be represented by a + d, where a is the atmospheric pressure, and d the difference between the two levels of mercury, both expressed in millimeters of mercury. The volume of the gas is proportional to the height of the gas column, h. To demonstrate Boyle's law, all that need be shown is that the product (a + d)h is a constant. This has been done with sample data presented in Table 1. In this example, atmospheric pressure was assumed to be 760 mm Hg, although for greater accuracy atmospheric pressure could have been measured.

Table 1. Sample Results of Procedure A

Difference in Hg levels (mm)	Height of gas column (mm)	Total pressure of gas[a] (mm Hg)	Product of pressure and height[b]
32	239	792	189,000
146	211	906	191,000
372	167	1132	189,000
512	149	1272	190,000
728	128	1488	190,000
947	114	1707	195,000

[a]Calculated by adding an atmospheric pressure of 760 mm Hg to the difference in the mercury levels.

[b]Calculated by multiplying the height of the gas column (in mm) by the total pressure of the gas (in mm Hg).

The data obtained in Procedure B are less precise than those from Procedure A, but their interpretation is similar. To illustrate Boyle's law with these data, one need show that the product of the volume multiplied by the pressure of the gas is constant. Data from a trial of Procedure B are presented in Table 2. Initially, the volume of the gas is 4 liters, neglecting the small volume of gas contained in the tubing, at which time its pressure is atmospheric pressure. As the gas is compressed into a volume of 3 liters, its pressure increases, supporting a column of mercury in the manometer. Then the pressure of the gas is the sum of the height of the column of mercury and atmospheric pressure. When the second flask is filled with water, the gas is compressed further, and its pressure increases. As the results in Table 2 show, the product of the volume multiplied by the pressure of the gas is virtually constant.

Table 2. Sample Results of Procedure B

Volume of gas (liters)	Difference in Hg levels (mm)	Total pressure of gas[a] (mm Hg)	Product of volume and pressure[b]
4.00	0	760	3040
3.00	258	1018	3050
2.00	775	1535	3070

[a]Calculated by adding an atmospheric pressure of 760 mm Hg to the difference in the mercury levels.

[b]Calculated by multiplying the volume of the gas (in liters) by the total pressure of the gas (in mm Hg).

REFERENCES

1. F. B. Dutton, *J. Chem. Educ.* 18:15 (1941).
2. J. R. Partington, *A Short History of Chemistry*, 3d ed., Macmillan and Co.: London (1957).
3. R. C. Weast, Ed., *Handbook of Chemistry and Physics*, 59th ed., CRC Press: Boca Raton, Florida (1978).

5.5

Boyle's Law and
the Mass of a Textbook

A number of books are placed, one by one, onto the plunger of a syringe mounted on a stand, and the volume of the gas in the syringe is recorded. The data illustrate Boyle's law and can be used to determine the mass of a book [1–6].

MATERIALS

2–3 drops glycerine

30-mL glass syringe

septum cap

clamp, with holder

ring stand

4 or more identical textbooks

PROCEDURE

Preparation

Assemble the apparatus as illustrated in Figure 1. Lubricate the syringe plunger with glycerine, and fill the syringe with 30 mL of air. Seal the tip of the syringe with the septum cap. Clamp the syringe vertically to the ring stand.

Presentation

Record the volume of gas in the syringe. Balance one textbook on the plunger of the syringe, gently press down on the book, and let the plunger spring back up. Record the volume of gas in the syringe and the number of books. Sequentially add a second, third, fourth, etc., textbook, and record the volume of gas and number of books for each step. Then, remove the books, one by one, and record the volume of gas in the syringe for each step. Average the volumes obtained as the books were removed with those obtained as the books were added.

The data obtained in this procedure can be used to illustrate Boyle's law or, alternatively, to determine the mass of one book, as described in the Discussion section.

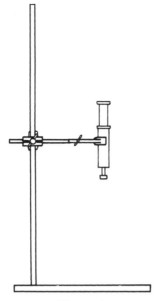

Figure 1.

DISCUSSION

According to Boyle's law, the volume of a gas is inversely proportional to its pressure, as represented by the equation

$$\frac{1}{V} = kP$$

where k is a proportionality constant. The data obtained from this demonstration can be used to illustrate Boyle's law. Sample data obtained by this procedure using copies of a typical chemistry textbook (having a mass of 2.0 kg) are presented in the table. When there are no books balanced on the plunger, the pressure of the gas in the syringe is atmospheric pressure, which may be represented by P_a. (When the data in the table were obtained, the atmospheric pressure was 751 torr.) When a book is placed on the plunger, the weight of the book exerts a pressure on the gas in the syringe. This pressure may be represented by B. Therefore, when there is one book on the plunger, the pressure of the gas in the syringe is $P_a + B$. When there are two books on the plunger, the pressure of the gas is $P_a + 2B$. In general, the pressure of the gas in the syringe is $P_a + nB$, where n is the number of books on the plunger.

Volume of Gas versus Number of Textbooks[a]

Number of books	Volume of gas (mL)
0	30.0
1	20.0
2	15.0
3	12.0
4	10.0

[a] The mass of each book is 2.0 kg.

For each number of books, n, a volume of gas, V_n, was measured. According to Boyle's law,

$$1/V_n = k(P_a + nB)$$

Multiplying through by k on the right side, this equation can be rewritten as

$$1/V_n = kBn + kP_a$$

which has the form of the general equation for a straight line (y = mx + b). According to Boyle's law, when $1/V_n$ is plotted versus n, the plot will be a straight line. The data from the table have been plotted in Figure 2, and the plot is indeed a straight line.

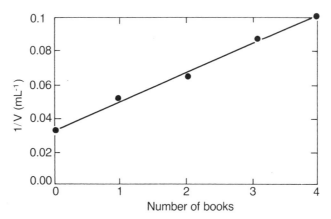

Figure 2. Plot of 1/V versus number of books.

From the plot of $1/V_n$ versus n, the mass of the textbook can be determined. The slope of the line is kB, the weight of the textbook multiplied by the proportionality constant, and the n = 0 intercept is kP_a, atmospheric pressure multiplied by the proportionality constant. From the n = 0 intercept and the value of atmospheric pressure, the value of the proportionality constant, k, can be determined, and from its value and the slope of the line, the value of B can also be determined.

The n = 0 intercept occurs at $1/V_n = 0.0333$ mL^{-1}, therefore, $kP_a = 0.0333$ mL^{-1}. Because the value of P_a is 751 torr, $k = 4.43 \times 10^{-5}$ mL^{-1}torr^{-1}. The slope of the line is 0.0167 mL^{-1}book^{-1}, which is kB. Therefore, B = 377 torr/book. Each book exerts a pressure of 377 torr on the gas in the syringe. In other words, each book exerts a pressure equivalent to a column of mercury 377 mm tall. Pressure is a force per area; the force in this case is produced by the weight of the book or by the weight of a column of mercury, and the area is the area of the plunger. The area of the plunger is equal to the volume of the syringe, 30.0 mL (30.0 cm^3), divided by the distance between the 0-mL and 30-mL marks on the syringe, in this case, 7.5 cm. Therefore, the area of the plunger is 30.0 cm^3/7.5 cm = 4.0 cm^2. The total volume of a mercury column 377 mm tall with a cross-sectional area of 4.0 cm^2 is 151 cm^3. Because the density of mercury is 13.6 g/cm^3, the mass of the mercury is 2050 g, which is also the mass of the book.

REFERENCES

1. G. Towe and G. R. Caughlan, *Amer. J. Phys.* 29:706 (1961).
2. D. A. Davenport, *J. Chem. Educ.* 39:252 (1962).
3. L. C. Grotz and J. E. Gauerke, *J. Chem. Educ.* 48:337 (1971).
4. W. J. Deal, *J. Chem. Educ.* 52:405 (1975).
5. D. W. Miller, *J. Chem. Educ.* 54:245 (1977).
6. D. A. Davenport, *J. Chem. Educ.* 56:322 (1979).

5.6

Thermal Expansion
of Gases

A flask is warmed and an egg is seated in the mouth of the flask. When the flask is chilled in an ice bath, the egg pops into the flask. The flask is then inverted so that the egg lodges in the neck of the flask. When the flask is warmed, the egg pops out [1]. When liquid nitrogen is poured over a balloon, the balloon shrinks [2]. Other procedures for demonstrating the thermal expansion of gases are also described [3].

MATERIALS FOR PROCEDURE A

petrolatum or stopcock grease

hard-boiled egg, shelled, slightly larger in diameter than mouth of Erlenmeyer flask

1-liter Erlenmeyer flask

stand, with clamp for flask

Bunsen burner

ice-water bath large enough to accommodate the Erlenmeyer flask

MATERIALS FOR PROCEDURE B

cylinder of helium gas, with valve

cylinder of oxygen gas, with valve

50 mL liquid nitrogen

2 round rubber balloons, with minimum inflated diameter of 25 cm

ca. 2 m string

shallow dish, with diameter larger than that of inflated balloons

insulated gloves

MATERIALS FOR PROCEDURE C

petrolatum or stopcock grease

hard-boiled egg, shelled, slightly larger in diameter than mouth of Erlenmeyer flask

1-liter Erlenmeyer flask

evaporating dish, at least 20 cm in diameter

stand, with clamp for flask

insulated gloves

50 mL liquid nitrogen

MATERIALS FOR PROCEDURE D

several pea-sized chunks dry ice (solid carbon dioxide), CO_2

2 round rubber balloons, with minimum inflated diameter of 25 cm

insulated gloves

10-cm length plastic tubing, with outside diameter of 20 mm (optional)

2 mL liquid nitrogen (optional)

PROCEDURE A

Preparation

Lightly grease the inside of the neck of the 1-liter Erlenmeyer flask with petrolatum (e.g., Vaseline) or stopcock grease. Clamp the flask onto the stand. Light the Bunsen burner and adjust it to produce a luminous flame.

Presentation

Gently warm the 1-liter Erlenmeyer flask with the Bunsen burner for about 1 minute. While the flask is warm, seat the egg, narrow end down, in the mouth of the flask. Unclamp the flask and immerse it in the ice-water bath. The egg will pop into the flask.

Grasp the flask by the neck and invert it so the egg lodges in its neck. Gently heat the side of the flask with the burner and rotate the flask to avoid scorching the egg. The egg will be forced from the flask.

PROCEDURE B

Preparation

Inflate one balloon with helium and the other with oxygen. Tether the helium-filled balloon to the bench with the string.

Presentation

Wearing insulated gloves, hold the helium-filled balloon over the dish and pour liquid nitrogen over it. The balloon will shrink to a small fraction of its original size. Release the balloon while it is contracted. It will not rise, but will stay in the dish. After the liquid nitrogen has evaporated, the balloon will warm and expand, and it will soon begin to rise.

Hold the oxygen-filled balloon over the dish and pour liquid nitrogen over it. It will shrink to even a smaller size than did the helium-filled balloon. Pour off the liquid nitrogen, place the collapsed balloon in the dish, and allow the balloon to return to its original size.

PROCEDURE C

Preparation

Lightly grease the inside of the neck of the 1-liter Erlenmeyer flask with petrolatum (e.g., Vaseline) or stopcock grease. Place the flask in the evaporating dish and clamp it to the stand. Seat the hard-boiled egg, narrow end down, in the mouth of the flask.

Presentation

Wearing insulated gloves, pour about 50 mL of liquid nitrogen into the evaporating dish. The egg will pop into the flask.

Unclamp the flask and pour some of the liquid nitrogen into the flask. Invert the flask so that the egg lodges in its neck. The egg will be driven from the flask by the evaporating and expanding nitrogen.

PROCEDURE D

Preparation

If liquid nitrogen will be used, prepare one of the balloons by inserting one end of the plastic tubing into its mouth.

Presentation

Wearing insulated gloves, place several pea-sized chunks of the dry ice in the other balloon and quickly seal it. As the carbon dioxide vaporizes, it will inflate the balloon.

Wearing insulated gloves, pour 2 mL of liquid nitrogen through the tubing inserted into the first balloon. Remove the tubing and seal the balloon. As the nitrogen vaporizes, it will inflate the balloon.

HAZARDS

Liquid nitrogen is extremely cold, having a boiling point of $-196°C$. Skin contact with liquid nitrogen or with an object chilled by liquid nitrogen can result in severe frostbite. Protective gloves should be worn whenever liquid nitrogen is handled.

Dry ice, solid carbon dioxide, has a temperature of $-77°C$, and handling it with bare hands can result in frostbite. Protective gloves should be worn whenever it is handled.

DISCUSSION

The temperature dependence of the volume of a fixed amount of gas at constant pressure was first reported by Jacques Alexandre Cesar Charles (1746–1823) in 1787. Similar investigations were also carried out by Joseph Louis Gay-Lussac in 1802. These investigators discovered that, at constant pressure, the volume of a gas is inversely proportional to its absolute temperature. The procedures in this demonstration can be used to demonstrate Charles's law qualitatively.

In Procedure A, the gas within the flask is warmed. The warm gas is then confined in the flask when the flask is sealed with the egg. As the gas cools, it contracts and draws the egg into the flask. It is likely that the pressure of the gas inside the flask is not constant during this process, because there is some friction to be overcome in moving the egg through the neck of the flask. Because the egg pops into the flask rather than gradually sliding in, the volume of the gas remains constant for awhile as it cools. Because the volume is constant and the gas is cooling, the pressure of the gas must be diminishing. Thus the pressure of the gas in the flask is lower than atmospheric pressure outside the flask, and the excess atmospheric pressure begins to push the egg into the flask. The reverse is accomplished by heating the cooled gases trapped within the flask when the egg is lodged in the neck of the inverted flask; the expanding gases push the egg from the flask.

In Procedure B, the volume of the gas in the balloon decreases as the gas is chilled, causing the balloon to shrink. This is closer to a constant-pressure process than is the process in Procedure A. The pressure of the gas within the balloon is greater than atmospheric pressure by an amount sufficient to distend the rubber of the balloon. Charles's law predicts a fourfold decrease in the volume of a gas as it is chilled from room temperature, 20°C (293K), to liquid nitrogen temperature, −196°C (77K). If the gas within the balloon condenses, as oxygen gas does at −183°C, the volume change is much greater. The density of liquid oxygen at −183°C is 1.149 g/mL, and that of the gas at 0°C is 1.429 g/liter [4]. Therefore, the volume of the balloon can shrink by a factor of more than 700.

The magnitude of the expansion of a substance as it goes from a condensed phase to the gas phase can be illustrated with Procedures C and D. In Procedure C, the egg is forced from the flask by nitrogen gas as it vaporizes from the liquid. In Procedure D, a balloon is inflated by the gas produced as dry ice or liquid nitrogen vaporizes.

REFERENCES

1. L. A. Ford, *Chemical Magic*, T. S. Denison and Company: Minneapolis, Minnesota (1959).
2. R. Barnard, *J. Chem. Educ.* 41:A139 (1964).
3. P. G. Marlowe, *J. Chem. Educ.* 57:307 (1980).
4. R. C. Weast, Ed., *CRC Handbook of Chemistry and Physics*, 59th ed., CRC Press: Boca Raton, Florida (1978).

5.7

Charles's Law

The volume and the temperature of a fixed amount of gas at constant pressure are recorded as the gas is heated or cooled. The data illustrate Charles's law and can be used to estimate the value of the absolute zero temperature on the Celsius scale [1, 2].

MATERIALS FOR PROCEDURE A

syringe needle, ca. 18 gauge

1-holed rubber stopper to fit 250-mL Erlenmeyer flask

250-mL Erlenmeyer flask

thermometer, $-10°C$ to $+110°C$

100-mL graduated cylinder

stand, with 2 clamps and 1 ring

ice-water bath, sufficiently large to contain Erlenmeyer flask

30-mL glass syringe (lubricated with a drop of glycerine)

heat gun or hand-held hair dryer (optional)

MATERIALS FOR PROCEDURE B

ca. 400 mL tap water

200 mL ice

150 mL 2-propanol (isopropyl alcohol), $CH_3CHOHCH_3$

ca. 100 g dry ice (solid carbon dioxide), CO_2, in small chunks

10-mL glass syringe

plastic syringe cap

3 250-mL beakers

ring stand or tripod, with wire gauze

thermometer, $-50°C$ to $+110°C$

Bunsen burner

insulated gloves

MATERIALS FOR PROCEDURE C

1-liter beaker

stand, with clamp, ring, and wire gauze

1-liter glass cylinder

glass tubing, ca. 60 cm in length, with outside diameter of 7 mm

glass-working torch

1-holed rubber stopper to fit 125-mL Erlenmeyer flask

125-mL Erlenmeyer flask

thermometer, $-10°C$ to $+110°C$

Bunsen burner

50-mL gas-measuring buret

PROCEDURE A [1]

Preparation

Carefully insert the syringe needle through the body of the rubber stopper, so the tip of the needle is inside the Erlenmeyer flask when the stopper is seated in its mouth. Insert the thermometer through the hole in the rubber stopper. Adjust the position of the thermometer, so that its bulb is at the center of the flask when the stopper is seated in the mouth of the flask.

The volume of the 250-mL Erlenmeyer flask when the stopper assembly is in place can be determined to the nearest milliliter by filling the flask with water, stoppering it with the prepared stopper, then removing the stopper and measuring the volume of the water with the 100-mL graduated cylinder. Dry the flask.

Firmly seat the stopper in the mouth of the flask. The stopper must fit tightly to prevent any gas leakage from the flask. Clamp the flask to the stand. Position the ice-water bath to immerse as much of the flask as possible, and support the bath on a ring attached to the stand. Wait for the temperature of the gas in the flask to level off at its lowest value.

Depress the plunger until there is no air in the syringe, and attach the syringe to the top of the needle. Clamp the body of the syringe to the stand. The assembly is depicted in Figure 1.

Presentation

Remove the ice bath and allow the flask to warm gradually. As the gas in the flask warms, it will expand into the syringe, raising the plunger. For each 5°C increase in temperature, record both the temperature and volume of gas in the syringe. If a heat gun is available, the flask can be warmed and the measurements recorded for temperatures above room temperature. The volume of the gas in the syringe is equal to the increase in the volume of the gas as it warms. The total volume of the gas is determined by adding these increases to the initial volume of the gas in the flask. The data obtained

Figure 1.

by this procedure can be used to illustrate Charles's law, as described in the Discussion section.

PROCEDURE B [1]

Preparation

Fill the 10-mL syringe with 6 mL of air and seal the syringe with a plastic syringe cap. Pour 200 mL of water into the first 250-mL beaker, and place the beaker on a wire gauze on a ring stand or tripod. Heat the water to about 70°C with the Bunsen burner. Fill the second beaker with ice and add water until the beaker is full. Pour 150 mL of isopropyl alcohol into the third beaker, and, wearing insulated gloves, gradually stir in small chunks of dry ice until the beaker contains 200 mL of slush.

Presentation

Record the volume of air in the syringe and the temperature of the air around the syringe. Immerse the syringe in the hot water. After about 1 minute, record the volume of the gas in the syringe and the temperature of the water. Remove the syringe from the hot water and immerse it in the ice bath. After about 1 minute, record the volume of the gas in the syringe and the temperature of the ice bath. Remove the syringe from the ice bath and immerse it in the dry ice–isopropyl alcohol bath. Record the volume of the gas in the syringe and the temperature of the bath. The data obtained in this procedure can be used to illustrate Charles's law and to estimate the value of absolute zero temperature, as described in the Discussion section.

PROCEDURE C [2]

Preparation

Assemble the following apparatus as illustrated in Figure 2. Place the 1-liter beaker atop the wire gauze on the ring attached to the stand. Place the 1-liter glass cylinder next to the stand. Bend the piece of 7-mm glass tubing as shown in Figure 2. Insert the short bend of the tube through the rubber stopper, and place the long U-bend of the tube in the cylinder, as shown in Figure 2. Clamp the flask onto the stand so as much as possible of the flask is inside the beaker. Fill the beaker with water and immerse the thermometer in the water. Position the Bunsen burner under the beaker. Fill the cylinder with water to just below the opening of the glass tube.

If you use a standard buret, be sure its stopcock is closed. Lower the buret over the free end of the tube in the cylinder. Seat the stopper containing the other end of the tube in the mouth of the Erlenmeyer flask.

Presentation

Raise the buret so the level of liquid within it is the same as the level of the water in the cylinder. Record the volume reading on the buret and the temperature of the water bath. Make all future volume readings in this fashion, with the liquid levels equalized.

Light the Bunsen burner and heat the water bath. Stir the water in the bath. As the gas in the flask warms, it expands into the buret. Record the temperature and the volume of the gas in the buret for every 5°C increase in the temperature of the bath.

The volume of the gas at each temperature reading is the sum of the volume of the flask, the volume of the glass tube, and the volume of gas in the buret. The volume of

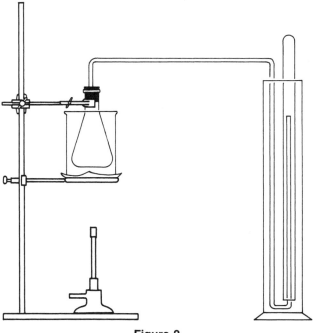

Figure 2.

the flask and tube can be determined by filling them with water and measuring the volume of the water using a 100-mL graduated cylinder. The data obtained by this procedure can be used to illustrate Charles's law and to estimate the value of absolute zero temperature, as described in the Discussion section.

HAZARDS

Solid carbon dioxide, dry ice, has a temperature of −78°C and can cause frostbite. Thermal protection in the form of gloves or a towel should be used when handling dry ice.

Because acetone and 2-propanol are flammable, they should be used away from open flames.

Syringe needles are sharp and potential sources of puncture wounds.

DISCUSSION

The temperature dependence of the volume of a fixed quantity of gas at constant pressure was first reported by Jacques Alexandre Cesar Charles (1746–1823) in 1787. This work was repeated by Joseph Louis Gay-Lussac in 1802 and is attributed at times to him as well as to Charles.

The data obtained from each of the procedures of this demonstration can illustrate Charles's law when they are plotted as the total volume of the gas versus temperature. Such a plot shows a linear relationship between temperature and volume of the gas, in agreement with Charles's law, which states that the volume of a gas is proportional to its absolute temperature. The slope of the line corresponds to the coefficient of thermal expansion of the gas. Most gases expand by 1/273 of their volume at 0°C for each Celsius-degree increase in their temperature. They contract by the same fraction for each Celsius-degree decrease in temperature. Therefore, the slope of a plot of volume versus temperature yields the coefficient of expansion of 1/273, or $0.00366°C^{-1}$. The temperature at the zero-volume intercept is called the absolute zero of temperature. Extrapolating the plot of volume versus temperature to zero volume yields a temperature of −273°C, the value of absolute zero on the Celsius scale.

REFERENCES

1. D. A. Davenport, *J. Chem. Educ.* 39:252 (1962).
2. D. T. Haworth, *J. Chem. Educ.* 44:353 (1967).

5.8

Determination of Absolute Zero

Air within a filter flask attached to a manometer is heated with a heat gun. The temperature of the air and the corresponding pressure are recorded periodically as the gas heats and then cools. The data collected can be used to determine a value for absolute zero on the Celsius temperature scale. An alternate procedure uses the change in the volume of a gas with changing temperature to estimate the value of absolute zero [1].

MATERIALS FOR PROCEDURE A

thermometer, −10°C to +110°C

1-holed stopper to fit 1-liter filter flask

1-liter filter flask

stand, with clamp

rubber tubing, ca. 60 cm in length

mercury-filled U-tube manometer

heat gun or hand-held hair dryer

MATERIALS FOR PROCEDURE B

2.8 liters tap water

2 2-liter beakers

tripod or ring stand, with wire gauze

Bunsen burner

baby bottle, with nipple

tongs

thermometer, −10°C to +110°C

PROCEDURE A

Preparation

Assemble the apparatus as illustrated in the figure. Carefully insert the thermometer into the 1-holed stopper, and seat the stopper in the mouth of the 1-liter filter flask. Adjust the position of the thermometer so its bulb is at the center of the flask. Clamp the flask onto the stand. Using the rubber tubing, attach the side arm of the filter flask to the U-tube manometer.

33

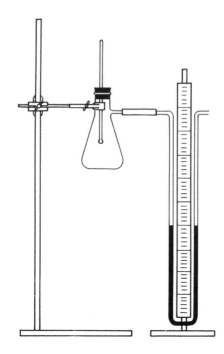

Presentation

Warm the entire outside surface of the filter flask with the heat gun. As the temperature of the gas increases, periodically record the temperature and the pressure of the gas. Once the temperature of the gas is over 110°C, stop heating the flask and allow it to cool. Record more temperature-pressure data as the gas inside the flask cools. The data obtained in this procedure can be used to determine a value for absolute zero on the Celsius scale, as described in the Discussion.

PROCEDURE B [1]

Preparation

Pour 1.4 liters of water into one of the 2-liter beakers. Set the beaker on a wire gauze on a tripod or ring stand, and heat the water to boiling with the Bunsen burner. While the water is heating, pour 1.4 liters of water into the other 2-liter beaker. Attach a nipple to a dry baby bottle.

Presentation

Holding the baby bottle with tongs, immerse it in the boiling water. Use the thermometer to measure the temperature of the water. After the bottle has been immersed for about 1 minute, quickly remove it from the hot water, invert it, and insert it, nipple down, into the second beaker of water. As the gas in the bottle cools, it will contract and draw water through the nipple into the bottle. After the bottle has been immersed for about a minute, use the thermometer to measure the temperature of the water. Lift the bottle from the water, still holding it inverted, and measure the volume of the gas in the bottle by using the graduations on the bottle. This is the volume of gas at the lower temperature. At the higher temperature, the gas filled the entire bottle, whose volume equals the sum of the volume of gas plus the volume of liquid it now contains. With the bottle upright, measure the volume of the liquid in the bottle using the bottle's graduations. Add this volume to the volume of the gas to get the volume of the gas at the higher temperature. These two temperature-volume data can be used to estimate the value of the absolute zero of temperature on the Celsius scale, as described in the Discussion section.

HAZARDS

Mercury is extremely toxic and should be handled with care to avoid prolonged or repeated exposure to the liquid or vapor. Continued exposure to the vapor may result in severe nervous disturbance, insomnia, and depression. Continued skin contact also can cause these effects as well as dermatitis and kidney damage. Mercury should be handled only in well-ventilated areas. Mercury spills should be cleaned up immediately by using a capillary attached to a trap and an aspirator. Small amounts of mercury in inaccessible places should be treated with zinc dust to form a nonvolatile amalgam.

DISCUSSION

The data collected in Procedure A can be used to illustrate the proportionality of temperature and pressure and to determine the value of absolute zero on the Celsius scale. When the pressure is plotted versus the temperature of the gas, the points lie in a straight line. The zero-pressure intercept of this line provides an estimate for the value of absolute zero on the Celsius scale.

There are a number of commercial devices which can be used in Procedure A of this demonstration. Most are composed of a metal sphere attached to a pressure gauge. The sphere can be heated or chilled in a bath, and by recording the temperature of the bath and the corresponding pressure of the gas, data similar to those from the procedure described above can be obtained. To use one of these commercial devices, follow the directions which accompany it.

The pressure of an ideal gas is directly proportional to its temperature. This can be represented by the equation

$$T = kP$$

where k is a proportionality constant. This equation applies only if the temperature is measured on an absolute temperature scale (i.e., one on which the value of absolute zero temperature is zero). The Celsius scale is obviously not an absolute temperature scale, because Celsius temperatures below zero are possible (and, in some climates, common!). The Celsius scale can be converted to an absolute scale by adding to it a number that increases the value of absolute zero to zero. The value of absolute zero on the Celsius scale is $-273.15°C$ [2]. By adding 273.15 to temperatures on the Celsius scale, an absolute temperature scale is obtained. This scale is called the Kelvin scale. On the Kelvin scale, absolute zero is 0K, and 0°C is 273.15K. If temperatures are expressed in Celsius, then the equation expressing the relationship between temperature and pressure of an ideal gas is

$$t + a = kP$$

or

$$t = kP - a$$

where t is the Celsius temperature, and a is the constant which converts the Celsius scale to an absolute one, namely 273.15. This equation indicates that the plot of the Celsius temperature of a gas versus its pressure is a straight line, with a slope of k and an intercept of $-a$.

Procedure B cannot be used to illustrate Charles's law, because only two points are obtained. However, if Charles's law behavior is assumed for the gas, an estimate of the value of absolute zero temperature can be obtained from a plot of the volume versus the temperature extrapolated to zero volume. The "charm" of this procedure is that absolute zero can be estimated by an experiment that could be carried out in the kitchen.

REFERENCES

1. D. A. Davenport, *J. Chem. Educ.* 46:878 (1969).
2. R. C. Weast, Ed., *CRC Handbook of Chemistry and Physics*, 59th ed., CRC Press: Boca Raton, Florida (1978).

5.9

Dependence of Pressure on the Amount of Gas

The pressure and mass of a gas in a container are measured for various amounts of gas. A plot of the pressure versus the mass illustrates that the pressure of a gas at constant volume and absolute temperature is proportional to the mass of the gas.

MATERIALS FOR PROCEDURE A

mass-of-air globe (e.g., Fisher #S41430 or Sargent-Welch #1514)

vacuum pump, with vacuum tubing

balance, with 0.01-g sensitivity

vacuum/pressure gauge (e.g., Fisher #S41374 or Sargent-Welch #S-39718)

bicycle pump

barometer (optional)

thermometer, $-10°C$ to $+110°C$ (optional)

MATERIALS FOR PROCEDURE B

hacksaw

file

empty butane-lighter refill can, with adapters

tire-pressure gauge

compressed gas cylinder of oxygen (O_2), nitrogen (N_2), or argon (Ar), with two-stage regulator

15-cm length heavy-walled rubber tubing

hose clamp (or 10 cm copper wire, 18 gauge)

vacuum pump

balance, with 1-mg sensitivity

barometer (optional)

thermometer, $-10°C$ to $+110°C$ (optional)

PROCEDURE A

Preparation and Presentation

Connect the mass-of-air globe to the vacuum pump and evacuate it. Weigh the globe to within 10 mg. Record the mass. Open the stopcock for a fraction of a second to readmit some air. Reweigh the globe and record the mass. Connect the globe to the vacuum/pressure gauge and record the pressure of the air in the globe.

Repeat the above procedure a number of times, each time admitting slightly more air into the globe after it has been evacuated.

Attach the globe to the bicycle pump and pump some air into it. Weigh the globe and record the mass. Connect the globe to the vacuum/pressure gauge and record the pressure in the globe.

Repeat the above procedure a number of times, each time pumping more air into the globe until the pressure exceeds the capacity of the gauge. Record atmospheric pressure from the barometer and room temperature from the thermometer.

The data obtained by this procedure can be used to illustrate that the pressure of a gas at constant temperature and volume is proportional to the mass of the gas, as described in the Discussion section. The atmospheric pressure can be determined.

PROCEDURE B [1]

Preparation

Using the hacksaw and file, modify one of the adapters that comes with the butane-lighter refill can to fit the tire-pressure gauge. Attach the adapter to the *empty* butane-lighter refill can.

Adjust the second stage of the two-stage regulator attached to the compressed gas cylinder to a pressure of about 60 psi. Connect the 15-cm piece of heavy-walled rubber tubing to the regulator and fasten it with the hose clamp.

Presentation

Insert the adapter into the vacuum tubing connected to the vacuum pump. Turn on the pump and depress the valve into the refill can to evacuate the can. After the can has been evacuated (about 20 seconds), release the valve and detach the adapter from the vacuum tubing. The can will automatically seal when the valve is released. Weigh the evacuated can to the nearest milligram and record the mass.

Open the valve on the compressed gas cylinder to flush air from the attached rubber tubing. Insert the adapter on the evacuated refill can into the rubber tubing and depress the valve on the can to fill the can with gas. After about 20 seconds, release the valve and remove the adapter from the tubing. Close the valve on the compressed gas cylinder.

Weigh the can and record the mass. Use the tire-pressure gauge to measure the pressure of the gas in the can. Record the pressure. Depress the valve to allow some gas to escape.

Repeat the steps of the previous paragraph a number of times until the pressure of the gas in the can falls below the minimum measurable with the pressure gauge, about

10 psi. Depress the valve on the can for about 30 seconds to allow the pressure within the can to equalize the atmospheric pressure. Record the mass of the can one last time. Record atmospheric pressure from the barometer and room temperature from the thermometer.

The data obtained in this procedure can be used to illustrate that the pressure of a gas at constant volume and temperature is directly proportional to the mass of the gas, as described in the Discussion section. The atmospheric pressure can be determined, and, if the demonstration is repeated with a second gas, the ratio of the molecular weights of the gases can also be determined.

DISCUSSION

The data obtained in both procedures of this demonstration can be used to illustrate that the pressure of a gas at constant volume and temperature is directly proportional to the mass of the gas. This can most readily be done by plotting the pressure of the gas in the container versus the mass of the gas. The mass of the gas is simply the difference between the weight of the filled container and that of the "empty," evacuated container. Both pressure gauges used in these procedures register the difference between atmospheric pressure and the pressure being measured. Therefore, the pressure of the gas in the can is equal to the sum of the pressure reading from the gauge plus atmospheric pressure.

If the atmospheric pressure has not been measured, then its value can be determined from the mass-pressure data collected in the demonstration. The pressure of the gas in the container, P, is equal to the pressure reading from the gauge, G, plus atmospheric pressure, A:

$$P = G + A$$

Substituting P from this equation for P in the ideal gas law equation,

$$PV = nRT$$

gives

$$(G + A)V = nRT$$

The number of moles of gas, n, is equal to the mass of the gas, m, divided by the molar mass of the gas, M.

$$n = \frac{m}{M}$$

Therefore,

$$(G + A)V = \frac{m}{M}RT$$

Rearranging this gives

$$G = \frac{mRT}{MV} - A$$

If the pressure reading from the gauge, G, is plotted versus the mass of the gas, m, then the plot will be a straight line with a slope of RT/MV and an m = 0 intercept of −A. Thus, the plot can be extrapolated to give an estimate of the atmospheric pressure, A.

If Procedure B is performed using two or more different gases, then the ratios of their molar masses can be determined. The data obtained with each gas can be plotted, pressure versus mass, and the slopes of the resulting lines are equal to RT/MV, where R is the gas constant, T the absolute temperature of the gas, M its molar mass, and V the volume of the container. The ratio of the slopes for two different gases is

$$\frac{R_1 T_1 / M_1 V_1}{R_2 T_2 / M_2 V_2}$$

where the subscript 1 refers to one gas and 2 to the other. Because $R_1 = R_2$, $V_1 = V_2$, and $T_1 = T_2$, the ratio of the slopes is equal to the inverse of the ratio of the molar masses.

$$\frac{slope_1}{slope_2} = \frac{M_2}{M_1}$$

If one of the gases is treated as a "known" and the other as an "unknown," the molar mass of the unknown can be determined from the ratio of the slopes and the molar mass of the known.

REFERENCE

1. G. M. Bodner, D. A. Davenport, and L. J. Magginnis, *J. Chem. Educ.* (in press).

5.10

Dalton's Law
of Partial Pressures

A sample of one gas is injected into a filter flask connected to a U-tube manometer, and the height of the mercury column is measured. The pressure is released, and the measurement is repeated with a different-sized sample of a second gas and then a third gas. Finally, identical samples of all three gases are injected into the filter flask, and the total pressure is measured. The data collected can be used to illustrate Dalton's law of partial pressures [1].

MATERIALS

1 bag each of helium (He), oxygen (O_2), and nitrogen (N_2) gases (See the introduction to this chapter, p. 4, for a description of these bags.)

5-cm length glass tubing, with outside diameter of 20 mm

solid rubber stopper to fit 500-mL filter flask

cork borer

septum cap to fit 20-mm glass tubing

10-cm length copper wire, 18 gauge

500-mL filter flask

5-cm length rubber tubing to fit tubulation of filter flask

mercury-filled open-armed U-tube manometer, with arms extending at least 12 cm above equilibrium level of mercury

30-mL syringe, with needle

PROCEDURE [1]

Preparation

Assemble the apparatus as illustrated in the figure. Bore a hole in the rubber stopper large enough to accommodate the 20-mm glass tube. Insert the glass tube into the hole in the stopper. Attach the rubber septum cap to the glass tube and fasten it by wrapping copper wire around the septum. Seat the stopper in the mouth of the 500-mL filter flask.

Connect the tubulation of the filter flask to the U-tube manometer with the rubber tubing.

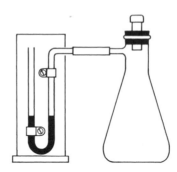

Presentation

Insert the free syringe needle through the septum cap to equalize the pressure inside the flask with the external atmospheric pressure. Remove the needle from the septum and attach it to the syringe. With the syringe, draw 25 mL of helium from the plastic bag. (See the introduction to the chapter, p. 4, for a description of this procedure.) Insert the needle of the syringe through the rubber septum and inject the helium into the flask. Record the difference in the heights of the two mercury columns in the manometer.

Repeat the procedure in the preceding paragraph using 15 mL of oxygen gas, and again with 30 mL of nitrogen gas. Then equalize the pressure in the flask and sequentially add 25 mL of helium, 15 mL of oxygen, and 30 mL of nitrogen. Record the difference in the levels of the mercury in the manometer. This last difference in levels will equal the sum of the differences produced by each gas individually, within the accuracy of the measurements.

HAZARDS

The pressure within the flask will rise to about 15% over atmospheric pressure. Therefore, the U-tube manometer must be capable of registering this pressure without spilling mercury. The distance from the top of the open arm of the manometer to the mercury level at atmospheric pressure should be at least 15% of 760 mm, namely, 120 mm. A manometer having a mercury reservoir at the top of its open tube is recommended to reduce the chances of mercury spillage.

Mercury is extremely toxic and should be handled with care to avoid prolonged or repeated exposure to the liquid or vapor. Continued exposure to the vapor may result in severe nervous disturbance, insomnia, and depression. Continued skin contact also can cause these effects as well as dermatitis and kidney damage. Mercury should be handled only in well-ventilated areas. Mercury spills should be cleaned up immediately by using a capillary attached to a trap and an aspirator. Small amounts of mercury in inaccessible placed should be treated with zinc dust to form a nonvolatile amalgam.

Syringe needles are sharp and potential sources of puncture wounds.

DISCUSSION

Among the numerous scientific contributions of John Dalton (1766–1844), we must include the law of partial pressures, first published in 1801, which states that the pressure of a mixture of gases is the sum of the pressures of the individual components.

$$P_{total} = P_1 + P_2 + P_3 + \ldots P_n$$

The data collected in this demonstration illustrate this relationship. Injecting 25 mL of He into the flask will increase the pressure of the gases in the flask by a measured amount, P_{He}. Likewise, injecting 15 mL of O_2 results in an increase of P_{O_2}, and the injection of 30 mL of N_2 produces an increase of P_{N_2}. When all three of these gases are injected at once, the pressure increase in the flask is $P_{He} + P_{O_2} + P_{N_2}$, illustrating that the total pressure is the sum of the individual pressures of these gases.

Three different gases and three different volumes of these gases are used in this demonstration. This need not be the case, although the clarity of the demonstration is increased by this procedure. If identical volumes of the three gases are used, then the pressure changes will be the same for all three. While this shows that the pressure change is independent of the identity of the gas, this may be an interfering factor in the interpretation of this demonstration in terms of Dalton's law. The additivity of the partial pressures is somewhat obscured when all three are the same. Using the same gas for all three samples obscures the identification of a particular pressure with a particular gas; that is, it obscures the "partial" nature of the pressure, because the mixture of two samples contains only one gas.

The pressure of a gas satisfies the definition of a colligative property, namely, one that depends on the number of molecules present and not on their nature or identity [2]. Thus, the pressure of a mixture of helium, oxygen, and nitrogen gases is equal to the pressure that would be exerted by the same number of molecules, an equimolar amount, of any *one* of these gases.

The concept of the mole allows us to count the number of atoms or molecules in a known *mass* of a *substance*. Concentration units such as molarity enable us to do the same for a known *volume* of a *solution*. Avogadro's law provides a link between the number of atoms or molecules and the *volume* of a *gas*. Dalton's law of partial pressures provides the final link between the number of atoms or molecules and the *pressure* of a *gas*.

Of course, Dalton's law of partial pressures holds only if the gases do not react. Results the same as those obtained here could not be obtained if we mixed, for example, $NH_3(g)$ and $HCl(g)$, or $NO(g)$ and $O_2(g)$.

REFERENCES

1. I. Talesnick, *Demonstrations in Chemistry*, Faculty of Education, Queen's University: Kingston, Ontario (undated).
2. S. Glasstone, *Textbook of Physical Chemistry*, 2d ed., Van Nostrand: New York (1946).

5.11

Avogadro's Hypothesis

A flask is weighed, filled with a measured volume of gas, and reweighed. This is repeated with several different gases. If the molecular masses of the gases are known, then these data can illustrate Avogadro's hypothesis [1]. An alternative procedure uses a syringe in place of the flask [2].

MATERIALS FOR PROCEDURE A

at least 3 cylinders of various compressed gases, such as oxygen (O_2), nitrogen (N_2), helium (He), methane (CH_4), hydrogen (H_2), carbon dioxide (CO_2), butane (C_4H_{10}), sulfur hexafluoride (SF_6), or argon (Ar)

2-way vacuum stopcock

1-holed rubber stopper to fit 125-mL flask

125-mL thick-walled Erlenmeyer or round-bottomed flask

vacuum pump, with hose

MATERIALS FOR PROCEDURE B

at least 3 bags of various gases, such as oxygen (O_2), nitrogen (N_2), helium (He), methane (CH_4), hydrogen (H_2), carbon dioxide (CO_2), butane (C_4H_{10}), sulfur hexafluoride (SF_6), or argon (Ar) (See the introduction to this chapter, p. 4, for a description of these bags.)

50-mL plastic syringe

iron nail, at least 5 cm in length

Bunsen burner

syringe cap

syringe needle

PROCEDURE A [1]

Preparation

Insert one arm of the stopcock through the 1-holed stopper and seat the stopper in the mouth of the Erlenmeyer or round-bottomed flask.

Presentation

Connect the free arm of the stopcock to the hose of the vacuum pump. Evacuate the flask by turning on the pump and opening the stopcock. After the flask is evacuated (about 30 seconds), close the stopcock and detach it from the hose. On a balance, weigh the flask to the nearest milligram.

Open the stopcock, loosen the stopper, and connect the stopper to a cylinder of oxygen. Pass oxygen gas through the flask for 15–30 seconds. Reseat the stopper, close the stopcock, and detach the flask from the cylinder. Weigh the flask to the nearest milligram.

Repeat the procedure in the preceding paragraph using other gases, such as nitrogen, helium, methane, hydrogen, carbon dioxide, butane, sulfur hexafluoride, and argon.

Subtract the mass of the "empty" flask from the masses of the flask containing each of the gases used. The resulting masses of the gases will be in the ratio of their molecular masses.

PROCEDURE B [2]

Preparation

Adjust the plunger of the plastic syringe to read a volume of 50 mL. Heat the nail with the Bunsen burner and force the hot nail through the plunger where the plunger meets the barrel, as shown in Figure 1. Remove the nail and push the plunger all the way into the syringe. Seal the syringe with the syringe cap.

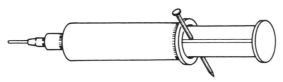

Figure 1.

Presentation

With the syringe tightly sealed, draw the plunger back until the nail can be inserted into the hole in the plunger. The syringe will be "empty." Weigh the empty syringe to the nearest milligram.

Remove the nail, attach the syringe needle, insert the needle through the septum of the bag of oxygen gas. Withdraw a 50-mL sample of oxygen. Remove the needle from the syringe, and replace it with the syringe cap. Insert the nail in the hole in the plunger. Weigh the syringe to the nearest milligram.

Repeat the procedure in the preceding paragraph with samples of other gases. Subtract the mass of the "empty" syringe from the masses of the syringe with gases. The resulting masses of the gases will be in the ratios of their molecular masses. Typical results of this procedure are given in the table.

Typical Results of Procedure B: Masses of 50-mL Samples of Various Gases

Gas	Molar mass (g/mole)	Calculated mass[a] (g)	Measured mass (g)	Number of molecules[b]
H_2	2.0	0.004	0.005	1×10^{21}
He	4.0	0.008	0.009	1×10^{21}
CH_4	16.0	0.032	0.041[c]	1.5×10^{21}
N_2	28.0	0.057	0.055	1.2×10^{21}
air	28.8	0.059	0.056	1.2×10^{21}
O_2	32.0	0.065	0.061	1.2×10^{21}
Ar	39.9	0.081	0.081	1.2×10^{21}
CO_2	44.0	0.089	0.088	1.2×10^{21}
C_4H_{10}	58.0	0.118	0.111	1.15×10^{21}
CCl_2F_2	120.9	0.245	0.228	1.14×10^{21}

[a] At 1 atm and 20°C, as calculated using the ideal gas law.

[b] As calculated from the measured mass, assuming 6.023×10^{23} molecules in one molar mass of gas.

[c] Discrepancy between calculated and measured masses may be due to the use of natural gas as the source of methane.

HAZARDS

The flask, like any evacuated container, must be regarded as an implosion hazard. Wrapping the flask with tape is recommended to prevent injury from flying glass shards should the flask implode.

Syringe needles are sharp and potential sources of puncture wounds.

DISCUSSION

In 1811, Avogadro proposed his hypothesis, which stated that equal volumes of different gases at the same temperature and pressure contain equal numbers of molecules. The samples of gas used in each procedure of this demonstration have the same volume and are at the same temperature and pressure. The results of this demonstration, combined with the known molecular masses of each of the gases, can illustrate Avogadro's hypothesis. For example, the molecular mass of oxygen is 16 times that of hydrogen, as determined through other independent experimental methods. Because the measured mass of a sample of oxygen gas is about 16 times the measured mass of an equivolume sample of hydrogen at the same temperature and pressure, the two samples of gas must contain the same number of molecules.

According to Avogadro's hypothesis, the samples of gas used in this demonstration contain the same numbers of molecules. Because the masses of these samples are different, the masses of the individual molecules of each gas must be different. Furthermore, the ratio of the masses of any two samples of gas must be the ratio of the masses of their molecules. Therefore, this demonstration allows the determination of the relative molecular masses of the gases used.

When the measured masses of the samples of gas are plotted versus their known molecular masses, a straight-line correlation is found. Figure 2 contains such a plot of

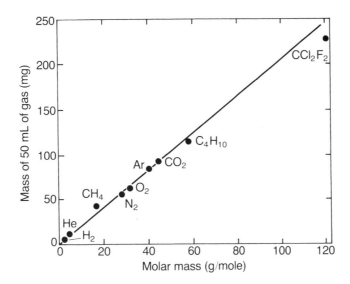

Figure 2. Plot of measured mass of 50 mL of gas versus molar mass of the gas.

the data in the table. This plot can be used to determine the molecular mass of a gas when its molecular mass is "unknown." One of the samples used may remain unidentified during the presentation of the demonstration, and its molecular mass determined by interpolation from the results with the identified samples.

REFERENCES

1. W. J. Deal, *J. Chem. Educ.* 52:405 (1975).
2. I. Talesnick, *Sci. Teacher* 44:47 (1977).

5.12

Determination of the Molecular Mass of the Gas from a Butane Lighter

The molar mass of the gas in a butane lighter is determined [1].

MATERIALS

1 liter tap water

butane-fueled lighter (or a canister of gas for refilling such a lighter)

60-cm length glass tubing, with outside diameter of 6 mm

glass-working torch

40-cm length plastic or rubber tubing to fit glass tubing

1-liter glass cylinder

50-mL gas-measuring tube (eudiometer)

thermometer, $-10°C$ to $+110°C$, with wire hook

pinch clamp

balance, with 1-mg sensitivity

barometer

PROCEDURE

Preparation

Bend the glass tubing into a gas-delivery tube as illustrated in the figure. Attach the plastic tubing to the gas-delivery tube.

Fill the 1-liter glass cylinder with tap water. Fill the gas-measuring tube with tap water, and cover its opening with a finger so that no air is in the tube. Invert the tube and lower it into the cylinder of water, being careful to admit no air into the tube. Suspend the thermometer in the cylinder by hanging it from the rim.

Securely connect the free end of the plastic tubing to the opening of the butane lighter so there is no gas leakage when the lighter valve is opened.

Presentation

Insert the gas-delivery tube about halfway down into the water-filled cylinder. Open the valve of the lighter to show the bubbles of gas escaping from the opening of the gas-delivery tube and to flush the tubing with gas. Leaving the open end of the tube

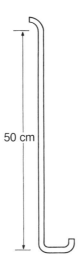

50 cm

under water, use a pinch clamp to seal the connected plastic tubing as close to the lighter as possible.

Detach the lighter from the tubing and weigh it to the nearest milligram. Record the mass and reattach the lighter to the tubing.

Push the open end of the gas-delivery tube to the bottom of the cylinder and position the mouth of the gas-measuring tube over the end of the gas-delivery tube. Open the valve of the lighter and allow 35–45 mL of the gas to collect in the gas-measuring tube. Make sure that all of the bubbles of gas released are collected in the gas-measuring tube. Close the lighter valve, detach the lighter from the tubing, and reweigh it. Record the mass of the lighter, along with the temperature of the water in the cylinder. To equalize the pressure of the gas in the gas-measuring tube with atmospheric pressure, adjust the gas-measuring tube so the levels of water inside and outside the tube are the same. Then, record the volume of the gas in the tube. Use a barometer to measure atmospheric pressure.

From the data obtained in this procedure, the molar mass of the gas from the lighter can be calculated, as described in the Discussion section.

HAZARDS

Butane is flammable, and therefore, the demonstration should be performed away from open flames.

DISCUSSION

The data collected in this demonstration can be used to determine the molar mass of the gas from the lighter. This can be done by assuming that the gas behaves as an ideal gas. According to the ideal gas equation,

$$PV = nRT$$

where P is the pressure of the gas,

V its volume,

n the number of moles of gas,

T its absolute temperature, and

R the gas constant.

The number of moles of a gas is equal to the mass of the gas, m, divided by its molar mass, M; that is, n = m/M. Therefore,

$$PV = \frac{m}{M}RT$$

and

$$M = \frac{mRT}{PV}$$

This last equation may be used to calculate the molar mass of the gas. If the value of R = 0.0821 liter·atm/mol·K is used, then the volume must be expressed in liters, pressure in atmospheres, temperature in Kelvin. If the mass of gas is expressed in grams, then the molar mass of the gas will be expressed in grams per mole. The mass of the gas collected in the tube is given by the difference between the masses of the lighter before and after the gas was delivered. The volume of the gas is the volume measured with the gas-measuring tube. The temperature of the gas is assumed to be that of the water it was bubbled through. The pressure of the gas inside the gas-measuring tube is the same as atmospheric pressure. However, the tube contains not only gas from the lighter, but also water vapor. The pressure of the gas from the lighter is the difference between the measured total pressure and the vapor pressure of water at the measured temperature. This vapor pressure may be found tabulated in many reference works, for example, the *CRC Handbook of Chemistry and Physics*.

The molecular formula of butane is C_4H_{10}. This corresponds to a molar mass of 58 g/mole. The molar mass calculated from the data obtained in this demonstration may be less than this, as low as 45 g/mole. This may be because the gas in the lighter is not pure butane, but a mixture containing lighter gases. When pure butane (e.g., from a gas cylinder) is used in this demonstration, the calculated molecular mass is 57 ± 2 g/mole.

REFERENCE

1. D. A. Davenport, *J. Chem. Educ.* 53:306 (1976).

5.13

Determination of the Molecular Mass of a Volatile Liquid: The Dumas Method

A small quantity of a liquid with a boiling point below 100°C is placed in a previously weighed flask, and the flask is covered with aluminum foil in which a pinhole has been made. The flask is immersed in a boiling-water bath until all of the liquid has vaporized. The flask is removed from the bath, allowed to cool to room temperature, and weighed. The flask is filled to the brim with water, and the volume of the water is measured in a graduated cylinder. From the data obtained and the temperature of the water bath and atmospheric pressure, the molar mass of the gas is determined [1, 2].

MATERIALS

750 mL tap water

10 drops concentrated (12M) hydrochloric acid, HCl

2 mL of a liquid having a boiling point below 100°C, such as acetone (CH_3COCH_3), ethanol (CH_3CH_2OH), ether ($CH_3CH_2OCH_2CH_3$), 2-propanol ($CH_3CHOHCH_3$), hexane (C_6H_{14}), or dichloromethane (CH_2Cl_2)

1-liter beaker

hot plate

boiling chips

15-cm test tube

ring stand, with clamp and holder

thermometer, $-10°C$ to $+110°C$

250-mL Erlenmeyer flask

aluminum foil, 5 cm × 5 cm

pin

balance, with 1-mg sensitivity

100-mL graduated cylinder

barometer

PROCEDURE

Preparation

Pour 750 mL of water into the 1-liter beaker and place the beaker on the hot plate. Add a few boiling chips to the beaker to prevent bumping, and add a few drops of 12M hydrochloric acid to prevent scale formation when the water is boiled. Clamp the test tube over the beaker. Place the thermometer in the beaker. Heat the water to boiling on the hot plate.

Presentation

Cover the 250-mL Erlenmeyer flask with the piece of aluminum foil. Make a small pinhole in the center of the foil. Weigh the flask to the nearest milligram. Remove the foil and add 2 mL of a volatile liquid to the flask. Replace the aluminum foil.

Immerse the flask in the boiling water so that as much as possible of its surface is covered without allowing any water to enter the flask. Adjust the test tube and clamp to keep the flask submerged. The volatile liquid in the flask will vaporize and expel all of the air from the flask. Schlieren lines will be visible above the pinhole as the vapors escape. Record the temperature of the water bath.

When all of the liquid has vaporized, quickly remove the flask from the bath. The vapor will condense on the walls of the flask almost immediately. Do not reimmerse the flask once it has been removed; to do so would produce unreliable results. Allow the flask to return to room temperature. Dry the outside of the flask, and weigh it with its contents and its aluminum cap. Drain the few remaining drops of liquid from the flask and fill the flask to the brim with water. Measure the volume of the water in the flask to the nearest milliliter with the 100-mL graduated cylinder. Record the barometric pressure. From the data collected, the molar mass of the vapors, and therefore of the liquid, can be determined, as described in the Discussion section.

HAZARDS

Because many of the liquids that can be used are flammable, the demonstrations should be performed away from open flames.

Dichloromethane vapor is narcotic in high concentrations and is irritating to the eyes.

Concentrated hydrochloric acid can produce skin burns, eye irritation, and irritation of the respiratory system.

DISPOSAL

The liquid remaining in the flask should be disposed of in a waste-solvent receptacle or allowed to evaporate in a hood.

DISCUSSION

If ideal gas behavior is assumed, the number of moles of vapor in the flask is given by

$$n = \frac{PV}{RT}$$

where P is the pressure of the vapor,
  V its volume,
  T its absolute temperature, and
  R the gas constant.

The vapors fill the flask when the flask is in the water bath. At that time, the volume of the vapor is equal to the volume of the flask, the temperature of the vapor is equal to the temperature of the water bath, and the pressure of the vapor is equal to pressure at the opening of the flask, namely, atmospheric pressure. Because all of these have been measured, the number of moles of vapor can be calculated.

The mass of the vapor in the flask is equal to the mass of the liquid which condensed from these vapors after the flask was removed from the water bath. The approximate value of this mass can be found by subtracting the mass of the "empty" flask from the combined mass of the flask and liquid, which was measured after the flask was removed from the bath. Thus, the mass of a certain amount of vapor and the number of moles in that amount of vapor are determined. From these, the mass of the vapor per mole can be calculated by dividing the mass by the number of moles.

The approximate mass of the liquid obtained by subtracting the mass of the "empty" flask from the combined mass of the flask and liquid will give a reasonably accurate value for the molar mass of the liquid. This mass is "approximate" because, when the flask is weighed the second time, it contains a small amount of liquid and the vapor of that liquid. When the flask is weighed the first time, it contains only air. Therefore, the flask and its contents displace more air the second time it is weighed than the first time. This results in a greater buoying effect because of the displaced air.

This buoying effect can be compensated for, if the additional volume of displaced air is known. The mass of the displaced air must be added to the "approximate" mass of the liquid to obtain its true mass. The additional volume of displaced air is equal to the volume of the condensed liquid in the flask plus the partial volume of the vapor in the flask. In general, the volume of the liquid is very small, and its effect is negligible; however, it may be measured if desired. The partial volume of the vapor is a more significant factor, especially if at room temperature the vapor pressure of the liquid is high. The partial volume of the vapor is equal to the volume of the flask, multiplied by the vapor pressure of the liquid, divided by the total pressure of the gas in the flask.

$$V_v = V_f \frac{P_v}{P_t}$$

If the identity of the liquid is known, its vapor pressure may be found in a reference, such as the *CRC Handbook of Chemistry and Physics*. The mass of the air displaced by the vapor of the liquid can be calculated using the ideal gas equation.

$$m = \frac{PV_v}{29RT}$$

In this equation, 29 is the average molar mass of air, in g/mol. This mass must be added to the mass of the "approximate" mass of the condensed liquid to correct for the buoyancy of the air displaced by the vapor of the liquid in the flask.

Commercial "Dumas flasks" are available from a number of sources. These flasks have a small opening at the end of a narrow tubular neck. They may be used in place of the Erlenmeyer flask and aluminum foil described in the procedure.

REFERENCES

1. J. B. Dumas and P. Boullay, *Annales de Chimie* 37:15 (1828).
2. D. P. Shoemaker and C. W. Garland, *Experiments in Physical Chemistry*, McGraw-Hill: New York (1962).

5.14

Flow of Gases Through a Porous Cup

Samples of gas are allowed to flow through a porous cup. The relative rates of flow cause liquid to be forced from a flask or air to be drawn into it [1, 2].

MATERIALS

300 mL tap water

3–5 drops food coloring (or other water-soluble dye)

cylinder of helium, with valve and tubing (or cylinder of hydrogen, or natural gas tap)

ca. 5-g chunk dry ice (solid carbon dioxide), CO_2

> *or*
>
> 10 g sodium bicarbonate (baking soda), $NaHCO_3$
>
> 10 mL vinegar (5% aqueous acetic acid, $HC_2H_3O_2$)

25-cm length glass tubing, with outside diameter of 7 mm

glass-working torch

2-holed rubber stopper to fit 500-mL Erlenmeyer flask

500-mL Erlenmeyer flask

5-cm length glass tubing, with outside diameter of 7 mm

10-cm length glass tubing, with outside diameter of 7 mm

1-holed rubber stopper to fit porous ceramic cup

porous ceramic cup, dry (e.g., Coors #60495 or as supplied in Sargent-Welch #0537)

30-cm length rubber tubing to fit 7-mm glass tubing

600-mL beaker

insulated gloves

PROCEDURE

Preparation

Form a 25-cm piece of glass tubing into a nozzle, as illustrated in Figure 1. Make a 45° bend about 5 cm from one end of the tubing, and constrict the opening in this end to about 0.5 mm.

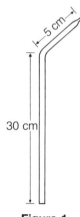

Figure 1.

Assemble the apparatus as illustrated in Figure 2. Insert the unconstricted end of the glass nozzle through one of the holes in the 2-holed rubber stopper so that it is within 1 cm of the bottom of the flask when the stopper is seated in the mouth of the flask. Insert a 5-cm piece of glass tubing through the other hole in the stopper and a 10-cm piece through the hole of the 1-holed stopper. Seat the 1-holed stopper in the porous ceramic cup. Connect the two short pieces of glass tubing with the 30-cm piece of rubber tubing.

Pour 300 mL of tap water into the flask and color it with a few drops of food coloring or other water-soluble dye, such as methylene blue. Seat the 2-holed stopper in the flask.

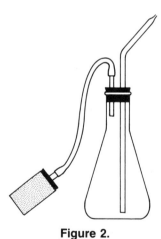

Figure 2.

Presentation

Invert a 600-mL beaker and fill it with helium, hydrogen, or natural gas (methane) by directing a gentle flow of the gas upward into the beaker for 10–15 seconds. Insert the porous cup into the inverted gas-filled beaker. Water from the flask will rise in the nozzle and squirt out its tip. Remove the cup from the beaker and the squirting stops.

Replace the cup and the squirting resumes. Turn the beaker upright for several seconds, then lower the cup into it. There will be no squirting.

Wearing gloves, place a small chunk of dry ice in the upright beaker, and allow about 30 seconds for carbon dioxide gas to fill the beaker. Alternatively, place about 10 g of sodium bicarbonate in the beaker and add about 10 mL of vinegar; the reaction between these two will fill the beaker with CO_2 gas. Lower the porous cup into the beaker. Bubbles of air will be drawn into the flask through the nozzle. Remove the cup from the beaker and the bubbling stops. Replace the cup in the beaker and the bubbling resumes.

HAZARDS

Solid carbon dioxide, dry ice, has a temperature of $-78°C$ and can cause frostbite. Thermal protection in the form of gloves or a towel should be used when handling dry ice.

Because water will be ejected from the nozzle for a considerable distance, care should be taken to avoid damage to the surroundings from the water jet.

DISPOSAL

Solutions should be flushed down the drain with water.

DISCUSSION

This demonstration illustrates qualitatively that low-molecular-weight gases flow more rapidly through a porous cup than do heavy gases. When the cup is inserted into the beaker of light gas (helium, hydrogen, or methane), the light gas flows into the cup more rapidly than the air within the cup flows out. Therefore, the pressure of the gases inside the cup and inside the flask increases. This increased pressure causes the liquid in the flask to rise in the tube and to be forced out through the nozzle. The lighter the gas, the greater the force on the liquid. Therefore, the jet of water produced when hydrogen or helium is used will travel farther than the one produced when methane is used.

When the porous cup is lowered into the beaker of carbon dioxide, the air in the cup flows outward more rapidly than the carbon dioxide flows inward. Therefore, the pressure of the gases in the cup and in the flask decreases. The pressure of the atmosphere than forces air into the flask to replace that flowing out through the cup.

The fountain effect produced with a light gas and the bubbling produced with carbon dioxide do not continue indefinitely for as long as the cup is held in the beaker. When the cup has been held in the beaker of helium for several minutes, the jet of water will gradually diminish and stop. The flow of gas into the cup is rapid initially, while the flow out is slow, producing a large pressure difference and a strong jet of water. The flow of helium into the cup increases the concentration of helium in the cup, and as the concentration of helium within the cup increases, the rate of flow out of the cup also increases. This results in a diminished pressure differential and in a smaller jet. Eventually, the rates of flow into and out of the cup become equal; when no pressure differential exists, the jet stops.

According to the kinetic molecular theory of gases, molecules at the same temperature have the same average kinetic energy, $\bar{e} = \frac{1}{2}mu^2$, where u is their average velocity. Therefore, for two gases at the same temperature, $m_1u_1^2 = m_2u_2^2$, or

$$\frac{u_1}{u_2} = \sqrt{\frac{m_2}{m_1}}$$

The ratio of the average speed of hydrogen molecules, with a molecular mass of 2.0, to that of the molecules in air, with an average molecular mass of 29, is $\sqrt{29/2}$, which is 3.8. Thus, when the ceramic cup containing air is placed in the beaker of hydrogen, the hydrogen flows into the cup nearly four times faster than the air flows out.

REFERENCES

1. H. N. Alyea and F. B. Dutton, Eds., *Tested Demonstrations in Chemistry*, 6th ed., Journal of Chemical Education: Easton, Pennsylvania (1965).
2. K. D. Schlecht, *J. Chem. Educ.* 61:251 (1984).

5.15

Ratio of Diffusion Coefficients: The Ammonium Chloride Ring

Ammonia vapor is introduced at one end of an air-filled tube, and hydrogen chloride gas at the other. After about 20 minutes, a ring of solid ammonium chloride forms inside the tube, nearer the end at which the hydrogen chloride was introduced [1].

MATERIALS

20 mL concentrated (12M) hydrochloric acid, HCl

20 mL 6M aqueous ammonia, NH_3 (To prepare 1.0 liter of stock solution, pour 400 mL of concentrated [15M] NH_3 into 400 mL of distilled water, and dilute the resulting solution to 1.0 liter with distilled water.)

cork borer

2 solid rubber stoppers (#3) to fit 22-mm glass tubing

2 cotton balls

1-m length glass tubing, with outside diameter of 22 mm

2 stands, with clamps

2 50-mL beakers

meter stick (optional)

PROCEDURE

Preparation

Use a cork borer to hollow the narrow ends of two rubber stoppers to a depth of about 1 cm, as shown in the figure. Place a small cotton ball in the hollow of each stopper.

Label one end of the glass tube "HCl" and the other "NH_3." Clamp the tube horizontally between two stands.

Presentation

Dip one of the stoppers holding a cotton ball into a 50-mL beaker containing the 12M HCl and the other into a 50-mL beaker containing the 6M NH_3. Simultaneously insert the two stoppers into the appropriate ends of the glass tube. After 15–20 minutes, a ring of white solid (ammonium chloride) will form on the inside of the tube. The ring will be closer to the HCl end of the tube than to the NH_3 end.

If the distance of the ring from each end of the tube is to be measured, it should be done soon after the ring forms, because the ring will drift with time.

HAZARDS

Concentrated hydrochloric acid can cause severe burns. Hydrochloric acid vapors are extremely irritating to the skin, eyes, and respiratory system.

Concentrated aqueous ammonia solution can cause burns and is irritating to the skin, eyes, and respiratory system.

DISPOSAL

The glass tube should be rinsed with water and the rinse flushed down the drain with water.

DISCUSSION

Diffusion is the process by which two gases mix in a system of uniform pressure. If the concentration is not uniform in a mixture of two gases, the gases diffuse into one another until the composition is uniform. According to Graham's law of diffusion, the rate of diffusion of a gas is inversely proportional to its molecular mass [2]. Therefore, the ratio of the diffusion rates of two gases is equal to the square root of the inverse of the ratio of their molecular masses.

$$\frac{\text{rate}(1)}{\text{rate}(2)} = \sqrt{\frac{m_2}{m_1}}$$

The molecular mass of ammonia is 17 amu, and that of hydrogen chloride is 36.5 amu. Therefore, the calculated ratio of the rate of diffusion of NH_3 to that of HCl is $\sqrt{36.5/17}$, which is 1.46. This suggests that NH_3 molecules should travel 1.46 times as far as HCl molecules in a given amount of time. From this, the ring of ammonium chloride should form 1.46 times as far from the NH_3 end of the tube as from the HCl end. However, the ratio most often obtained in this demonstration is 1.27 [1]. This ratio is the ratio of the diffusion coefficients of NH_3 and HCl, not the ratio of the rates of diffusion.

The redistribution of the molecules in a nonuniform mixture of two gases is accomplished through a net flow of molecules from one location to another. *Flow* is defined as the amount of material transported in unit time through a unit of area perpendicular to the direction of the flow. Therefore, the flow of a gas may have the units of mole/cm^2 · sec, and is represented by the differential equation

$$J = -D\frac{dc}{dz}$$

where J is the flow,

D is the diffusion coefficient of the gas,

c is the concentration of the gas molecules, and

z is the distance.

The minus sign is used because the flow is in the opposite direction from the increasing concentration gradient, dc/dz.

A number of factors in addition to the concentration gradient affect the flow rate of a gas, that is, affect the value of the diffusion coefficient. One is the average speed of its molecules. It would seem at first that the greater the average speed of its molecules, the faster a gas flows. However, the faster its molecules move, the more often they collide with other molecules, and these collisions retard the flow of the gas. Thus the diffusion coefficient depends upon the average speed of the molecules and on their mean free path (i.e., the distance between collisions). The mean free path of a molecule depends upon the size of the molecules around it, that is, on the collisional cross section of the molecules: large molecules have larger cross sections than small molecules. The average speed of gas molecules depends upon their mass and on the temperature of the gas. The diffusion coefficient of gas 1 into gas 2 is given by the equation

$$D_{12} = K \sqrt{\frac{1}{m_1} + \frac{1}{m_2}} \frac{\sqrt{T}}{S_{12}}$$

where m_1 and m_2 are the molecular masses of the gases,

S_{12} is the collisional cross section between molecules 1 and 2, and

K is a constant [1].

The reason that the results obtained from this demonstration are not what might be predicted on the basis of Graham's law of diffusion is that the process occurring in the tube is not a simple diffusion. In this demonstration there are three gases involved—ammonia, hydrogen chloride, and air. Ammonia is flowing through air, and hydrogen chloride is flowing through air. The ratio of the distance travelled by NH_3 to that travelled by HCl in this demonstration is the ratio of the flow rate of NH_3 through air to that of HCl through air. The ratio of these flow rates is equal to the ratio of the diffusion coefficients of each of these gases into air. This ratio is given by the following equation [1]:

$$\frac{D_{12}}{D_{13}} = \left(\frac{M_3}{M_2}\right)^{\frac{1}{2}} \left(\frac{(M_1 + M_2)}{(M_1 + M_3)}\right)^{\frac{1}{2}} \frac{S_{13}}{S_{12}}$$

In this equation,

D_{12} = diffusion coefficient of NH_3 into air;

D_{13} = diffusion coefficient of HCl into air;

M_1 = average molecular mass of air;

M_2 = molecular mass of NH_3;

M_3 = molecular mass of HCl;

S_{12} = cross-section for air-NH_3 molecular collisions;

S_{13} = cross-section for air-HCl molecular collisions.

Using an average molecular mass for air of 29.0 and a ratio of S_{13}/S_{12} of 1.04, the above equation gives a ratio of D_{12}/D_{13} equal to 1.28 [1]. This ratio is in very close agreement with the results of this demonstration.

REFERENCES

1. E. A. Mason and B. Kronstadt, *J. Chem. Educ.* 44:740 (1967).
2. T. Graham, *Phil. Mag.* 2:175, 269, and 351 (1833).

5.16

Molecular Collisions: The Diffusion of Bromine Vapor

Bromine vapor is allowed to fill two containers, one containing air and the other evacuated. The bromine fills the first container only gradually, while it fills the second almost instantaneously, illustrating the effect of molecular collisions on the rate of gas diffusion [1, 2]. The same effect is shown by the reverse process of condensing bromine vapor into the solid.

MATERIALS FOR PROCEDURE A

2–3 mL bromine

2 test tubes, 10 mm × 75 mm (or smaller)

2 forceps

glass-working torch

gloves, plastic or rubber

either

> 2 ignition test tubes, 25 mm × 200 mm
>
> solid rubber stopper to fit ignition tube
>
> 1-holed rubber stopper to fit ignition tube
>
> stopcock

or

> special evacuable tube (See Procedure A for description and materials needed.)
>
> gloves, plastic or rubber
>
> 1-liter glass cylinder
>
> glass plate to cover cylinder
>
> stopcock lubricant

vacuum pump, with hose

MATERIALS FOR PROCEDURE B

200 mL acetone, CH_3COCH_3

100 g dry ice (solid carbon dioxide), CO_2

63

600-mL beaker

insulated gloves

pair of bromine diffusion tubes (e.g., Sargent-Welch #4424)

   or

10 mL bromine

glass-working torch

2 35-cm lengths heavy-walled glass tubing, with outside diameter of ca.
   25 mm

2 10-cm lengths heavy-walled glass tubing, with outside diameter of ca.
   10 mm

gloves, plastic or rubber

vacuum pump, with hose

PROCEDURE A

Preparation

Working in a hood, seal two samples of liquid bromine in ampules. The ampules
can be made from two small test tubes by holding the two ends of a tube with two pairs
of forceps, heating the tube about 1 cm from its mouth with a glass-working torch until
the glass is soft, and pulling the forceps in opposite directions to form a constriction
in the tube (see Figure 1). Allow the tubes to cool before proceeding. Wearing gloves,

Figure 1.

fill the tubes one-third full with liquid bromine. Quickly heat the constriction in each
tube to soften the glass, and then pull the open end of the test tube with forceps to seal
the ampule. Do not thicken the glass at the seal, because the ampule must break open
easily during the demonstration.

The demonstration can be presented using either a specially constructed apparatus, described in the next paragraph, or thick-walled ignition test tubes. If the test
tubes are to be used, then carefully place one ampule apiece into each of two test tubes.
Seal one of the tubes with the solid rubber stopper. Insert one arm of the stopcock
through the 1-holed stopper and seal the other tube with this stopper. Attach the free
end of the stopcock to a vacuum pump, open the stopcock, evacuate the tube, then
close the stopcock.

For large rooms, a special apparatus can be constructed for this demonstration. It
is assembled from the parts shown in Figure 2. Part A is made of glass tubing with an
outside diameter of 23 mm. Seal closed one end of a 40-cm piece of this glass tubing,
and about 10 cm from the open end, seal on a 6-cm piece of the tubing. Seal two outer
standard-taper joints (19/38 or 24/40) onto the two openings in this tube assembly. Part
B is constructed by sealing an inner standard-taper joint onto one end of a 15-cm piece

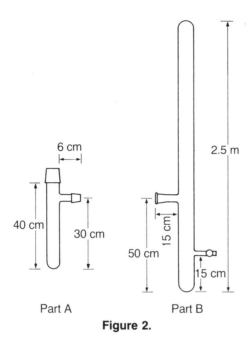

6 cm

2.5 m

40 cm

30 cm

15 cm

50 cm

15 cm

Part A

Part B

Figure 2.

of glass tubing with an outside diameter of 23 mm. The other end of the tubing is sealed onto a 2.5-m piece of glass tubing with an outside diameter of 45 mm, about 50 cm from one end. Seal a stopcock onto the 2.5-m glass tube about 15 cm from its end. Seal both ends of this tube closed and mount it on a white board about 2 cm thick, 7–10 cm wide, and 2.5 m long.

If the special apparatus is to be used, carefully place an ampule of bromine in Part A through its end opening and seal it with a greased, glass, standard-taper stopper. Lubricate the standard-taper joint on the short arm of Part A and insert it into the taper joint on Part B. Attach the free end of the stopcock to a vacuum pump, open the stopcock, and evacuate the entire apparatus. Close the stopcock.

Presentation

If the two ignition tubes are being used, pick up the tube containing the solid stopper (the tube containing air at atmospheric pressure) and give it a quick shake to break the ampule of bromine. The reddish brown vapor of bromine will fill the tube gradually, requiring 5–10 minutes to produce a uniform intensity throughout the tube. Then pick up the evacuated tube and shake it to break the ampule of bromine it contains. The reddish brown bromine vapors will fill the tube almost instantaneously.

If the special apparatus is being used, follow the procedure in this paragraph. Wearing gloves, break open an ampule of bromine and drop it into a 1-liter glass cylinder. Cover the cylinder with a glass plate. Reddish brown bromine vapor will gradually fill the cylinder from the bottom, requiring about an hour to be visible halfway up the cylinder. Break the ampule of bromine in the special apparatus by grasping Part A and quickly rotating it at the joint it shares with Part B, so that the ampule falls onto the glass stopper. The reddish brown bromine vapors will fill the apparatus virtually instantaneously.

PROCEDURE B

Preparation

Pour 200 mL of acetone into the 600-mL beaker. Wearing insulated gloves, gradually add about 100 g of dry ice chunks, until a slush forms in the beaker.

If diffusion tubes are not available, they can be constructed as described in this paragraph. With a glass-working torch, seal closed one end of each of two 35-cm pieces of heavy-walled glass tubing with an outside diameter of about 25 mm. Seal a 10-cm piece of small-diameter (ca. 10 mm) heavy-walled glass tubing onto the open end of each tube. Wearing gloves, place about 5 mL of bromine in each tube. Seal one of the tubes closed by heating the small-diameter tubing near the center. (This tube contains bromine with air at approximately atmospheric pressure.) Place the sealed end of the other tube in the dry ice–acetone bath. After the bromine has solidified, attach the open end of the narrow tube to the vacuum pump and evacuate the tube. Seal this tube closed by heating the small-diameter tubing at the center until it collapses. (This tube contains bromine without air.) Allow the tube to warm to room temperature before using it.

Presentation

Place one end of each tube in the dry ice–acetone bath. The intensity of the color of the bromine vapor in the tube containing only bromine will fade uniformly as the bromine solidifies at the bottom of the tube. The intensity of the color in the other tube will fade initially only near the bottom of the tube, and a gradient in the intensity of the color will develop. Eventually, the color in this second tube will fade to the same extent as that in the first tube, but it will take many times longer.

HAZARDS

Wear gloves while handling bromine containers. Bromine is a very strong oxidizing agent, and skin contact with the liquid can result in severe burns. Because bromine vaporizes readily at room temperature to yield toxic fumes, adequate ventilation must be provided.

The evacuated container must be regarded as an implosion hazard. A grid-like wrapping of transparent tape, which will leave the contents visible, is recommended for the container. The pressure difference between the inside and the outside of the container can approach 760 torr.

DISPOSAL

To remove bromine from the special apparatus, draw the vapor out by connecting the stopcock to a water aspirator. It may be necessary to fill the apparatus with air and reconnect it to the aspirator several times to draw out all of the bromine.

To dispose of the bromine in the 1-liter cylinder, allow the bromine to disperse in a hood. Alternatively, slowly pour a concentrated sodium thiosulfate ($Na_2S_2O_3$) solution into the cylinder until all of the bromine has been reduced. **Caution! The reaction**

between bromine and thiosulfate is exothermic, and the mixture will become hot. Then, flush the resulting, cooled solution down the drain with water.

The dry ice–acetone bath should be allowed to warm to room temperature. The acetone can then be recovered for reuse.

DISCUSSION

In each procedure of this demonstration, the two samples of bromine vapor are at the same temperature. Therefore, according to the kinetic molecular theory of gases, the molecules have the same average kinetic energy and the same average velocity. The difference in the rates at which the bromine vapors fill the containers is a consequence of a difference in the number of collisions they undergo with other molecules in the containers. The bromine molecules in the air-filled container undergo more frequent collisions than the bromine molecules in the container from which the air molecules were removed. In addition, the distances that the bromine molecules travel between collisions (their mean free paths) are shorter for the molecules mixed with air.

The mean free path of a molecule is equal to its average velocity divided by the frequency of its collisions. The frequency of collisions, z, between molecules in a sample of gas is given by the equation

$$z = \sqrt{2}\pi\theta^2\bar{v}N$$

where θ is the molecular diameter,
$\bar{v}$ is the average molecular speed, and
N is the number of molecules per cubic centimeter [3].

Therefore, the mean free path of a molecule is

$$a = \frac{1}{\sqrt{2}\pi\theta^2 N}$$

In a mixture of bromine and air, the average molecular diameter falls between the diameter of bromine (0.43 nm) and air (0.36 nm). The density of a gas at 20°C and 1 atm, calculated from the ideal gas law equation, is

$$\frac{n}{V} = \frac{P}{RT} = \frac{1 \text{ atm}}{(82.1 \text{ mL}\cdot\text{atm/mol}\cdot\text{K})(293\text{K})}$$

$$= 4.2 \times 10^{-5} \text{ mol/mL}$$
$$= 2.5 \times 10^{19} \text{ molecules/mL}$$

Therefore, the mean free path of bromine molecules in a mixture with air at 20°C and 1 atm is 5.6×10^{-6} cm.

A good vacuum pump can produce a pressure as low as 0.001 torr (1.3×10^{-6} atm). This may be the pressure within the evacuated tube before the bromine has vaporized. Under these conditions, the number of molecules per cubic centimeter is 3.3×10^{-3}. Therefore, the mean free path of the molecules is 4.3 cm. So, when the bromine begins to vaporize, the molecules can travel a million times farther in the evacuated tube than in the tube containing air. However, as the bromine vaporizes, the number of molecules per cubic centimeter increases, and their mean free path decreases. Once equilibrium between bromine liquid and vapor has been established at 20°C, the pressure of bromine is equal to the vapor pressure of bromine, 173 torr [4]. At this time, the mean free path of the bromine molecules is 2.1×10^{-5} cm.

REFERENCES

1. S. B. Arenson, *J. Chem. Educ.* 17:469 (1940).
2. J. E. Johnston, *J. Chem. Educ.* 47:A439 (1970).
3. J. O. Hirschfelder, C. F. Curtis, and R. B. Bird, *Molecular Theory of Gases and Liquids*, Interscience Publishers, John Wiley and Sons: New York (1964).
4. *International Critical Tables*, Vol. 3, McGraw-Hill: New York (1928).

5.17

Graham's Law of Diffusion

A volume of gas contained in a buret tube flows into the atmosphere through a porous glass frit, while air flows in the opposite direction. From the initial volume of gas and the final volume of air in the tube, the ratio of the flow rates of the two gases can be determined.

MATERIALS

500 mL tap water

2 or more cylinders of compressed, water-inert gases, such as helium (He), methane (CH_4), or argon (Ar), with valves

Büchner funnel, 3 cm in diameter, having a fritted glass disk with pores of 1 micron in diameter

25-mL buret body (or 25-mL Mohr pipette)

silicone caulking

50-cm length glass tubing, with outside diameter of 7 mm

10-cm length glass tubing, with outside diameter of 7 mm

glass-working torch

1-holed rubber stopper to fit mouth of Büchner funnel

ca. 60 cm vacuum tubing

water aspirator

1-liter glass cylinder

ca. 60-cm length rubber tubing to fit 7-mm glass tubing

filter paper to fit Büchner funnel

10-mL graduated cylinder

PROCEDURE

Preparation

Fabricate a diffusion tube by sealing a small (3-cm diameter), ultrafine porosity, fritted glass Büchner funnel onto the end of the body of a 25-mL buret or onto one end of a 25-mL Mohr pipette. (This can be done using silicone caulking.) Form a 50-cm length of glass tubing into a gas-delivery tube, as shown in the figure. Insert a 10-cm glass tube through the 1-holed rubber stopper, and connect the glass tube to a water aspirator with vacuum tubing.

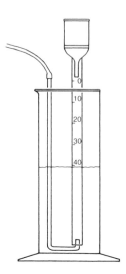

Assemble the apparatus shown in the figure. Pour 500 mL of tap water into the 1-liter glass cylinder. With rubber tubing, connect the gas-delivery tube to the valve of the compressed gas cylinder, and insert the delivery tube into the glass cylinder. Rest the diffusion tube in the cylinder. Place a damp piece of filter paper in the funnel to saturate the air above the frit with water vapor.

Presentation

Seat the stopper in the top of the funnel. Turn on the aspirator and draw water up the diffusion tube until it is just below the 0-mL mark. Read the water level. If the gas to be used has a molecular mass below the average molecular mass of air (29 g/mole), transfer 20–25 mL of gas from the cylinder into the diffusion tube. If the gas has a higher molecular mass, then use about 10 mL of the gas. Adjust the height of the diffusion tube in the cylinder so the levels of water inside and outside the tube are the same and record the volume. Continually adjust the height of the buret to maintain the equality of the water levels. Gases that are lighter than air flow from the tube more quickly than air flows in, and the volume of gas in the tube decreases. Gases that are heavier than air flow out more slowly than air flows in, and the volume of gas in the tube increases. If the diffusion coefficient of the gas is to be calculated from the results of this demonstration, record the time required for the volume of gas in the tube to change by 1 mL. Eventually, most of the original gas will have left the tube, and the volume of air in it will no longer change. Record the final volume of gas in the diffusion tube. Repeat the procedure with other gases. Measure the volume of the dead space (V_0), the volume between the glass frit and the zero mark of the buret, by inverting the diffusion tube, filling this space with water, and pouring the water into a 10-mL graduated cylinder. Graham's law of diffusion can be illustrated with the initial and final volume of gas, as described in the Discussion section.

DISCUSSION

The apparatus used in this demonstration is closely related to the apparatus first used by Graham to study the diffusion of gases [1]. Gases flow through the frit in both directions—the gas in the diffusion tube flows out, and air flows in. These gases may flow at different rates. For example, the gas in the tube may flow out of the tube more rapidly than air flows into the tube. In this case, the volume of gas in the tube will decrease. As the gas flows out and air flows in, the composition of the gas within the tube becomes closer to that of air, and the difference in the diffusion rates decreases. Eventually, all of the original gas in the tube will have flowed out and been replaced, at least partially, by air. At this point, the volume of air in the tube will remain constant. The volume of the air within the tube is less than the volume of the gas initially introduced.

The difference (ΔV) between the volume of gas introduced into the tube (V_{gas}) and the final volume of air in the tube is related to the ratio of the diffusion rate of the gas (J_{gas}) and to that of air (J_{air}) by the following equation [2]:

$$\frac{J_{gas}}{J_{air}} = \frac{V_{gas}}{V_{gas} + \Delta V}$$

The volume of the gas initially introduced can be found by adding the volume of the dead space, V_0, to the initial volume reading from the buret. Data obtained by this technique are presented in the table [2, 3]. The ratios of the rates of flow are in close agreement with the square root of the inverse of the ratio of their molecular masses.

Ratios of Flow Rates at Constant Pressure [2]

Gas	J_{gas}/J_{air}	$(m_{air}/m_{gas})^{1/2}$ [a]
He	2.7	2.7
CH_4	1.3	1.3
Ar	0.86	0.85

[a] The average molecular mass of air is taken as 29 g/mole.

In order for this technique to work properly, several precautions must be taken. The diameter of the pores in the Büchner funnel must be about 1 micron. If they are much larger, flow will be too rapid, and an accurate initial volume reading will be difficult to obtain. The gas should be introduced quickly and its initial volume measured immediately, because it will begin flowing out of the diffusion tube as soon as it is introduced. It is important to keep the levels of water inside and outside the tube equal at all times, in order to prevent a pressure difference between the gas in the tube and the air outside. The flow rates of the gases are inversely proportional to the square roots of their molecular masses only at constant pressure.

REFERENCES

1. T. Graham, *Phil. Mag.* 2:175, 269, and 351 (1833).
2. E. A. Mason and B. Kronstad, *J. Chem. Educ.* 44:740 (1967).
3. R. B. Evans, III, L. D. Love, and E. A. Mason, *J. Chem. Educ.* 46:423 (1969).

5.18

Graham's Law of Effusion

The time required for a fixed volume of a number of gases to effuse into a vacuum is measured. These times are proportional to the molecular masses of the gases [1].

MATERIALS

1 or more bags of the following gases: hydrogen (H_2), helium (He), methane (CH_4), oxygen (O_2), and carbon dioxide (CO_2) (See the introduction to this chapter, p. 4, for a description of these bags.)

3–4 drops glycerine

100-mL glass syringe

solid rubber stopper to fit 4-liter filter flask

#10 cork borer (19 mm in diameter)

10-cm length glass tubing, with outside diameter of 19 mm

septum cap to fit 19-mm glass tubing

copper wire, 18 gauge

4-liter filter flask

stand, with clamp to hold flask

vacuum pump, with hose

syringe needle, 27 gauge, cut to a length of ca. 6 mm

stopwatch

PROCEDURE

Preparation

Lubricate the syringe with a few drops of glycerine. Bore a hole in the stopper to accommodate the 19-mm tube, and insert the tube through the stopper. Attach a rubber septum to the tube and secure the septum with copper wire. Seat the stopper in the mouth of the 4-liter filter flask and clamp the flask to the stand. Use vacuum hose to connect the tubulation of the flask to the vacuum pump.

Presentation

Turn on the vacuum pump and evacuate the flask. Attach the needle to the syringe and fill the syringe with 100 mL of air. Carefully insert the needle of the syringe through the rubber septum, as shown in Figure 1. (The 27-gauge needle is very fragile and easily bent.) Use a stopwatch to measure how long it takes for 90 mL of air to effuse through the needle into the evacuated flask.

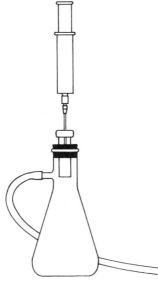

Figure 1.

Repeat the process with 100-mL samples of other gases. Prepare these samples by inserting the syringe needle through the septum of each of the bags of gas and drawing 100 mL of gas into the syringe.

HAZARDS

The filter flask is subject to implosion when evacuated. Therefore, it should be wrapped with netting or tape.

Corrosive or toxic gases should not be used, because they will be released into the room after passing through the vacuum pump.

Syringe needles are sharp and potential sources of puncture wounds.

DISCUSSION

Two distinct laws of gas behavior were advanced by Thomas Graham (1806–1869). Graham's law of diffusion was published in 1833, and his law of effusion was published 13 years later. Diffusion occurs when two gases mix without a pressure differential. Effusion occurs when a gas escapes through a small hole in a thin-walled container into a vacuum. This demonstration involves effusion rather than diffusion.

The table lists typical effusion times for 90 mL of various gases. Also listed are

Effusion Times of Various Gases [1]

Gas	t (sec)	t^2 (sec^2)	Molar mass (g/mole)
H_2	7.2	52	2.0
He	11.2	125	4.0
air	21.7	471	29.0
O_2	22.5	506	32.0
CO_2	26.2	686	44.0

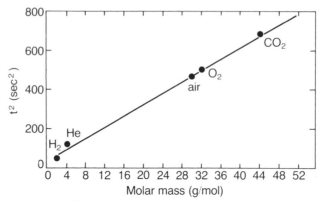

Figure 2. Plot of t^2 versus molar mass.

the squares of the effusion times and the molecular masses of the gases. In Figure 2, the square of the effusion time is plotted versus the molar mass of the gases. This shows that the square of the effusion time is proportional to the molar mass. It follows then that the effusion time is proportional to the square root of the molar mass. Because the effusion time is inversely proportional to the rate of effusion, the rate of effusion is inversely proportional to the square root of the molar mass of the gas.

If the flow of a gas is to be defined as effusion, at least three criteria must be satisfied [2]. First, the average distance a molecule travels between collisions must be at least ten times the radius of the hole through which the gas is flowing. Second, the pressure of the gas must be small enough that a molecule collides with the walls of the container more frequently than with other molecules. Therefore, each molecule escapes from the container independently of any other, and there is no bulk flow of gas through the hole. Third, the thickness of the wall of the container must be a small fraction of the diameter of the container, so that each molecule that enters the hole will succeed in escaping from the container. Although the procedure employed in this demonstration does not strictly meet all of these criteria, the results are in excellent agreement with Graham's law of effusion, which states that the rate of effusion of a gas is inversely proportional to its molar mass. If a longer needle is used, the third criterion will be violated more significantly, and the results of the demonstration will not be in such close agreement with Graham's law.

REFERENCES

1. D. A. Davenport, *J. Chem. Educ.* 39:252 (1962).
2. W. J. Deal, *J. Chem. Educ.* 45:676 (1975).

5.19

Liquid-Vapor Equilibrium

Sealed tubes containing bromine liquid and vapor are placed in baths at different temperatures. The reddish brown color of the bromine vapor is most intense in the warmest tube and least intense in the coldest tube.

MATERIALS

350 mL tap water

250 mL crushed ice

3 evacuated bromine diffusion tubes (e.g., as available in Sargent-Welch #4424)

   or

   15 mL bromine

   200 mL acetone, CH_3COCH_3

   100 g dry ice (solid carbon dioxide), CO_2

   glass-working torch

   3 35-cm lengths heavy-walled glass tubing, with outside diameter of ca. 25 mm

   3 10-cm lengths heavy-walled glass tubing, with outside diameter of ca. 10 mm

   gloves, plastic or rubber

   600-mL beaker

   insulated gloves

   vacuum pump, with hose

2 500-mL beakers

hot plate (or Bunsen burner and ring stand with wire gauze)

thermometer, $-10°C$ to $+110°C$

PROCEDURE

Preparation

   If evacuated diffusion tubes are not available, they can be constructed as described in this paragraph. With a glass-working torch, seal closed one end of each of three 35-cm lengths of heavy-walled glass tubing with an outside diameter of about 25 mm. Seal a 10-cm length of small-diameter (ca. 10 mm), heavy-walled glass tubing onto the open end of each tube. Wearing gloves, place about 5 mL of bromine in each tube. Pour 200 mL of acetone into the 600-mL beaker. Wearing insulated gloves, gradually add about 100 g of dry ice chunks, until a slush forms in the beaker. Immerse the sealed end of one of the tubes in the dry ice–acetone bath. After the bromine has solidified, attach the narrow tube at its opening to the vacuum pump and evacuate the tube. Seal this tube by heating the small-diameter tubing at the center until it collapses. Repeat this process with the other two tubes. Allow the tubes to warm to room temperature before using them.

Prepare a warm water bath by heating 300 mL of tap water to 50°C in a 500-mL beaker. Prepare an ice bath by placing 250 mL of crushed ice in a 500-mL beaker and adding 50 mL of cold tap water.

Presentation

Place one end of one tube in the ice bath. Place the end of another tube in the warm water bath. The remaining tube should be left at room temperature. The reddish brown color of the bromine vapor will be most intense in the tube immersed in the warm water bath and least intense in the tube immersed in the ice bath.

HAZARDS

Wear gloves while handling bromine containers. Bromine is a very strong oxidizing agent, and contact between the skin and liquid bromine can cause severe burns. Because bromine vaporizes readily at room temperature to yield toxic fumes, adequate ventilation must be provided.

The boiling point of bromine is 59°C, and the vapor pressure of bromine is over 1 atm above this temperature. Therefore, the tubes should not be immersed in baths above this temperature, to avoid creating high pressures within the tubes.

DISPOSAL

To dispose of any excess bromine, allow the bromine to disperse in a hood. Alternatively, slowly pour a concentrated sodium thiosulfate ($Na_2S_2O_3$) solution into the cylinder until all of the bromine has been reduced. **Caution! The reaction between bromine and thiosulfate is exothermic, and the mixture will become hot.** Then, flush the resulting, cooled solution down the drain with water.

The dry ice–acetone bath should be allowed to warm to room temperature. The acetone can then be recovered for reuse.

DISCUSSION

Bromine is a reddish brown, volatile, diatomic liquid with a suffocating odor. Its melting point is −7.25°C, and its normal boiling point is 59.47°C [1].

Bromine is suitable for demonstrating the effect of temperature on the vapor pressure of a liquid. The intensity of the deep reddish brown color of the vapor is a direct indication of the vapor pressure. The table gives the temperature dependence of the vapor pressure of bromine [2, 3].

Temperature Dependence of the Vapor Pressure of Bromine

Temperature (°C)	Pressure (torr)	Temperature (°C)	Pressure (torr)
−7.3	44.4	30	264
−5.0	50.5	35	324
0.0	65.9	40	392
5.0	85.3	45	472
10	109	50	564
15	138	55	670
20	173	58.78	760
25	214		

In this demonstration the reddish brown color of the vapor in a sealed vessel intensifies as the vessel is heated. A similar observation is made in Demonstration 6.17, "The Equilibrium between Nitrogen Dioxide and Dinitrogen Tetroxide." In both cases, temperature changes shift an equilibrium. In this demonstration, the equilibrium is between two phases of a pure substance. In Demonstration 6.17, the equilibrium is a chemical reaction involving NO_2 and N_2O_4.

REFERENCES

1. M. Windholz, Ed., *The Merck Index*, 9th ed., Merck and Co.: Rahway, New Jersey (1976).
2. E. W. Washburn, Ed., *International Critical Tables*, Vol. 2, McGraw-Hill: New York (1928).
3. J. Fischer and J. Bingle, *J. Am. Chem. Soc.* 77:6511 (1955).

5.20

Solid-Vapor Equilibrium

Sealed tubes containing iodine solid and vapor are placed in baths at different temperatures. The violet color of the iodine vapor is most intense in the warmest tube and least intense in the coldest tube.

MATERIALS

350 mL tap water

250 mL crushed ice

3 evacuated iodine diffusion tubes (e.g., as available in Sargent-Welch #4425)

> *or*

> 15 g iodine

> 200 mL acetone, CH_3COCH_3

> 100 g dry ice (solid carbon dioxide), CO_2

> glass-working torch

> 3 35-cm lengths heavy-walled glass tubing, with outside diameter of ca. 25 mm

> 3 10-cm lengths heavy-walled glass tubing, with outside diameter of ca. 10 mm

> gloves, plastic or rubber

> 600-mL beaker

> insulated gloves

> vacuum pump, with hose

2 500-mL beakers

hot plate (or Bunsen burner and ring stand with wire gauze)

PROCEDURE

Preparation

If evacuated diffusion tubes are not available, they can be constructed as described in this paragraph. With a glass-working torch, seal closed one end of each of three 35-cm pieces of heavy-walled glass tubing with an outside diameter of about 25 mm. Seal a 10-cm piece of small-diameter (ca. 10 mm), heavy-walled glass tubing onto the open end of each tube. Wearing gloves, place about 5 g of iodine in each tube. Pour 200 mL of acetone into the 600-mL beaker. Wearing insulated gloves, gradually add about 100 g of dry ice chunks, until a slush forms in the beaker. Immerse the sealed end of one of the tubes in the dry ice–acetone bath. Attach the narrow tube at its opening to the vacuum pump and evacuate the tube. Seal this tube by heating the small-diameter tubing at the center until it collapses. Repeat this process with the other two tubes. Allow the tubes to warm to room temperature before using them.

Prepare a hot water bath by heating 300 mL of tap water to boiling in a 500-mL beaker. Prepare an ice bath by placing 250 mL of crushed ice in a 500-mL beaker and adding 50 mL of cold tap water.

Presentation

Place one end of one tube in the ice bath. Place the end of another tube in the hot water bath. The remaining tube should be left at room temperature. The violet color of the iodine vapor will be most intense in the tube immersed in the hot water bath and least intense in the tube immersed in the ice bath.

HAZARDS

Because iodine, a moderately strong oxidizing agent, vaporizes readily at room temperature to yield toxic fumes, adequate ventilation must be provided. Wear gloves while handling iodine containers.

DISPOSAL

To dispose of any excess iodine, allow the iodine to disperse in a hood. Alternatively, pour a concentrated sodium thiosulfate ($Na_2S_2O_3$) solution into the iodine container until all of the iodine has been reduced. **Caution! The reaction between iodine and thiosulfate is exothermic, and the mixture will become hot.** Then, flush the resulting, cooled solution down the drain with water.

The dry ice–acetone bath should be allowed to warm to room temperature. The acetone can then be recovered for reuse.

DISCUSSION

Iodine is a blue-black, crystalline solid with a metallic luster. Its melting point is 113.60°C, and its normal boiling point is 185.24°C [1]. Solid iodine vaporizes readily at room temperature to a corrosive, violet gas with a sharp odor. This process, in which a solid vaporizes directly into a gas without passing through the liquid phase, is called sublimation. (Another substance with which sublimation is readily observed is solid carbon dioxide, dry ice.) The vapor pressure of iodine at 25°C is only 0.305 torr [1], but the intensity of its color makes the vapor easily visible.

Iodine is suitable for demonstrating the effect of temperature on the vapor pres-

Temperature Dependence of the Vapor Pressure of Iodine

Temperature (°C)	Pressure (torr)	Temperature (°C)	Pressure (torr)
0	0.0299	60	4.31
10	0.0808	70	8.22
20	0.202	80	15.1
30	0.471	90	26.8
40	1.03	100	45.5
50	2.16		

sure of a solid. The intensity of the violet color of the vapor is a direct indication of the vapor pressure. The table gives the temperature dependence of the vapor pressure of iodine [2].

REFERENCES

1. M. Windholz, Ed., *The Merck Index*, 9th ed., Merck and Co.: Rahway, New Jersey (1976).
2. E. W. Washburn, Ed., *International Critical Tables*, Vol. 3, McGraw-Hill: New York (1928).

5.21

Boiling Liquids
at Reduced Pressure

Water is made to boil at 50–60°C under reduced pressure created by an aspirator or by the condensation of water vapor. (Acetone may be made to boil at room temperature. This is described in Demonstration 1.1 of Volume 1 in this series.)

MATERIALS FOR PROCEDURE A

ca. 500 mL tap water

thermometer, −10°C to +110°C

2-holed rubber stopper to fit boiling flask

3-way stopcock

1-liter, thick-walled, round-bottomed boiling flask

boiling chips

tripod

Bunsen burner

stand, with clamp for flask

water aspirator, with trap and vacuum tubing

MATERIALS FOR PROCEDURE B

ca. 500 mL tap water

ice cube (or cold-water tap)

10-cm length of glass tubing to fit through rubber stopper

1-holed rubber stopper to fit boiling flask

10-cm length of rubber tubing to fit glass tubing

pinch clamp

1-liter, thick-walled, round-bottomed boiling flask

Bunsen burner

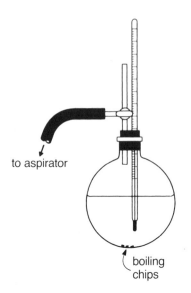

to aspirator

boiling
chips

PROCEDURE A

Preparation

Insert the thermometer through one hole of the 2-holed rubber stopper and insert one arm of the 3-way stopcock through the other hole. Fill the flask approximately half full with tap water. Add several boiling chips, and seat the stopper in the mouth of the flask. Adjust the thermometer so that its bulb is below the surface of the water (see figure). Place the flask on the tripod, open the stopcock, and heat the flask with the Bunsen burner until the temperature of the water is 50–60°C. Remove the flask from the tripod and clamp it to the stand.

Presentation

Attach the vacuum tubing from the trap to one of the open arms of the stopcock. Connect the trap to the aspirator. Turn the aspirator to maximum water flow, and adjust the stopcock so that air is drawn from the flask. When the pressure inside the flask is low enough, the water begins to boil. Read the temperature of the boiling water from the thermometer. As the water boils, its temperature decreases. Eventually the water will cool sufficiently so that it ceases to boil.

Before disconnecting the assembly, turn the stopcock to allow air into the flask, and turn off the water aspirator.

PROCEDURE B [1]

Preparation

Insert the glass tube through the rubber stopper. Attach the rubber tubing to the glass tube, and seal the rubber tubing with the pinch clamp.

Fill the round-bottomed flask approximately one-third full with tap water.

Presentation

Heat the water in the flask to boiling with a Bunsen burner. Allow the water to cool for 1 minute. Seal the flask tightly with the rubber stopper, and hold an ice cube on its inverted bottom. (Alternatively, it may be inverted under cold running water.) The water in the flask will boil for several seconds.

HAZARDS

The flask may implode if the pressure inside is lowered sufficiently. This risk can be minimized by using a thick-walled boiling flask rather than a standard round-bottomed flask.

DISPOSAL

To disassemble the flask used in Procedure B, remove the pinch clamp from the rubber tubing to release the partial vacuum in the flask. Then remove the stopper from the flask.

DISCUSSION

This demonstration shows that a liquid can boil below its normal boiling point. The boiling point of a liquid is the temperature at which the vapor pressure of the liquid is equal to the external pressure. When the external pressure above a liquid is 1 atm (760 torr), the liquid boils at its normal boiling point. When the external pressure above a liquid is reduced, the vapor pressure needed to induce boiling is also reduced, and the boiling point of the liquid decreases. The temperature dependence of the vapor pressure of water is illustrated by the data in the table.

In Procedure A, the pressure above the water in the flask is lowered by an aspirator. In Procedure B, the pressure is lowered by cooling the air in the flask and by

Temperature Dependence of the Vapor Pressure of Water [2]

Temperature (°C)	Pressure (torr)	Temperature (°C)	Pressure (torr)
0	4.579	55	118.04
5	6.543	60	149.38
10	9.209	65	187.54
15	12.788	70	233.7
20	17.535	75	289.1
25	23.756	80	355.1
30	31.824	85	433.6
35	42.175	90	525.76
40	55.324	95	633.90
45	71.88	100	760.00
50	92.51		

condensing some of the water vapor in the flask. (The reduction in pressure caused by condensing water vapor is dramatically demonstrated by the collapse of a can in Demonstration 5.1.) In both procedures of this demonstration, the pressure is reduced sufficiently to cause the water to boil. As the water boils, heat is lost because of the heat of vaporization of water, which is 46.78 kJ/mole [2]. In Procedure A, the water will eventually cool to a temperature at which the vapor pressure is less than the external pressure produced by the aspirator, and the water will stop boiling. (The cooling of water by evaporation is the basis of the cryophorus demonstration, in which the water is made to freeze. The apparatus required for the cryophorus demonstration is commercially available, e.g., Sargent-Welch #1667, or it can be fabricated easily.) In Procedure B, as the water boils, the pressure in the flask increases to the point where it is greater than the vapor pressure of the liquid, and the liquid stops boiling. The flask in Procedure B should be no more than half full of liquid water, to provide space for the vapor produced by boiling. If too little space is provided, the vapor pressure of the liquid water may be reached before boiling is apparent. Demonstration 1.1, "Evaporation as an Endothermic Process," in Volume 1 of this series, uses an apparatus very similar to that in Procedure A of this demonstration to highlight the loss of heat by a boiling liquid.

REFERENCES

1. A. Joseph, P. F. Brandwein, E. Morholt, H. Pollack, and J. F. Castka, *A Sourcebook for the Physical Sciences*, Harcourt, Brace, and World: New York (1961).
2. J. A. Dean, Ed., *Lange's Handbook of Chemistry*, 12th ed., McGraw-Hill: New York (1979).

5.22

Vapor Pressure

The temperature dependence of the vapor pressure of acetone is demonstrated by heating and cooling the liquid in a filter flask connected to a U-tube manometer [1].

MATERIALS

500 g crushed ice

250 mL cold tap water

40 mL acetone, CH_3COCH_3

3 rubber stoppers, one apiece to fit each of the three filter flasks

#10 cork borer (19 mm in diameter)

3 5-cm lengths of glass or plastic tubing, with outside diameter of 19 mm

3 rubber septa to fit 19-mm tubing

20 cm copper wire, 20–24 gauge

250-mL filter flask

500-mL filter flask

1-liter filter flask

2-liter beaker

mercury-filled U-tube manometer

20-cm length rubber tubing

syringe needle

10-mL syringe, with needle

heat gun or hand-held hair dryer

PROCEDURE

Preparation

Bore a hole large enough to pass the 19-mm tubing through each rubber stopper. Insert a 5-cm length of the tubing through each stopper. Attach a rubber septum to each piece of tubing and fasten it to the tubing with copper wire. Seat each stopper in the mouth of its flask.

Prepare an ice bath by placing 500 g of crushed ice and 250 mL of cold tap water in the 2-liter beaker.

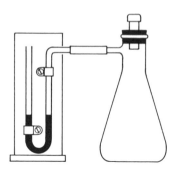

Presentation

Connect the 500-mL flask to the manometer with the 20-cm length of rubber tubing, as shown in the figure. Insert the separate syringe needle through the rubber septum to relieve any excess pressure.

With the syringe, inject 10 mL of acetone into the flask. As the acetone vaporizes, the pressure in the flask will increase and cause a change in the levels of mercury in the manometer. Once the mercury levels have stopped changing, record the pressure increase in the flask. This is the vapor pressure of acetone at room temperature.

Repeat the above steps using the 250-mL and 1-liter flasks. In each case, the mercury levels will change by the same amount, demonstrating that the vapor pressure is independent of the volume of the container.

If all of the acetone in the 500-mL flask has evaporated, add another 10 mL. Reconnect the stoppered 500-mL flask to the manometer. To demonstrate the effect of temperature on the vapor pressure of acetone, warm the flask gently with a heat gun or hair dryer. As the temperature increases, the manometer will register increasing pressure in the flask. Caution! Do not heat the flask strongly—the pressure may exceed the capacity of the manometer, resulting in a mercury spill. Chill the flask by immersing it in the ice bath. The manometer will show a large decrease in pressure.

HAZARDS

Because acetone is flammable, this demonstration should be performed away from flames.

Mercury is extremely toxic and should be handled with care to avoid prolonged or repeated exposure to the liquid or vapor. Continued exposure to the vapor may result in severe nervous disturbance, insomnia, and depression. Continued skin contact also can cause these effects as well as dermatitis and kidney damage. Mercury should be handled only in well-ventilated areas. Mercury spills should be cleaned up immediately by using a capillary attached to a trap and an aspirator. Small amounts of mercury in inaccessible places should be treated with zinc dust to form a nonvolatile amalgam.

Syringe needles are sharp and potential sources of puncture wounds.

DISPOSAL

The acetone should be poured into a suitable container for used solvents or flushed down the drain with water.

DISCUSSION

This demonstration illustrates that the vapor pressure of a volatile liquid is independent of the volume of its container. This is worth demonstrating, because many students, perhaps erroneously applying Boyle's law to vapor pressure, expect the vapor pressure to vary with volume. In order to observe the independence of vapor pressure from volume, it is necessary for liquid to be present in the flask, because the vapor pressure is established only when there is an equilibrium between the liquid and the vapor phases. In addition, it is necessary to allow sufficient time for this equilibrium to be established, as indicated by the pressure within the flask remaining constant.

This demonstration also illustrates the dependence of the vapor pressure on temperature. As the temperature increases, the vapor pressure increases; as the temperature decreases, so does the vapor pressure. Indeed, when the flask is immersed in the ice bath, acetone's vapor pressure is significantly lower than at room temperature, as the data in the table indicate.

Temperature Dependence of the Vapor Pressure of Acetone [2]

Temperature (°C)	Pressure (atm)	Temperature (°C)	Pressure (atm)
−59.4	0.001	78.6	2.0
−31.1	0.013	113.0	5.0
−9.4	0.053	144.5	10.0
7.7	0.13	181.0	20.0
39.5	0.53	214.5	40.0
56.5	1.0		

REFERENCES

1. I. Talesnick, *Demonstrations in Chemistry*, Faculty of Education, Queen's University: Kingston, Ontario (undated).
2. R. C. Weast, Ed., *CRC Handbook of Chemistry and Physics*, 63d ed., CRC Press: Boca Raton, Florida (1982).

5.23

Relative Velocity
of Sound Propagation:
Musical Molecular Weights

The pitch of the sound of an organ pipe increases as the molar mass of the gas passed through it decreases, and the pitch decreases as the molar mass increases [1].

MATERIALS

several cylinders of compressed gases, such as hydrogen (H_2), helium (He), methane (CH_4), nitrogen (N_2), oxygen (O_2), carbon dioxide (CO_2), difluorochloromethane (CHF_2Cl), difluorodichloromethane (CF_2Cl_2), perfluoroethane (C_2F_6), and sulfur hexafluoride (SF_6)

organ pipe (or whistle), with a natural pitch below 1000 Hz†

ca. 60-cm rubber hose

PROCEDURE

Sound the organ pipe by blowing air through it. This can be accomplished by blowing into the base of the pipe with your mouth, or by connecting the rubber hose to the base of the pipe and blowing into the hose.

Attach the rubber hose to the base of the pipe and attach the other end to the valve of one of the cylinders of gas. Sound the pipe by opening the valve of the cylinder and allowing the gas to flush the pipe until the pitch of the sound stabilizes. (Use the lowest exit pressure of gas that sounds the pipe. If the pressure is high, the organ pipe may sound at an overtone.) If the molar mass of the gas is smaller than the average molar mass of air (29 g/mol), then the pitch of the pipe will be higher than when air is passed through it. If the molar mass of the gas is greater, then the pitch will be lower.

Repeat the presentation with the cylinders containing other gases.

DISCUSSION

As this demonstration illustrates, the pitch of an organ pipe (or other wind instrument) depends upon the molar mass of the gas forced through it. The lighter the gas, the higher the pitch. The relationship between the molar mass of the gas and the per-

---

† Contact a musical instrument vendor to locate a pipe organ manufacturer from whom such a pipe may be available.

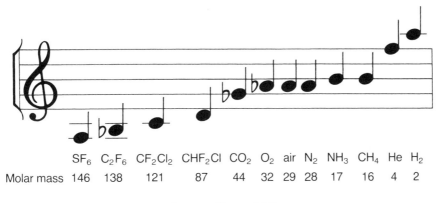

	SF$_6$	C$_2$F$_6$	CF$_2$Cl$_2$	CHF$_2$Cl	CO$_2$	O$_2$	air	N$_2$	NH$_3$	CH$_4$	He	H$_2$
Molar mass	146	138	121	87	44	32	29	28	17	16	4	2

Source: Figure 4 [1]

ceived pitch of the sound for an A-880 organ pipe is given in the figure. The natural frequency of the pipe used in this demonstration should be below 1000 Hz, because a pipe with a higher pitch will resonate at an inaudible frequency when hydrogen or helium is forced through it.

The perceived pitch of a sound is determined by the frequency of the vibrations in the medium. The higher the frequency, the higher the pitch. The major factor affecting the resonance frequency of the pipe is the speed of sound in the gas, which is determined by the average speed of the gas molecules [1]. The product of the frequency of a wave multiplied by its wavelength is equal to the speed of the wave. In the organ pipe, the wavelength is determined by the size of the pipe, and is constant. Therefore, as the frequency changes, so must the speed of the wave (sound). The table shows the speed of sound in various gases at 0°C. However, any attempt to derive the speed of sound or the molecular weight of the gas from the pitch of the organ pipe will be unsuccessful, perhaps because of incomplete filling of the pipe by the gas.

Speed of Sound in Various Gases [2]

Gas	Speed (m/s)
H$_2$	1269.5
He	970
N$_2$	337
air	331.45
O$_2$	317.2

REFERENCES

1. D. A. Davenport, M. Howe-Grant, and V. Srinivasan, *J. Chem. Educ.* 56:523 (1979).
2. R. S. Berry, S. A. Rice, and J. Ross, *Physical Chemistry*, Interscience Publishers, John Wiley and Sons: New York (1980).

5.24

Electrical Conductivity of Gases

A gas glows under high voltage when its pressure is reduced below atmospheric pressure [1].

MATERIALS

2 nails, ca. 10 cm long

1-holed rubber stopper to fit 25-mm tubing

solid rubber stopper to fit 25-mm tubing

right-angle glass bend, with outside diameter of 7 mm and length of each arm ca. 5 cm

30–120-cm length glass tubing, with outside diameter of 25 mm

1 or 2 stands, with clamps and clamp holders

vacuum pump (or water aspirator), with vacuum tubing

either

induction coil (e.g., Sargent-Welch #2392B)

6–12-volt DC source (e.g., automobile battery)

4 insulated cables, with clips on both ends

or

Tesla coil

2 insulated cables, with clips on both ends

1 or more cylinders of compressed, nonflammable gas, such as neon (Ne), carbon dioxide (CO_2), or argon (Ar), with valve (optional)

rubber tubing to attach to valve of compressed gas cylinder (optional)

PROCEDURE

Preparation

Assemble the gas discharge tube as illustrated in the figure. Drill a 3-mm (1/8-inch) hole through each of the stoppers. Insert one nail through each of the new holes. Insert the right-angle bend through the original hole in the 1-holed stopper. Seat the stoppers in the open ends of the glass tube and clamp the glass tube to the stands. Connect the free end of the right-angle bend to the vacuum pump with vacuum tubing.

If an induction coil is to be used, attach the primary terminals to the low-voltage DC source. Using cables with clips, attach one of the secondary terminals to one nail and the other terminal to the other nail.

If a Tesla coil is to be used, use a cable with clips to attach one of the nails to a ground, such as a water pipe, and another cable to attach the other nail to the tip of the Tesla coil.

Presentation

Darken the room lights, and turn on the vacuum pump and the high-voltage source (i.e., the induction coil or Tesla coil). A violet glow will appear inside the tube as the air is pumped from the tube. As the pressure in the tube drops, the intensity of the glow will increase and then decrease. (The decrease in intensity may not appear if a water aspirator is used, because the vacuum it produces is not as great as that obtained with a vacuum pump.) Turn off the high-voltage source and disconnect it from the nails. Disconnect the vacuum tubing from the glass bend and turn off the pump.

If a compressed gas cylinder is available, connect the free end of the glass bend to the cylinder and loosen the stopper at the opposite end of the assembly. Open the valve on the cylinder and flush gas through the assembly for 20–30 seconds. Close the valve and reseat the stopper. Reconnect the glass bend to the vacuum pump and the nails to the high-voltage source. Turn on the high-voltage source and the vacuum pump. When the glow appears, the color is characteristic of the gas used and may be different from that obtained with air.

Repeat the steps in the previous paragraph using other nonflammable gases.

HAZARDS

The induction coil can produce painful electrical shocks if both terminals are touched simultaneously.

The Tesla coil can produce small burns to the skin if its tip is touched.

DISCUSSION

The glow observed in this demonstration was first discovered by the English physician William Watson in 1752 [1]. In an attempt to determine whether air is a conductor of electricity, he pumped the air from a glass bulb containing a pair of electrodes. As the air was pumped out, the amount of electricity conducted between the electrodes increased, and the bulb was filled with a violet light.

In this demonstration, when the pressure of the gas in the tube is high, about 100 torr or more, no electric current flows between the electrodes and no glow is observed. When the pressure is reduced sufficiently, a current eventually flows while sparks jump between the electrodes. As more gas is removed, the appearance of the tube changes. At lower pressure, a steady glow will appear throughout the tube. As the pressure is lowered further, a dark space will appear a short distance from the cathode (the Faraday dark space), and eventually, alternating disks of light and dark will appear.

The processes occurring within the tube are complicated and depend on a number of factors, including the pressure of the gas in the tube, the voltage applied to the electrodes, and the frequency of the voltage oscillations. However, a simple explanation is possible for some of the many phenomena that can be observed during this demonstration.

In an electric field, such as exists between the electrodes in this demonstration, a free electron will be accelerated toward the anode, the positive electrode. A free electron may be produced spontaneously in the gas by ionizing cosmic radiation or at the cathode by thermal ejection. As the electron accelerates toward the anode, its speed and, therefore, its kinetic energy increase. If the concentration of gas molecules in the tube is high, the electron cannot move far before it strikes a gas molecule. If the electron has not moved far, it has not gained much energy from the electric field, and the collision will have little effect on the gas molecule. As the pressure of the gas in the tube is reduced, the concentration of gas molecules decreases, and a free electron can travel farther before colliding with a gas molecule. Therefore, when it strikes a gas molecule, the electron may have enough energy to excite one of the molecule's electrons from a low-energy orbital to a higher-energy orbital. When the excited molecule returns to its ground state, it emits a photon of light. The energy of this photon and, therefore, its wavelength and the perceived color of the emitted light are characteristic of the molecule.

As even more gas is pumped from the tube, the electrons travel farther between collisions and become more energetic. When a highly energetic electron strikes a molecule, an electron may be ejected from the molecule, leaving a positively charged ion. The ejected electron and the positive ion will be accelerated in opposite directions by the electric field. Thus, at low gas pressure, there is a stream of electrons traveling from cathode to anode, and a stream of ions travelling from anode to cathode. These ions and electrons collide and lose energy in the form of light, but the wavelength of the light is not dependent on the gas, because the energy of the colliding particles varies continuously. Therefore, the color of the emitted light changes as the pressure is reduced, and the color at low pressure is independent of the gas in the tube [2].

REFERENCES

1. C. L. Stong, *Sci. Am.* 198(2):112 (1958).
2. F. A. Maxfield and R. R. Benedict, *Theory of Gaseous Conduction and Electronics*, McGraw-Hill: New York (1941).

5.25

Superheated Steam [†]

The water vapor from boiling water passes through a copper coil that is heated in a burner flame. The temperature of the steam is measured. The steam is capable of charring paper [1].

MATERIALS

200 mL tap water

60-cm length copper tubing, with outside diameter of 9 mm (3/8 inch)

500-mL Erlenmeyer flask

1-holed rubber stopper to fit Erlenmeyer flask

2 ring stands, with clamps

hot plate or Bunsen burner

2 Meker burners

thermometer, $-5°C$ to $+360°C$

2 sheets white paper

matches

PROCEDURE

Preparation

Arrange the apparatus as illustrated in the figure. Prepare the copper coil by making a right-angle bend 5 cm from one end; bend the other end into a coil with an inside diameter of 7 cm. Crimp the end of the tubing coming from the coil, leaving a 3-mm opening. Securely clamp both the neck of the flask and the opening of the copper tubing.

Presentation

Pour 200 mL of water into the flask, reattach it to the apparatus, and heat it. When the water is boiling vigorously, measure the temperature of the steam escaping from the end of the copper tubing. Hold a piece of white paper at the opening of the tubing. The paper may become dampened.

[†] We wish to thank Ronald I. Perkins of Greenwich High School, Connecticut, for suggesting this procedure while he was an instructor in the 1983 University of Wisconsin System Workshop for Science and Science Education Faculty.

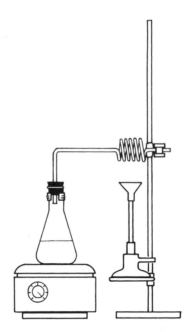

While the water is boiling, heat the coiled portion of the copper tubing with the two Meker burners. The copper will darken as oxide forms on its surface. With the thermometer, measure the temperature of the steam as it escapes from the copper tubing. Then, hold another piece of paper at the opening of the tubing. This time the superheated steam will char the paper. Hold the head of a match near the opening; the match will ignite.

HAZARDS

The superheated steam produced in this experiment can reach temperatures over 180°C and can cause burns to skin exposed to the vapors escaping from the copper coil. The apparatus should be allowed to cool before it is handled.

DISCUSSION

Steam is the word used to denote water in its vapor state. Everyone is familiar with the transition of water from liquid to vapor that occurs when water boils. However, many people identify steam with the small clouds that form over a cup of hot water or the jet of mist that escapes from a boiling teakettle. Actually these manifestations are tiny droplets of liquid water suspended in air. Steam is a colorless gas, invisible in air.

The presence of steam can be detected by its temperature, as illustrated in this demonstration. The standard state of water at 25°C is liquid, and the vapor pressure of water at this temperature is 23.8 torr. Therefore, any water vapor with a pressure above 23.8 torr must be at a temperature above 25°C. The temperature of the water vapor escaping from the copper coil is significantly above room temperature. The presence of the water vapor can be detected by condensing it on some cool surface. If this surface is paper, the vapor may wet it through. Water vapor can be heated in the copper coil well above the temperature of boiling water. This is called superheated steam. Superheated steam can be hot enough to char the paper. Usually, the paper will not ignite, because the water vapor which is charring the paper is also flushing oxygen away from the paper. However, if the superheated steam is hot enough, a match can be readily ignited, because the match head contains its own oxidizer.

When water vapor at the boiling point (100°C) transfers heat to other matter, such as paper, it loses heat energy. Removing heat energy from steam at the boiling point causes it to condense. Thus, the paper becomes wet when it is exposed to steam at 100°C. The energy released by condensing steam is represented by the heat of vaporization of water at 100°C, 40.65 kJ/mol [2]. When the paper is exposed to superheated steam, the paper chars because its temperature is raised above the decomposition temperature of the organic materials of which it is made. The heat released by the cooling of steam above 100°C is represented by the specific heat of steam, 35 J/mol-K [3]. These data indicate that far more heat is liberated by the condensation of a mole of water than by the cooling of a mole of steam by several hundred degrees. It is the heat liberated by the condensation of steam that makes steam far more capable of producing severe burns than is hot water.

REFERENCES

1. R. E. Stowe, *J. Chem. Educ.* 37:A547 (1960).
2. M. Windholz, Ed., *The Merck Index*, 9th ed., Merck and Co.: Rahway, New Jersey (1976).
3. R. C. Weast, Ed., *CRC Handbook of Chemistry and Physics*, 59th ed., CRC Press: Boca Raton, Florida (1978).

5.26

Kinetic Molecular Theory Simulator

The behavior of gas molecules as described by the kinetic molecular theory of gases is simulated with small, hard balls and a vibrating stage [1–4].

MATERIALS

6–12 steel balls, ca. 3 mm in diameter (e.g., the ball bearings of a bicycle)

bar magnet, ca. 5 cm in length

matchbox

molecular motion simulator (e.g., Sargent-Welch #1710M)

overhead projector

12–24 plastic balls, ca. 3 mm in diameter

cork chip, ca. 8–10 mm in diameter

4–6 plastic balls, 5–6 mm in diameter

wooden bar to fit securely between the sides of the square frame of the molecular motion simulator

PROCEDURE

Preparation

Magnetize the 6–12 steel balls by storing them in contact with the bar magnet for several days or by rubbing them over the magnet for several minutes.

Make a scoop from the matchbox by removing one end of the box.

Mount the simulator on an overhead projector and carefully level the apparatus.

Presentation

Turn on the overhead projector and perform one or more of the following steps:
1. Place 6–12 plastic balls of the same diameter on the glass plate of the simulator, and turn on the vibrator. The motion of the balls illustrates several features of the kinetic molecular theory of gases:
    a. the constant, random motion of gas molecules,
    b. the small diameter of the gas molecules compared with the distances between the molecules,
    c. the collisions of gas molecules with each other and with the sides of the container,
    d. the variations in the velocities of the molecules,

  e. the elastic nature of the collisions between gas molecules, and

  f. the rarity of three-body collisions.

2. Drop a cork chip several times the size of a ball onto the glass plate. The motion of the cork chip as it is bombarded by the balls illustrates Brownian motion. Remove the cork chip.

3. Double the number of balls on the glass plate. The motion of the balls will illustrate that, as the concentration of gas molecules increases,

  a. the number of collisions between gas molecules increases, and

  b. the mean free path of the molecules between collisions decreases.

 Use the matchbox scoop to remove half of the balls.

4. Add 4–6 plastic balls of larger diameter to the glass plate. The larger balls have greater mass. The more massive balls will illustrate that

  a. on average, the more massive molecules move more slowly than the lighter molecules, but

  b. at any given moment, some larger molecules are moving faster than some smaller molecules.

5. Increase the speed of the vibrator. This increases the average kinetic energy of the balls and simulates an increase in the temperature of a gas. The change in the motion of the balls illustrates the change in the behavior of gas molecules as their temperature increases, namely, that

  a. the average speed of the molecules increases,

  b. the frequency of collisions between molecules increases, and

  c. the mean free path of the molecules decreases

6. Turn off the vibrator, remove the large balls, and move all of the small balls to one side of the glass plate. Place the wooden bar across the center of the glass plate, securely wedged between the sides of the square frame, with all of the balls on one side (see Figure 1). Adjust the speed of the vibrator so that it is possible to count the

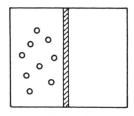

Figure 1.

collisions of the balls with each wall. Ask a student to count the number of times one of the walls is struck by the balls in a 1-minute period. Ask three other students to do the same with the other three walls. Record the total number of collisions. Without changing the speed of the vibrator, quickly pluck the wooden bar from the apparatus, doubling the size of the container. Ask the students to repeat their count. The behavior of the balls will illustrate that

  a. the average speed of the molecules is independent of the size of the container, and

  b. the frequency of the collisions with the sides of the container decreases as the volume of the container increases.

7. Replace the plastic balls on the glass plate with the magnetized steel balls. Adjust the vibrator to a sufficiently high speed to keep the balls in motion. Then, gradually

decrease the speed of the vibrator, simulating a decrease in the temperature of a gas. As the speed decreases, the balls will begin to clump together. This illustrates the effect of intermolecular forces on the behavior of gas molecules; that is, these forces cause the gas to condense into a liquid before the temperature has reached absolute zero.

DISCUSSION

The kinetic molecular theory of gases was developed from ideas of D. Bernoulli (1738), J. J. Waterston (1845), K. A. Kronig (1856), and R. Clausius (1857). However, the precise mathematical form of the theory can be attributed to the work of J. Clerk Maxwell (1860) and L. Boltzmann (1868).

The postulates of the kinetic theory of gases may be summarized as follows [5]:

(a) Gases are composed of tiny particles whose diameters are much smaller than the distances between the particles.

(b) These particles obey Newton's first law of motion; that is, they move in straight lines until they collide either with each other or with a wall of the container.

(c) The collisions of a gas particle with other gas particles and with the walls of its container are elastic; that is, kinetic energy is conserved in these collisions.

This theoretical model of a gas can be illustrated by the molecular motion simulator as described in this demonstration. The kinetic molecular model of a gas consists of a large number of moving particles, and the motion of these particles is random, or chaotic. This random, or chaotic, motion is illustrated in step 1 of the procedure, in which the balls rolling on a two-dimensional surface represent the gas particles moving through three-dimensional space. As the simultator shows, the particles move at various speeds, and the speed of a given particle changes as it collides with other particles and with the walls of the container.

The simulator is deficient in one major respect. The balls lose kinetic energy through friction with the glass surface, while, according to the kinetic theory of gases, the particles of a gas lose no kinetic energy between collisions. Thus, the simulator must use the vibrator to maintain the motion of the balls, replacing the kinetic energy lost through friction. Energy must be fed into the simulator continuously to keep the balls moving, while a gas needs no such energy input to maintain the motion of its particles. In addition, the collisions of the balls with each other and with the frame of the simulator are not perfectly elastic; some kinetic energy is lost. However, the collisions are sufficiently close to being elastic that the deviations from elasticity are not apparent when the balls are moving quickly.

When examined under a microscope, smoke appears to be a collection of tiny particles. These particles move, not in straight lines, but in rapid series of lurches from one direction to another, a sort of jiggling motion. This motion is called Brownian motion and can be explained in terms of the kinetic theory of gases. The smoke particles are much larger and more massive than the particles of the gas in which they are suspended. The motion of the smoke particles is produced through their constant bombardment by the tiny, invisible gas molecules. This effect is demonstrated by the motion of the cork chip in step 2 of this demonstration.

When a particle of gas collides with a wall of its container and rebounds, it exerts a force on the wall at the moment of collision. This force divided by the area of the wall is equal to the pressure exerted on the wall by the particle at the moment of impact. If the particle is moving back and forth between the sides of the container, then the instan-

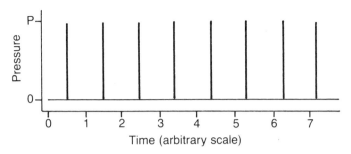

Figure 2. Pressure exerted on the wall of a container of volume V by a single particle in the container, as a function of time.

taneous pressure exerted by the particle would be as shown in Figure 2. Actual pressure gauges are incapable of registering the impact of a singe molecule, but instead register the average effect of the impacts of many molecules. As the concentration of molecules is increased, the frequency of their collisions with the walls increases. This is illustrated in step 3 of this demonstration. The increase in the number of collisions of the particles with the walls of the container leads to an increase in the average force exerted on the walls of the container and, therefore, to an increase in the pressure.

The kinetic theory of gases leads to a linear correlation between the average kinetic energy of gas molecules and the temperature of the gas. The temperature of a gas is directly proportional to the average kinetic energy of the gas molecules. The kinetic energy of a gas molecule is proportional to its mass multiplied by the square of its velocity. Therefore, molecules with greater mass move more slowly than molecules of lesser mass at the same temperature. This is illustrated in step 4 of this demonstration. A simulation of the effect of a temperature increase on the motion of gas particles is presented in step 5.

The instantaneous pressure exerted on one wall of the container by a single particle travelling from side to side is depicted in Figure 2. If the size of the container were increased, then the molecule would strike the wall less frequently, but with the same force at each impact. The pressure on one wall would be as shown in Figure 3. Because pressure gauges register time-averaged pressures, the pressure registered in the case of Figure 3 would be less than that in the case of Figure 2. Thus, increasing the volume of a container results in a lower pressure. The decrease in the frequency of collisions with the walls of a container when the volume of the container is increased is illustrated in step 6 of this demonstration.

The postulates of the kinetic molecular theory of gases lead to the ideal gas law as

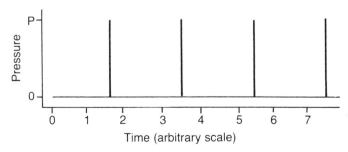

Figure 3. Pressure exerted on the wall of a container of volume 2V by a single particle in the container, as a function of time.

the equation of state for gases. While the ideal gas law is an accurate description of the behavior of gases under conditions of low pressure and high temperature, it inadequately describes the behavior of matter under other circumstances. Perhaps the most obvious deviation from ideality in a gas is its condensation at sufficiently low temperatures. The ideal gas law does not account for condensation at any temperature. Therefore, the gas molecules must have properties not described by the kinetic molecular theory. One of these properties is intermolecular attraction. There are forces of attraction between molecules of all gases. These forces are simulated in step 7 by using magnetized balls in the molecular motion stimulator. When the kinetic energy of the balls is low enough, they begin to stick together, as gas molecules stick together in the condensation process.

REFERENCES

1. B. A. Fiekers and G. S. Gibson, *J. Chem. Educ.* 22:305 (1945).
2. R. H. Woolley and D. McLachlan, Jr., *J. Chem. Educ.* 27:187 (1950).
3. W. H. Slabaugh, *J. Chem. Educ.* 30:68 (1953).
4. R. C. Plumb, *J. Chem. Educ.* 43:648 (1966).
5. G. W. Castellan, *Physical Chemistry*, Addison-Wesley Publishing Co.: Reading, Massachusetts (1964).

6

Chemical Behavior of Gases

Rodney Schreiner, Bassam Z. Shakhashiri,
Glen E. Dirreen, and Lenard J. Magginnis

The early development of modern chemistry relied heavily on the observation of chemical properties of gases. This introduction is intended to highlight this reliance, but a significant number of other important influences are not mentioned here. For a full picture of the development of modern chemistry, teachers should consult the literature of the history of chemistry, such as references 1–3.

It was through the study of gases, the simplest of the states of matter, that chemical research was turned in a fruitful direction during the 18th century. Hales's development of the pneumatic trough in the 1720s introduced an apparatus which made it possible for 18th-century scientists to generate and manipulate gases experimentally [4]. The properties of gases which these scientists investigated led to the first free flights by man in hot air balloons and soon thereafter in hydrogen balloons [5]. In addition, this research on gases occupies a strategic position in the early development of modern chemistry. It was by studying the properties of gases that the early chemists—particularly Black, Cavendish, Priestley, and Lavoisier—laid the foundations for modern chemistry. All four of these chemists began their careers in the age when the phlogiston theory explained combustion as the escape of a substance called phlogiston from a burning material. The first three of these chemists discovered new gases with novel properties, but interpreted their results mainly in terms of the phlogiston theory. It was Lavoisier who had the insight to organize these discoveries into a system of chemistry without phlogiston, a system which introduced the modern chemical elements. The atomic theory of John Dalton also rested on discoveries made during the study of gases. Both the identification of the chemical elements and the atomic theory owe their existence to investigations of chemical properties of gases.

Early in the 18th century, a popular area of investigation for chemists of the day was the production of "airs" (as gases were called then). The development of a method for handling samples of air over water allowed the collection and measurement of the volume of gases. This led to interest in chemical reactions which produced gases. Stephen Hales (1677–1761) described the production of a number of gases from a variety of substances. He heated a variety of substances in a gun barrel and collected the evolved gases by the displacement of water. Among the substances he used were coal (which liberated methane), saltpeter (which produced oxygen), and a mixture of iron filings and dilute sulfuric acid (which liberated hydrogen). Perhaps under the influence of Newton, nearly all of Hales's work was quantitative, attempting to determine the amount of "air" which could be liberated from various substances. He failed to distinguish among the various gases he undoubtedly produced; they were all "air" to Hales.

Joseph Black (1728–1799) also studied the production of gas, but his interests grew out of his medical training. He was interested in the antacid properties of magnesia (magnesium carbonate), so he treated magnesia with acid and collected the gas which was liberated. He called this gas "fixed air," because it seemed to be trapped, or fixed, in the magnesia. Because he wanted to understand how magnesia functioned as an antacid, he studied the properties of fixed air carefully. He found that this gas was not identical with air and that it had properties unique to itself. Unlike ordinary air, fixed air would extinguish a burning candle that was inserted into it, and fixed air would produce a precipitate when it was bubbled through limewater. The formation of this precipitate appeared to be a characteristic of fixed air, and Black used it as a test for fixed air. With this test he demonstrated that fixed air was also produced in fermentation, in respiration, in the combustion of charcoal, and in the decomposition of limestone by heat. Indeed, when limestone was heated, fixed air was liberated and a residue of quicklime remained. The quicklime dissolved in water to form limewater, and when fixed air was bubbled through limewater, limestone was regenerated. Based on these results, Black concluded that

$$\text{limestone} = \text{quicklime} + \text{fixed air}$$

Black widened his studies to include the properties of alkalis and their reactions with acids and with each other. He found his chemical investigations so rewarding that he accepted the first professorship in chemistry at Glasgow and later moved to Edinburgh.

Henry Cavendish (1731–1810) also studied the properties of gases, including fixed air. He was particularly interested in what he called "factitious air"—that is, any gas that is contained in other bodies and can be liberated from them by chemical means. Cavendish produced fixed air by another method of Black, treating magnesia or limestone with acid. Extrapolating from this method, Cavendish treated other materials with acid, and discovered another factitious air produced by the action of acid on iron. This air was certainly different from fixed air, because when a flame was brought to it, rather than extinguishing the flame as fixed air would, this new air exploded! Not surprisingly, "inflammable air" was the name given by Cavendish to this new air. The inflammable air produced by the action of different acids on a variety of metals was identical, and it had a density far less than that of ordinary air. Cavendish interpreted these results in terms of the phlogiston theory. He considered a metal to be compounded of its calx and phlogiston and tentatively suggested that inflammable air was pure phlogiston. The acid liberated the phlogiston from the metal without flame, thereby producing an air that still possessed its flammability.

Perhaps the most thorough investigator of the chemical properties of gases was Joseph Priestley (1733–1804), who within the span of a dozen years discovered as many different gases. Considering that he had little formal scientific training, this was a remarkable achievement. Priestley's successes rested in some measure on his substitution of mercury for water in the pneumatic trough, which allowed him to collect gases that were soluble in water. Among the gases he discovered were ammonia ("alkaline air"), hydrogen chloride ("marine acid air"), nitrogen oxide ("nitrous air"), dinitrogen oxide ("diminished nitrous air"), nitrogen dioxide ("nitrous vapors"), sulfur dioxide ("vitriolic acid air"), and nitrogen ("phlogisticated air"). However, Priestley's most important discovery was the gas liberated when the red calx of mercury was heated. He found that a candle burned brilliantly when it was placed in this gas, and that a mouse could survive far longer in a sealed container of this gas than in the same amount of ordinary air. Being a confirmed phlogistonist, Priestley interpreted these observations in terms of the phlogiston theory, which held that when a substance burns, phlogiston escapes from it and enters the air. Because substances burned more intensely in Priestley's new gas, it appeared to have a greater capacity to absorb phlogiston than did ordinary air—that is, it appeared to be less saturated with phlogiston than was ordinary air. Therefore, Priestley called the new gas "dephlogisticated air." Even though most of Priestley's experimental results were accurate, he was misled in his explanations by his devotion to the phlogiston theory.

Although schooled as a lawyer, Antoine Lavoisier (1743–1794) had excellent training in science, and he was aware of most of the work being done by his contemporaries throughout Europe. He, like Hales before him, had a predilection for the quantitative and repeated experiments of Cavendish and Priestley with an eye to the mass balance. He determined that both sulfur and phosphorus gained weight when they burned in air. In addition, when phosphorus was burned in air within a bell jar over mercury, the volume of the air decreased by one-fifth, which was indicated by the rise of mercury into the jar. The white product of the combustion increased in weight by an amount equal to the weight of the air consumed in the combustion. Elaborating on experiments performed by Priestley, Lavoisier also examined what happened when the red calx of mercury was heated. Heating this red calx produced mercury and a colorless gas, and the mass of the mercury and gas produced was the same as that of the original calx. Lavoisier believed that the explanation for these observations offered by the phlogiston theory—that phlogiston was liberated from a substance when it burned, and that it combined with a calx when the calx was transformed to a metal—was inadequate. He suspected that the true situation was just the opposite. He suggested, in his hypothesis of combustion, that when a substance burns it combines with air or some part of it. Based on observations of Priestley's "dephlogisticated air," Lavoisier identified it as the component of air involved in combustion, and called it "oxygene" (oxygen). On the other hand, when a calx is transformed into its metal, the calx loses oxygen to the air, which accounts for its loss in weight.

With Lavoisier's theory of combustion, there was no need for the concept of phlogiston. Not only could the new theory account for all of the qualitative aspects of combustion and calx formation, but it could also explain the quantitative observations more readily than the phlogiston theory could. But of even greater significance for the development of modern chemistry were the implications of this new theory regarding the composition of substances. Metal calxes were now to be thought of as compounded from metals and oxygen. Were metals and oxygen themselves compounded of other substances? Lavoisier suggested that a substance which yields no other substance when subjected to the full force of chemical analysis should be considered an element. Based on this definition, Lavoisier identified 33 elements, most of which are still on the modern list of chemical elements. His interpretation of the term *element* and his identification of many elements provided the impetus for the rapid development of chemistry which occurred in the early decades of the 19th century.

Lavoisier's definition and identification of chemical elements spurred the development of chemical analysis, the determination of the elemental composition of materials. As the data from chemical analyses accumulated, several relationships in the composition of substances appeared. First, it seemed that some substances always contained the same weight proportion of elements. These substances were called compounds, and their definite composition was expressed as the law of constant composition. Second, it appeared that two different compounds may be composed of the same elements, but that the proportions of these elements differed between the compounds, and furthermore, the ratio of the weights of one element that combined with a fixed weight of the other element in these two compounds could be represented with small whole numbers. This relationship was expressed in the law of multiple proportions.

These two laws proved essential for the revival of the atomic theory. John Dalton (1766–1844) restated the atomic theory, which had been proposed and rejected repeatedly for over 2 millennia. However, Dalton attributed properties to atoms which would explain both the laws of constant composition and multiple proportions. He claimed that atoms are indivisible, that all atoms of a particular element have exactly the same mass, and that each element has a characteristic atomic mass. He proposed that chemical compounds are formed when atoms of the elements combine in definite small whole-number ratios, such as 1:1 or 1:2. Because all atoms of a particular element have the same mass, the fixed ratio of atoms in a compound dictates that its composition by weight be fixed. This accounts for the law of constant composition. The law of multiple proportions can also be explained: when two compounds of the same elements differ only in the combining ratios of the atoms, the ratios of their weights can be expressed in small whole numbers.

Because Dalton claimed that each element has a characteristic atomic mass, he attempted to determine the values of these masses. He realized that the absolute mass of each atom could not be determined, because he did not know how many atoms were contained in any given sample of an element. However, he could determine their relative masses. In order to do this, he needed two kinds of information: the relative weights of elements in compounds and the ratios of the atoms of the elements in those compounds. Chemical analyses provided the required information of the relative weights of elements in compounds. However, there was no experimental method to determine the ratio of atoms, so Dalton made a number of assumptions. He assumed that if only one compound of two elements were known, then the ratio of the atoms of its elements is 1:1. When two compounds were known, they were assumed to have 1:1 and 2:1 compositions. Selecting the lightest element known, hydrogen, and giving it a relative mass of 1, Dalton assigned atomic masses to a number of other elements. Using Lavoisier's analysis of water as 85% oxygen by weight, and assuming an atomic ratio of 1:1 (HO), Dalton calculated that an atom of oxygen is 5.66 times as heavy as an atom of hydrogen. When more accurate analyses were available, Dalton revised his relative atomic mass of oxygen to 7. This does not agree with the modern value of 16, because the chemical analysis was not very accurate, and, more important, because Dalton had assumed the wrong combining ratio of atoms in water. Dalton faced a dilemma: without the combining ratios it was impossible to determine the relative atomic masses, and without the relative atomic masses, the combining ratios could not be ascertained.

Another chemical property of gases was to prove of utmost importance in resolving this dilemma. Beginning in 1805 Joseph Louis Gay-Lussac (1778–1850) began investigating the combining volumes of reacting gases. Initially, he studied the reaction between hydrogen and oxygen. He discovered that the volume of hydrogen was always exactly twice the volume of oxygen involved when these volumes were measured at the same temperature and pressure. Gay-Lussac investigated similar studies of other reactions involving gases. In every case he found the volumes of reacting gases were always in ratios expressed by small whole numbers. He stated this in his law of combining volumes in 1808. Gay-Lussac was aware of Dalton's atomic theory and surmised that his discovery might have some bearing on this theory. However, he failed to suggest a specific connection between volume ratios and atomic ratios. This was done by Jöns Jakob Berzelius (1779–1848), who inferred from the law of combining volumes that equal volumes of gases must contain equal numbers of particles, or that the number of particles must be in small whole-number ratios. Because he had no evidence to the contrary, Berzelius adopted the former, simpler hypothesis. If this were the case, then the ratio of atoms in water would not be 1:1. Two volumes of hydrogen react with 1 volume of oxygen, indicating that water contains twice as many atoms of hydrogen as oxygen. Therefore, its formula is H_2O.

The situation is not quite as simple as this would suggest, as Dalton himself pointed out. If the volume of water vapor produced by 2 volumes of hydrogen and 1 volume of oxygen were measured, 2 volumes of water vapor would be found. This contradicts the hypothesis that equal volumes of gases contain equal numbers of atoms. How can a certain number of oxygen atoms produce twice as many particles of water? A solution to this problem was suggested by Amadeo Avogadro (1776–1856) in 1811. He proposed that the particles of elements in gases can be composed of more than one atom. Specifically, if hydrogen and oxygen particles were diatomic, then the reaction could be represented by the equation

$$2\,H_2 + O_2 \longrightarrow 2\,H_2O$$

This equation explains the observed volume relationships, if equal volumes of gases contain equal numbers of particles (molecules). Unfortunately, this suggestion by Avogadro was not widely accepted until 1860, by which time the problem of molecular formulas and atomic weights had developed into a completely chaotic situation. It was only then that Avogadro's hypothesis was again given serious consideration, and a table of atomic weights consistent with contemporary values was constructed.

There are a great number of chemical reactions that involve gases, and many of these reactions can be carried out in a classroom. Because of the constraints of space, it is impossible to include in this chapter all demonstrations that happen to involve a gas as a reactant or product. Some basis for the selection of the demonstrations is required. One of the criteria used in selecting the demonstrations in this chapter is that they illustrate chemical properties of gases that were significant in the development of chemistry. This allows the teacher to select demonstrations from this chapter which illustrate the fundamental principles of chemistry, whether or not they are presented in a historical context. Demonstrations 6.1, 6.6, 6.8, 6.12, 6.13, 6.15, 6.16, and 6.19 through 6.23 illustrate properties of gases that were significant in the development of modern chemistry. The remaining demonstrations present additional properties of these and other gases of general interest.

REFERENCES

1. E. Farber, *The Evolution of Chemistry*, Ronald Press Co.: New York (1952).
2. J. R. Partington, *A Short History of Chemistry*, 3d ed., Harper and Brothers: New York (1960).
3. A. J. Ihde, *The Development of Modern Chemistry*, Harper and Row: New York (1964); reprint ed., Dover Publications: New York (1984).
4. J. Parascandola and A. J. Ihde, *Isis* 60: Part 3, 351 (1969).
5. A. F. Scott, *Sci. Am.* 250:126 (1984).

6.1

Preparation and Properties of Carbon Dioxide

A gas is prepared by the reaction of marble chips with acid and is collected by the downward displacement of water. This gas does not support combustion; it forms a precipitate in limewater; and it is very soluble in aqueous sodium hydroxide. A large volume of this gas can be liberated from soda water, and when the gas is generated in a liquid containing bits of solid, the bits rise and sink periodically.

MATERIALS FOR PROCEDURE A

50 g marble chips, $CaCO_3$ (calcium carbonate)

100 mL tap water

2 mL acetone

25 mL saturated aqueous calcium hydroxide, $Ca(OH)_2$ (To prepare 1.0 liter of solution, boil 1.0 liter of distilled water, cover, and allow to cool overnight. Add 1.75 g of solid $Ca(OH)_2$. Shake well and allow to settle. If cloudy at time of use, filter.)

900 mL 1.0M aqueous sodium hydroxide, NaOH (To prepare 1.0 liter of 1.0M aqueous NaOH, dissolve 40 g of NaOH in 500 mL of distilled water and dilute the resulting solution to 1.0 liter.)

100 mL 6M hydrochloric acid, HCl (To prepare 1.0 liter of solution, add 500 mL of concentrated [12M] HCl to 300 mL of distilled water and dilute the resulting solution to 1.0 liter.)

2 right-angle glass bends, with outside diameter of 7 mm and length of each arm ca. 5 cm

2-holed rubber stopper to fit 250-mL Erlenmeyer flask

long-stemmed glass funnel (or thistle tube)

250-mL Erlenmeyer flask

2 90-cm lengths plastic or rubber tubing to fit 7-mm glass tubing

pan or dish, preferably transparent, with capacity of 4–10 liters

4 500-mL Erlenmeyer flasks

300-mL beaker (preferably tall form)

2 glass plates or watch glasses to cover 300-mL and 50-mL beakers

50-mL beaker

1-liter beaker

1-holed rubber stopper to fit 500-mL Erlenmeyer flask

4 solid rubber stoppers to fit 500-mL Erlenmeyer flasks

wooden splint

matches

dropping pipette

MATERIALS FOR PROCEDURE B

25 g sodium bicarbonate, $NaHCO_3$ (baking soda)

10 mL vinegar (5% w/w acetic acid, $HC_2H_3O_2$, in water)

candle, ca. 8 cm tall

matches

piece of cardboard, 4 cm $\times$ 4 cm

1-liter beaker

MATERIALS FOR PROCEDURE C

800 mL tap water

100 mL carbonated beverage, cold

baby bottle, with cap

latex pipette bulb

2 disks of 1/16-inch plastic, with diameter of 3.8 cm and 7-mm hole in the center

1-liter beaker

hot plate (or burner)

MATERIALS FOR PROCEDURE D

1500 mL tap water

100 g sodium bicarbonate, $NaHCO_3$ (baking soda)

10 p   , uncooked spaghetti

400 mL vinegar (5% w/w acetic acid, $HC_2H_3O_2$, in water)

2-liter beaker (or similar glass container)

stirring rod

PROCEDURE A

This procedure uses the original method of Joseph Black to synthesize carbon dioxide. It is not necessary to use this method of preparation; the apparatus can be used with other sources of CO_2 as well. For example, the generation flask can be filled with soda water and shaken to liberate the gas, as described in Procedure C.

Preparation

Assemble the apparatus as illustrated in Figure 1. Insert one of the right-angle bends through the 2-holed rubber stopper. Insert the stem of the funnel through the other hole in the stopper so that the tip of the stem is within 5 mm of the bottom of the 250-mL Erlenmeyer flask when the stopper is seated in the mouth of the flask. Attach the 90-cm length of plastic or rubber tubing to the free end of the glass bend. Place 50 g of marble chips (calcium carbonate) in the flask, and add 100 mL of water. Seat the stopper assembly in the mouth of the flask.

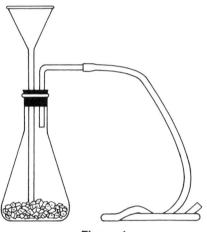

Figure 1.

Pour water into a pan or dish to a depth of 5 cm. Fill four 500-mL Erlenmeyer flasks to the brim with water. Cover the mouth of each flask with a hand while inverting it in the pan of water.

Pour 2 mL of acetone into the 300-mL beaker and cover the beaker with a glass plate or watch glass.

Pour 25 mL of aqueous calcium hydroxide (limewater) into a 50-mL beaker and cover the beaker with a glass plate.

Pour 900 mL of 1.0M aqueous sodium hydroxide into the 1-liter beaker. Insert the second right-angle glass bend into the 1-holed stopper. Attach the 90-cm length of plastic or rubber tubing to the free end of the glass bend.

Presentation

Pour 2–3 mL of 6M hydrochloric acid into the funnel. Fizzing will appear where the acid contacts the marble chips. Hold the open end of the rubber tubing under the water in the pan and observe the bubbles which emerge. Add more acid to the funnel to keep the bubbles forming at a moderate rate. After the bubbles have been forming for about 30 seconds, place the opening of the tubing under the mouth of one of the inverted Erlenmeyer flasks, and fill the flask with gas by the downward displacement of the water from the flask. When the flask is full of gas, seal it with a rubber stopper and remove it from the pan. Fill the other three flasks with gas in the same manner.

Light a wooden splint and insert the burning splint into one of the flasks of gas. The flame will be extinguished.

Ignite the acetone in the 300-mL beaker by holding a lighted match to the lip of the tilted beaker, and set the beaker of burning acetone on the table. Holding a second Erlenmeyer flask of gas no more than 10 cm above the mouth of the beaker, pour gas from the flask into the beaker. The flame will be extinguished. Pour the carbon dioxide from the beaker and relight the acetone to show that some flammable vapors remain.

Pour 25 mL of saturated aqueous calcium hydroxide into the third flask of gas. Swirl of solution around within the flask. A white precipitate will form in the solution. As the flask is swirled more, the white precipitate may redissolve.

Seat the 1-holed stopper assembly in the mouth of the last flask. Place the open end of the tubing in the beaker of 1.0M sodium hydroxide. Using a dropping pipette, withdraw several milliliters of the NaOH solution from the beaker. Quickly unstopper the flask, add the NaOH solution from the pipette, and restopper the flask. Swirl the solution around in the flask until the NaOH solution in the beaker begins to flow through the tubing into the flask. The solution will continue to flow slowly into the flask until the flask is nearly completely full.

PROCEDURE B[†]

Preparation

Light the candle and hold it horizontally over the card so that several drops of molten wax drop onto the center of the card. Quickly extinguish the candle and press its base into the soft wax on the card. When the wax hardens, the candle should be fixed to the card. Set the card and candle in the 1-liter beaker. Spread 25 g of sodium bicarbonate over the card on the bottom of the beaker.

Presentation

Light the candle. Slowly pour 3–4 mL of vinegar into the beaker. The vinegar will react with the sodium bicarbonate, causing effervescence. In a few seconds, the candle will be extinguished. Attempt to relight it; the match will be extinguished as it is lowered into the beaker. Once the fizzing has stopped, carefully tip the beaker to pour the carbon dioxide gas out of it. Then relight the candle; now it will burn. It can be extinguished again by pouring more vinegar into the beaker.

PROCEDURE C

Preparation

Remove the nipple from the cap of a baby bottle. Assemble the cap, latex pipette bulb, and plastic disks as illustrated in Figure 2.

Pour 800 mL of water into the 1-liter beaker and heat it to boiling on the hot plate.

† We wish to thank Ronald I. Perkins of Greenwich High School, Connecticut, for suggesting this procedure while he was a Fellow of the Institute for Chemical Education, University of Wisconsin–Madison.

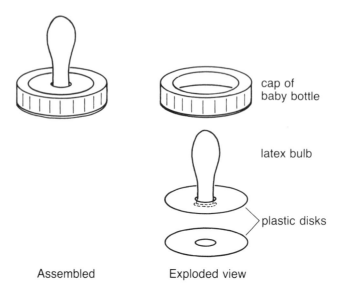

cap of
baby bottle

latex bulb

plastic disks

Assembled   Exploded view

Figure 2.

Presentation

Pour 100 mL of a cold carbonated beverage into the baby bottle and seal it with the cap assembly. Shake the bottle to induce fizzing in the liquid. When shaking no longer produces much effervescence, warm the bottle in a hot water bath. The gas given off by the carbonated beverage is enough to stretch the latex bulb to a nearly spherical shape.

PROCEDURE D [1]

Preparation

Pour 1500 mL of water into the 2-liter beaker. Add about 100 g of sodium bicarbonate to the water and stir the mixture until all of the solid has dissolved.

Presentation

Break 10 pieces of uncooked spaghetti into 2- to 3-cm pieces. Drop the pieces of spaghetti into the beaker and stir the mixture. Pour 300 mL of vinegar into the mixture and stir it for several seconds. Bubbles of CO_2 will form. In a minute or two, sufficient bubbles will cling to the pieces of spaghetti to cause them to rise to the surface. When they reach the surface, the bubbles escape, and the spaghetti falls back to the bottom. When the action slows, add another 100 mL of vinegar.

HAZARDS

Hydrochloric acid may cause severe burns. Hydrochloric acid vapors are extremely irritating to the skin, eyes, and respiratory system.

Saturated calcium hydroxide solution (limewater) is a base with a pH = 12.4. It can irritate the skin and eyes.

Sodium hydroxide can cause severe burns of the eyes and skin. Dust from solid sodium hydroxide is very caustic.

Acetone is very flammable. Care must be taken to prevent the diffusion of flammable vapors before igniting the acetone in the beaker.

DISPOSAL

The CO_2-generation flask should be filled with water and the resulting solution flushed down the drain. The marble chips may be collected and dried for reuse.

The contents of each of the 500-mL flasks should be flushed down the drain with water.

After removing the candle and card, the contents of the beaker in Procedure B should be flushed down the drain with water.

The contents of the beaker used in Procedure C should be flushed down the drain with water.

After removing the spaghetti, the contents of the beaker in Procedure D should be flushed down the drain with water.

DISCUSSION

Carbon dioxide is a colorless gas at 25°C and 1 atm. When chilled to −78.5°C at this pressure, it condenses to a white solid. This solid sublimes directly to the gas at atmospheric pressure. However, at a pressure of 5.2 atm, the solid melts to a colorless liquid at −56.6°C. The density of the gas is 1.977 g/liter at 0°C, and that of the solid is 1.56 g/cm^3 at −79°C [2].

Carbon dioxide was the first gas to be recognized as a substance distinct from air [3]. Joseph Black (1728–1799) was the first to realize this fact during the course of his studies of the effects of acids on magnesia alba ($MgCO_3$). His primary interest lay in discovering the exact nature of the action of magnesia alba in its common use as a stomach antacid, and he reported his preliminary results in his doctoral dissertation, *"De Humore Acido a Cibis orto, et Magnesia Alba"* ("On the Acid Humour Arising from Food, and Magnesia Alba" [4]). Subsequently, he extended his work to other carbonates. He found that these carbonates released an "air" when treated with acid. Moreover, this air was not the same as ordinary air, but had unique properties of its own. Limewater (aqueous $Ca(OH)_2$) quickly clouded when in contact with this air; strong alkalies of soda and potash (NaOH and KOH) were gradually neutralized by it; animals could not survive in it; and flames were extinguished in it. (The latter property is illustrated by Procedure C of this demonstration.) Because this air appeared to be trapped, or fixed, in the carbonates, he called it "fixed air."

Black also discovered that when chalk ($CaCO_3$) was heated, it lost weight and released fixed air, leaving behind quicklime (CaO). Quicklime dissolved in water to produce a solution of lime. This lime solution would react with fixed air, producing chalk. This cycle can be described by these modern chemical equations:

$$\underset{\text{chalk}}{CaCO_3} \xrightarrow{\text{heat}} \underset{\text{quicklime}}{CaO} + \underset{\text{fixed air}}{CO_2}$$

$$CaO + H_2O \longrightarrow \underset{\text{slaked lime}}{Ca(OH)_2}$$

$$Ca(OH)_2 + CO_2 \longrightarrow CaCO_3 + H_2O$$

By quantitative experiments, Black was able to describe the role of fixed air in this and other cycles that he studied. His work was instrumental in dispelling the notion that all gases were composed of the "element" air. By using the reaction of fixed air with slaked lime, Black developed a quantitative measure of its amount in a sample of gas. He was able to show that it was a component of ordinary air, and that exhaled air contained more fixed air than what was contained in ordinary air [5].

Procedure A of this demonstration uses the original method of Black to synthesize carbon dioxide. It is not necessary to use this method of preparation; the apparatus can be used with other sources of CO_2 as well. The generation flask can be filled with soda water (those of lemon-lime flavor seem most effective) and shaken to liberate the gas, as described in Procedure C. Carbonated water as a beverage is one of the earliest commercial uses of carbon dioxide, developed by Joseph Priestley himself [5]. The generator described in Procedure A can also be used with the reactants of Procedures B and D—sodium bicarbonate in place of marble chips and vinegar in place of hydrochloric acid. Alternatively, the flask can be filled with dry ice, and the gas produced by its sublimation can be collected as described in Procedure A. The flame and limewater tests can be used with these other sources to demonstrate that they do indeed produce carbon dioxide.

In addition to demonstrating the formation of a precipitate by the reaction of limewater with carbon dioxide, Procedure A also shows that this precipitate redissolves with excess carbon dioxide. (See the Discussion section of Demonstration 4.10 in Volume 1 of this series.)

$$CaCO_3(s) + CO_2(g) + H_2O(l) \longrightarrow Ca^{2+}(aq) + 2 HCO_3^-(aq)$$

The dissolution of carbon dioxide gas in alkali solutions, such as aqueous sodium hydroxide, can occur by two mechanisms [6, 7]. At a pH < 8, the predominant mechanism for the reaction is

$$CO_2(aq) + H_2O(l) \longrightarrow H_2CO_3(aq)$$

$$H_2CO_3(aq) + OH^-(aq) \longrightarrow HCO_3^-(aq) + H_2O(l)$$

At a pH > 10, the carbon dioxide reacts directly with the hydroxide ion, as indicated in the following mechanism:

$$CO_2(aq) + OH^-(aq) \longrightarrow HCO_3^-(aq)$$

$$HCO_3^-(aq) + OH^-(aq) \longrightarrow CO_3^{2-}(aq) + H_2O(l)$$

At pH values between 8 and 10, both mechanisms contribute significantly to the reaction. In this demonstration, the pH of the 1.0M NaOH solution is greater than 10, so the second mechanism is the dominant one in the reaction between the carbon dioxide and alkali. A more extensive discussion of the process of dissolution of carbon dioxide and water may be found in the Discussion section of Demonstration 4.10 in this series.

REFERENCES

1. D. Herbert, *Mr. Wizard's Supermarket Science*, Random House: New York (1980).
2. R. C. Weast, Ed., *CRC Handbook of Chemistry and Physics*, 59th ed., CRC Press: Boca Raton, Florida (1978).
3. A. F. Scott, *Sci. Am.* 250(l):126 (1984).
4. Translated by C. Brown, *J. Chem. Educ.* 12:225, 268 (1935).
5. A. J. Ihde, *The Development of Modern Chemistry*, Dover Publications: New York (1984).
6. D. M. Kern, *J. Chem. Educ.* 37:14 (1960).
7. D. A. Palmer and R. van Eldik, *Chem. Rev.* 83:651–731 (1983).

6.2

Reactions of Carbon Dioxide in Aqueous Solution

Solid carbon dioxide is dropped into tall cylinders filled with colored liquids. Over a period of several minutes the colors of the liquids change, some more than once [1]. The color of a liquid in a syringe changes as the plunger is drawn [2].

MATERIALS FOR PROCEDURE A

10 liters distilled water

10 mL thymolphthalein indicator solution (To prepare 100 mL of stock solution, dissolve 0.04 g of thymolphthalein in 50 mL of ethanol and dilute the resulting solution to 100 mL with distilled water.)

10 mL phenolphthalein indicator solution (To prepare 100 mL of stock solution, dissolve 0.05 g of phenolphthalein in 50 mL of ethanol and dilute the resulting solution to 100 mL with distilled water.)

10 mL phenolsulfonephthalein, phenol red, indicator solution (To prepare 100 mL of stock solution, dissolve 0.04 g of phenolsulfonephthalein in 11 mL of 0.1M sodium hydroxide and dilute the resulting solution to 100 mL with distilled water.)

10 mL 3′,3″-dibromothymolsulfonephthalein, bromothymol blue, indicator solution (To prepare 100 mL of stock solution, dissolve 0.04 g of 3′,3″-dibromothymolsulfonephthalein in 6.4 mL of 0.01M sodium hydroxide and dilute the resulting solution to 100 mL with distilled water.)

10 mL methyl red indicator solution (To prepare 100 mL of stock solution, dissolve 0.02 g of methyl red in 60 mL of 95% ethanol and dilute the resulting solution to 100 mL with distilled water.)

10 mL methylene blue indicator solution (To prepare 100 mL of stock solution, dissolve 0.01 g of methylene blue in 100 mL of distilled water.)

10 mL Yamada's universal indicator [3] (To prepare 200 mL of stock solution, dissolve 0.005 g thymol blue, 0.012 g methyl red, 0.060 g bromothymol blue, and 0.10 g phenolphthalein in 100 mL of ethanol. Add 0.01M sodium hydroxide until the solution is green, and dilute the resulting solution to 200 mL with distilled water.)

200 mL 5M aqueous ammonia, NH_3 (To prepare 1.0 liter of stock solution, pour 330 mL of concentrated [15M] NH_3 into 500 mL of distilled water, and dilute the resulting mixture to 1.0 liter with distilled water.)

or

200 mL 0.1M sodium hydroxide, NaOH (To prepare 1.0 liter of stock solution, dissolve 4.0 g NaOH in 500 mL of distilled water and dilute the resulting solution to 1.0 liter with distilled water.)

500 g dry ice, CO_2 (solid carbon dioxide)

20 mL 6M hydrochloric acid, HCl (To prepare 1.0 liter of stock solution, pour 500 mL of concentrated [12M] HCl into 300 mL of distilled water and allow the mixture to cool to room temperature. Dilute the resulting solution to 1.0 liter with distilled water.)

12 1-liter glass cylinders (or similar containers)

insulated gloves

MATERIALS FOR PROCEDURE B

25 mL club soda or other colorless carbonated beverage

10 drops methyl red indicator solution (For preparation, see Materials for Procedure A.)

50-mL plastic syringe

iron nail, ca. 7 cm in length

Bunsen burner

100-mL beaker

syringe cap

MATERIALS FOR PROCEDURE C

50 g dry ice, CO_2 (solid carbon dioxide) in chunks

5–20 liters hot tap water in pan or pail

insulated gloves

PROCEDURE A [1, 4, 5]

Preparation

Fill each of the 12 cylinders four-fifths full with distilled water and arrange the cylinders in pairs. To each of the first pair of cylinders add 5 mL of thymolphthalein indicator solution. To each cylinder of the second pair add 5 mL of phenolphthalein indicator. Add 5 mL of phenol red to the third pair, and 5 mL of bromothymol blue to the fourth pair. To each of the cylinders of the fifth pair add 5 mL of methyl red *and* 5 mL of methylene blue. Add 5 mL of universal indicator to each cylinder of the sixth pair.

Add enough 5M NH_3 solution or 0.1M NaOH solution to each of the first pair of cylinders (containing thymolphthalein) to produce a blue color in the solution (approximately 10 mL per liter of solution). Add the same amount of NH_3 or NaOH to each of the other ten cylinders.

Presentation

Wearing gloves, drop a lump of solid carbon dioxide into one cylinder of each pair. Leave the other cylinder unchanged for comparison. As the carbon dioxide dissolves, the pH decreases and the solutions change color. If ammonia is used to make the indicator solutions basic, the pH will not fall low enough to cause methyl red to change color. These color changes may be induced by adding 10 mL of 6M hydrochloric acid to the cylinders containing both methyl red and methylene blue and to the cylinder containing the universal indicator. When sodium hydroxide is used to make the indicator solutions basic, the color changes will occur more quickly than when ammonia is used, and the color changes for methyl red will be more nearly complete.

Commercial dry ice has a high density and sinks to the bottom of the solutions, so the bubbles of carbon dioxide stir the solution and saturate it rather rapidly. Solid carbon dioxide prepared from tank CO_2 has a lower density, which causes it to float. It saturates the solution more slowly and less uniformly, so the color changes occur only near the top of the cylinder. In this case the lower portion of the solution can serve as a reference for the original color.

PROCEDURE B [2]

Preparation

Adjust the plunger of the plastic syringe to read a volume of 50 mL. Heat the nail with the Bunsen burner and force the hot nail through the plunger where the plunger meets the barrel, as shown in the figure. Remove the nail and push the plunger all the way into the syringe.

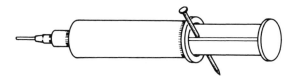

Presentation

Pour 25 mL of club soda into the 100-mL beaker and add 10 drops of methyl red indicator solution. The solution will be red. Draw 10 mL of this solution into the syringe, invert the syringe, and expel any air. Seal the syringe with the syringe cap.

Hold the syringe, pull back on the plunger to the 50-mL mark, and lock the plunger in position by inserting the nail through the hole in the plunger. Shake the syringe to cause fizzing in the solution. The liquid in the syringe will turn yellow. Release the plunger and shake the syringe to speed the redissolving of the carbon dioxide. The color of the solution will return to red. This process can be repeated several times, although it may be necessary to expel any gas from the syringe before the plunger is withdrawn again.

PROCEDURE C

Preparation and Presentation

Wearing gloves, drop about 50 g of dry ice chunks into the container of hot water. As the bubbles of cold CO_2 gas reach the humid air at the surface of the water, the water vapor condenses into a fog.

HAZARDS

Concentrated aqueous ammonia solution can cause burns and is irritating to the skin, eyes, and respiratory system.

Solid CO_2, dry ice, has a temperature of $-78°C$ and can cause frostbite. Thermal protection in the form of gloves or a towel should be used when handling dry ice.

DISPOSAL

Solutions should be discarded by flushing them down the drain with water.

Solid CO_2 should be allowed to sublime in a hood or other well-ventilated area.

DISCUSSION

The solubility of carbon dioxide in water at 1 atm of CO_2 as a function of temperature is given in Table 1 [6]. Dissolved carbon dioxide reacts with water according to the following equations:

$$CO_2(aq) + H_2O(l) \longrightarrow H_2CO_3(aq) \tag{1}$$

$$H_2CO_3(aq) + H_2O \longrightarrow H_3O^+(aq) + HCO_3^-(aq) \tag{2}$$

$$HCO_3^-(aq) + H_2O \longrightarrow H_3O^+(aq) + CO_3^{2-}(aq) \tag{3}$$

It is customary to treat this system as if all of the neutral, dissolved carbon dioxide were present as carbonic acid, H_2CO_3. When that is done, the value for the first dissociation constant (K_1) of carbonic acid is 4.5×10^{-7} [7]. This constant does not apply to any one of the above equations, although it has the form of the equilibrium con-

Table 1. Solubility of CO_2 in Water [6]

T (°C)	g CO_2/100 g H_2O	millimoles CO_2/liter of solution
0	0.3346	76.0
5	0.2774	63.0
10	0.2318	52.7
15	0.1970	44.8
20	0.1688	38.4
25	0.1449	32.9
30	0.1257	28.6
35	0.1105	25.1
40	0.0975	22.2

stant for equation 2, in which the concentration assigned to the H_2CO_3 is the sum of the concentrations of CO_2(aq) and H_2CO_3(aq).

The reaction of equation 1, the production of carbonic acid from aqueous carbon dioxide, is itself an equilibrium, and the ratio of CO_2(aq) to H_2CO_3(aq) is about 600:1 at 25°C [8]. This means that the concentration of H_2CO_3 is only 1/600 of that assumed in the tabulated value of its dissociation constant, 4.5×10^{-7}. Its true dissociation constant is $600 \times 4.5 \times 10^{-7}$, or 2.7×10^{-4}. Thus, carbonic acid is a much stronger acid than the tabulated value of K_1 would imply. In dealing with the calculations appropriate to this demonstration, the tabulated value of K_1 is more useful. However, in any discussion of the relative strengths of acids, the true dissociation constant is the more useful one.

The equilibrium constant for the second dissociation of carbonic acid applies to equation 3 and has the value 5.6×10^{-11} [7].

Because all of the solutions in Procedure A end up saturated with CO_2, the description of the equilibrium system will involve the use of K_1. The failure of the ammonia solutions to reach as low a pH as the sodium hydroxide solutions may be explained in terms of the buffering due to the bicarbonate ion formed in neutralizing the base.

$$[H^+] = \frac{4.5 \times 10^{-7} \times \text{solubility of } CO_2 \text{ in mol/liter}}{\text{initial moles of base/volume in liters}}$$

The expected pH of a pure solution of carbonic acid may be calculated in the traditional way.

The indicators used undergo the color changes at the pH's listed in Table 2. Yamada's universal indicator exhibits the color changes at the pH's given in Table 3. For

Table 2. Indicator Color Changes[a]

Indicator	Color change	pH range
thymolphthalein	blue to colorless	10.6–9.4
phenolphthalein	pink to colorless	10.0–8.2
thymol blue	blue to yellow	9.6–8.0
phenol red	red to yellow	8.0–6.6
bromothymol blue	blue to yellow	7.6–6.0
methyl red	yellow to red	6.0–4.8

[a] Methyl red modified with methylene blue changes from green to magenta in the pH range 6.6–4.8.

Table 3. Indicator Colors at Various pH Values

Indicator	10	9	8	7	6	5	4
universal indicator	violet	indigo	blue	green	yellow	orange	red
phenolphthalein	pink	light pink	colorless				
thymol blue	blue	green	yellow				
bromothymol blue	blue			green	yellow		
methyl red	yellow					orange	red

comparison, Table 3 also includes the colors of the component indicators at the various pH's. The color of the universal indicator is explainable in terms of the colors of the component indicators. However, there are significant differences in the concentrations and color intensities of the different components, so that the color of the universal indicator is not easily predicted in all cases.

In Procedure A, base is added to the cylinders initially to allow the pH to be established at approximately 11. The ammonia provides base by the reaction

$$NH_3(aq) + H_2O(l) \longrightarrow NH_4^+(aq) + OH^-(aq)$$

for which the standard equilibrium constant at 25°C is 1.8×10^{-5} [7].

The neutralization reactions of importance are those which provide bicarbonate ion.

$$H_2O(l) + NH_3(aq) + CO_2(aq) \longrightarrow NH_4^+(aq) + HCO_3^-(aq)$$

$$OH^-(aq) + CO_2(aq) \longrightarrow HCO_3^-(aq)$$

It takes a greater number of moles of ammonia than of sodium hydroxide to establish the pH at 11, so the ammonia solutions have a greater concentration of bicarbonate ion than do the sodium hydroxide solutions and are buffered at a higher pH at the end of Procedure A.

The process occurring in Procedure B is the reverse of that in Procedure A. In Procedure A, carbon dioxide gas dissolves to form a saturated aqueous solution, and the effect on the pH of the solution is observed. In Procedure B, carbon dioxide is removed from a solution, and the effect of that process on the pH of the solution is observed. Procedure B investigates the effect of pressure on the solubility of carbon dioxide. LeChatelier's principle suggests that a decrease in the pressure over a solution of carbon dioxide would result in a shift to the left of the equilibrium, represented by the following equation:

$$CO_2(g) + H_2O(l) \rightleftharpoons H^+(aq) + HCO_3^-(aq)$$

As the pressure is reduced, CO_2 bubbles out of the solution, the solution becomes less acidic, and the methyl red indicator turns from red to yellow. When the pressure returns to normal, the CO_2 redissolves, and the methyl red indicator returns to its acidic color, red.

REFERENCES

1. B. Z. Shakhashiri, G. E. Dirreen, and W. R. Cary, "Lecture Demonstrations," in *Source Book for Chemistry Teachers*, ed. W. T. Lippincott, Americal Chemical Society: Washington, D.C. (1981).
2. C. Zidick, *Chem 13 News*, May (1978).
3. L. S. Foster and I. J. Gruntfest, *J. Chem. Educ.* 14:274 (1937).

4. H. N. Alyea and F. B. Dutton, Eds., *Tested Demonstrations in Chemistry*, 6th ed., Journal of Chemical Education: Easton, Pennsylvania (1965).

5. B. Z. Shakhashiri, *Educ. in Chem.* (Britain) 18:16 (1981).

6. A. Seidell, *Solubilities of Inorganic and Metal Organic Compounds*, 3d ed., Van Nostrand: New York (1940).

7. R. C. Weast, Ed., *CRC Handbook of Chemistry and Physics*, 59th ed., CRC Press: Boca Raton, Florida (1978).

8. A. F. Trotman-Dickenson, Executive Ed., *Comprehensive Inorganic Chemistry*, Vol. 1, p. 1232, Pergamon Press: Oxford (1973).

6.3

Reaction Between Carbon Dioxide and Limewater

Carbon dioxide is bubbled through a clear, colorless solution in a glass vessel. A relatively dense precipitate forms fairly rapidly and then, upon continued bubbling, eventually redissolves. If the solution is then heated, the precipitate reforms. A person exhaling through a sample of the same solution produces the precipitate but is not able to redissolve it. This is described in Demonstration 4.10 in Volume 1 of this series.

6.4

Carbon Dioxide Equilibria and Reaction Rates: Carbonic Anhydrase-catalyzed Hydration †

A small amount of strong base is quickly added to a solution of ice-cold hydrochloric acid containing acid/base indicator and the time required for the solution to regain its acid indicator color is measured. The demonstration is repeated with acetic acid, with carbonic acid, and with carbonic acid to which a drop of blood has been added.

MATERIALS

80 mL distilled water

40 mL 0.1M hydrochloric acid, HCl (To prepare 1.0 liter of stock solution, add 8.3 mL of concentrated [12M] HCl to 500 mL of distilled water and dilute the resulting solution to 1.0 liter with distilled water.)

42 mL 0.1M acetic acid, $HC_2H_3O_2$ (To prepare 1.0 liter of stock solution, add 6.0 mL glacial [17M] $HC_2H_3O_2$ to 500 mL of distilled water and dilute the resulting solution to 1.0 liter with distilled water.)

5 mL 2M sodium hydroxide, NaOH, in a 10-mL graduated cylinder (To prepare 1.0 liter of stock solution, dissolve a total of 80 g NaOH, in 20-g increments, in 500 mL of distilled water and allow the solution to cool to room temperature. Dilute the resulting solution to 1.0 liter with distilled water.)

3.0 mL bromothymol blue (3',3''-dibromothymolsulfonephthalein) or phenol red (phenolsulfonephthalein) indicator solution (To prepare 250 mL of stock solution, dissolve 0.1 g of either indicator in 25 mL 0.01M sodium hydroxide and dilute the resulting solution to 250 mL with distilled water.)

80 mL 0.1M carbonic acid, H_2CO_3, carbon dioxide gas dissolved in water (Commercial seltzer water, or club soda without added citrate, that has been opened and kept capped in a refrigerator for about 2 days, with daily opening and reclosing of the bottle, is suitable. Once this initial "equilibration" of the dissolved carbon dioxide to about atmospheric pressure has occurred, the tightly capped and refrigerated bottle can be used successfully for at least 2 weeks.)

---

† This demonstration was developed by Professor Jerry A. Bell, Simmons College, in response to a discussion generated by observations made during demonstrations of gas/liquid equilibria carried out in a syringe by Clem Zidick, a high school teacher from Anchorage, Alaska, at the 1983 Camille and Henry Dreyfus Summer Institute for High School Chemistry Teachers, Princeton University, Princeton, New Jersey. Also see *Chem 13 News*, May 1978, and Procedure B of Demonstration 6.2.

several drops mammalian blood (Outdated whole human blood from a blood bank or a sample from a clinical laboratory [taken in a heparinized tube and kept refrigerated] are suitable sources. The blood may also be obtained from the demonstrator. With the sort of kit used by diabetics to obtain a drop of blood for self-testing, the demonstrator can painlessly prick a middle finger and collect a drop or two of blood in a long-tipped pipette. To assure that the procedure is sterile, all directions that come with such a kit must be followed.)

6 100-mL Erlenmeyer flasks

ice bath, sufficiently large to hold the containers of HCl, $HC_2H_3O_2$, and H_2CO_3

4 disposable pipettes

clock or watch, with second hand

PROCEDURE

Preparation

Pour 40 mL of distilled water into each of two of the 100-mL Erlenmeyer flasks. Add about 2 mL of 0.1M $HC_2H_3O_2$ to one of the flasks and about 1 mL of 2M NaOH to the other. Add 0.5 mL of indicator solution to each of the flasks. These will serve as references of the indicator color in acidic and basic solutions, respectively.

Place the containers of the stock solutions of HCl, $HC_2H_3O_2$, and H_2CO_3 in the ice bath, and allow sufficient time for them to become ice cold (at least 15 minutes).

Presentation

Pour about 40 mL of ice-cold 0.1M hydrochloric acid into one of the 100-mL Erlenmeyer flasks and add 0.5 mL of indicator. Either indicator will produce a yellow color. Use a disposable pipette to quickly inject about 1 mL of 2M NaOH solution into the acid solution, and time how long it takes for the basic color of the indicator (blue for bromothymol blue, red for phenol red) to disappear. Even without any agitation, the solution will return to its acidic color, yellow, in less than a second.

Repeat the procedure in the previous paragraph using 40 mL of ice-cold 0.1M acetic acid in place of hydrochloric acid. The results will be similar; no difference can be observed between the speed of the neutralization of a weak acid and that of a strong acid.

Pour about 40 mL of ice-cold 0.1M H_2CO_3 solution into another of the 100-mL Erlenmeyer flasks and add 0.5 mL of indicator solution. Use another disposable pipette to quickly inject about 1 mL of 2M NaOH solution, and time how long it takes for the basic color of the indicator to disappear. The solution will not return to the same yellow color it was initially, because the carbonic acid/bicarbonate buffer system that is formed keeps the pH between 6 and 7, producing a pale yellowish green with bromothymol blue or light orange with phenol red. The fading of the basic color is, however, still quite evident. The basic color of the solution will persist for about 25–35 seconds before fading.

Repeat the procedure of the previous paragraph, but add a drop of whole blood to the carbonic acid solution before adding NaOH. Rinse the pipette used to deliver the blood 2–3 times by drawing up solution from the reaction vessel and expelling it back into the flask. Then, add the NaOH solution and time the reaction. The basic color of the reaction will persist for about 5–10 seconds. Again, the final color of the solution is not that of the purely acidic form of the indicator and there is also the added reddish orange color imparted by the blood.

HAZARDS

Glacial acetic acid can cause severe burns, and the vapors are irritating to the eyes and respiratory system.

Sodium hydroxide can cause severe burns of the eyes and skin. Dust from solid sodium hydroxide is very caustic.

Be careful to use sterile techniques if you take samples of your own blood.

DISPOSAL

All solutions should be flushed down the drain with water.

DISCUSSION

The hydrochloric acid and acetic acid solutions are used in this demonstration to show that the neutralization of acids, whether strong or weak, is very rapid. This is an important point to establish, because students often think that a weak acid will give up its protons slowly enough to be observed on a time scale of seconds or minutes. There is no observable difference between the reaction carried out at room temperature or with the ice-cold acids, but the acids are used cold to be sure students do not think that temperature is the factor causing the slow change in the case of the carbonic acid solution.

Almost all the carbon dioxide in an aqueous solution of carbon dioxide is present as CO_2 molecules in equilibrium with a small amount of carbonic acid. The overall rate of reaction of hydroxide with this solution is slow because of the slow rate of the reaction forming carbonic acid from dissolved carbon dioxide:

$$CO_2(aq) + H_2O(l) \longrightarrow H_2CO_3(aq)$$

The rate equation for this reaction is

$$-d[CO_2]/dt = k_{CO_2}[CO_2]$$

where $k_{CO_2} = 0.03 \text{ s}^{-1}$ [1].

At a pH > 10, which corresponds to the solution formed in the first instant after base is added in this demonstration, the competing, faster reaction of hydroxide ion accounts for most of the reaction of carbon dioxide and yields (overall) carbonate ion.

$$CO_2(aq) + OH^-(aq) \longrightarrow HCO_3^-(aq)$$
$$HCO_3^-(aq) + OH^-(aq) \longrightarrow CO_3^{2-}(aq) + H_2O(l)$$

The rate of the slow first step for this reaction is given by

$$-d[CO_2]/dt = k_{OH^-}[OH^-][CO_2]$$

where $k_{OH^-} = 8500$ $M^{-1}s^{-1}$. The pH of the solution falls rapidly during this stage of the reaction. When the pH falls below about 9, the direct reaction between carbon dioxide and hydroxide becomes unimportant at 0°C. The carbonate which was produced from the hydroxide reacts with the carbonic acid which continues to be formed slowly from the remaining dissolved carbon dioxide. Thus, the pH continues to fall toward the equilibrium value for the bicarbonate/carbonic acid buffer system formed, but it falls much more slowly because the rate-determining step is now the slow hydration of the dissolved carbon dioxide. The rate constants for these reactions, as well as the important equilibrium constants, are given by Kern in his excellent review of the many decades of work on this complex system [1]. In the first paragraph of his review, Kern also gives directions for a demonstration, similar to that outlined above, for illustrating the slow rate of the hydration reaction. An interesting "double clock" experiment, in which both the rapid and slow stages of this overall process are studied, has been devised by Jones, Haggett, and Longridge [2]. These authors also suggest a demonstration of the slow stage that is very similar to that described here.

The indicator chosen to signal the "end point" of these neutralization reactions should have a color transition in the pH range between 6 and 7, because the final bicarbonate/carbonic acid buffer system has a pH near 6.4, the pK_a for carbonic acid (actually the final equilibrium mixture of dissolved carbon dioxide and carbonic acid). Bromothymol blue (pH range 6.0–7.6) and phenol red (6.6–8.0) are excellent, readily available candidates for this task. The final color for phenol red is a bit more difficult to distinguish in solutions containing some blood, but still gives quite satisfactory results.

Carbon dioxide is produced as a result of catabolism in our cells, carried away in the blood (mainly as bicarbonate), released as carbon dioxide gas in the lungs, and finally exhaled. Riddance of this end product of oxidation of our fuel molecules is important, because its buildup would quickly lead to acidosis (and death). However, the uncatalyzed dehydration reaction—

$$H_2CO_3(aq) \longrightarrow CO_2(aq) + H_2O(l)$$

—the reverse of the hydration reaction, is also very slow, with a rate constant of 20 s^{-1} [1]. If there were not some mechanism for speeding up these reactions, life, as we know it, would not be possible. Fortunately, blood contains an enzyme, carbonic anhydrase, that catalyzes these reactions.

Carbonic anhydrase is a relatively small metalloprotein, containing zinc, with a molecular weight of about 30,000 daltons. The zinc ion is essential to the enzyme's function; its removal causes the enzyme to lose its activity. The metalloenzyme can be reconstituted from the apoenzyme (protein without the metal ion) by adding zinc or other metal ions; no other metal ion works as well, although there are others that restore some activity [3]. The enzyme is inhibited by thiocyanate; the time for neutralization in the presence and absence of thiocyanate can be measured to test this assertion [4].

Carbonic anhydrase has the highest activity of any enzyme known, with a turnover number of about 600,000 per second [5]. This high activity is required if the dehydration of carbonic acid is to be rapid enough to eliminate most of it during the blood's short residence time (about 1 second) in the lungs. One can sense how efficient the catalyst is by observing that a drop of blood in about 40 mL of reaction mixture in this demonstration (a dilution of at least 500-fold) decreases the time required for the neutralization by a factor of 3 to 5.

REFERENCES

1. D. M. Kern, *J. Chem. Educ.* 37:14 (1960).
2. P. Jones, M. L. Haggett, and J. L. Longridge, *J. Chem. Educ.* 41:610 (1964).
3. R. S. McQuate, *J. Chem. Educ.* 54:645 (1977).
4. R. L. McGeachin, *J. Chem. Educ.* 32:191 (1955).
5. L. Stryer, *Biochemistry*, 2d ed., W. H. Freeman and Co.: San Francisco (1975).

6.5

Combustion of Magnesium in Carbon Dioxide

Ignition of magnesium and an oxidizing agent in a block of dry ice results in a brilliant flare of light and a black and white residue. This is described as Demonstration 1.37 in Volume 1 of this series.

6.6

Preparation and Properties of Hydrogen

A gas is prepared by the reaction of zinc with acid and is collected by the downward displacement of water. This gas is flammable.

MATERIALS

30 g mossy zinc

75 mL distilled water

100 mL 6M hydrochloric acid, HCl (To prepare 1.0 liter of solution, add 500 mL of concentrated [12M] HCl to 300 mL of distilled water and dilute the resulting solution to 1.0 liter.)

right-angle glass bend, with outside diameter of 7 mm and length of each arm ca. 5 cm

2-holed rubber stopper to fit 250-mL Erlenmeyer flask

long-stemmed glass funnel (or thistle tube)

90-cm length plastic or rubber tubing to fit 7-mm glass tubing

250-mL Erlenmeyer flask

pan or dish, preferably transparent, with capacity of 4–10 liters

500-mL Erlenmeyer flask

matches

15-cm candle

PROCEDURE

Preparation

Assemble the apparatus as illustrated in the figure. Insert one end of the right-angle bend through the 2-holed rubber stopper. Insert the stem of the funnel through the other hole in the stopper so that the tip of the stem is within 5 mm of the bottom of the 250-mL Erlenmeyer flask when the stopper is seated in the mouth of the flask. Attach the 90-cm length of plastic or rubber tubing to the free end of the glass bend. Place 30 g of mossy zinc in the flask, and add 75 mL of distilled water. Seat the stopper assembly in the mouth of the flask.

Pour water into the pan or dish to a depth of 10 cm. Fill the 500-mL Erlenmeyer

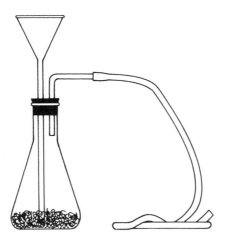

flask to the brim with water. Cover its mouth with the palm of a hand while inverting it in the pan of water.

Presentation

Pour 2–3 mL of 6M hydrochloric acid into the funnel. Fizzing will appear where the acid contacts the zinc. Hold the open end of the plastic or rubber tubing under the water in the pan and observe the bubbles which emerge. Add more acid to the funnel to keep the bubbles forming at a moderate rate. Flush the air from the generation flask and delivery tubing by allowing the bubbles to escape for at least 30 seconds. Then situate the opening of the tubing under the mouth of the inverted Erlenmeyer flask and fill the flask with gas by the downward displacement of the water from the flask.

Light the candle. Quickly lift the Erlenmeyer flask of gas from the pan of water, and, holding the flask upside down, insert the top of the burning candle into the flask. Observe that the candle's flame is extinguished inside the flask and that an almost invisible flame appears at the mouth of the flask. The flame at the mouth of the flask melts some of the wax at the sides of the candle, thereby giving evidence of its presence. Remove the candle from the flask and note that the candle relights and that the gas continues to burn at the mouth of the flask. This process of extinguishing and relighting the candle may be repeated several times. Set the flask down and watch as the flame gradually sinks into the flask and then is extinguished. Because the flame is not very luminous, it may be more easily observed in a darkened room.

HAZARDS

Hydrogen gas is very flammable and yields explosive mixtures with air. Therefore, the generated hydrogen should be allowed to sweep the air from the generation apparatus before the collection of gas is started, and care must be taken to avoid getting air into the collection flask before igniting the hydrogen.

Hydrochloric acid may cause severe burns. Hydrochloric acid vapors are extremely irritating to the skin, eyes, and respiratory system.

Because the hydrogen flame is not very luminous, it can be hard to see, and it is possible to be inadvertently burned by the flame.

The rim of the Erlenmeyer flask becomes hot as the hydrogen burns, so it should be allowed to cool before the flask is handled.

DISPOSAL

The hydrogen-generation flask should be filled with water and the resulting solution flushed down the drain. Any remaining zinc can be collected and dried for reuse.

DISCUSSION

At room temperature and atmospheric pressure, hydrogen is a colorless and odorless gas. At 1 atm its boiling point is $-252.87°C$, and its melting point is $-259.14°C$ [1].

Hydrogen was discovered and originally named "inflammable air" in 1766 by Henry Cavendish [2]. Cavendish followed the method of Joseph Black for producing "fixed air" by the action of acids on carbonates. Extrapolating from Black's method, Cavendish tried substituting other materials for the carbonates that Black had used. When he substituted such metals as zinc, iron, or tin, he again observed fizzing, and he was able to collect the resulting gas. He found this new gas to be much lighter than normal air. In addition, instead of extinguishing a flame as fixed air did, this gas exploded in air when a flame was brought near it. Hence came the name inflammable air for this new gas.

The reaction used in this demonstration is identical to one of the methods originally used by Cavendish. In modern terms, the reaction is

$$2\ HCl(aq)\ +\ Zn(s)\ \longrightarrow\ H_2(g)\ +\ ZnCl_2(aq)$$

Although the flammability of hydrogen was noted at its discovery, the nature of the products formed from the burning of hydrogen was not discovered until some years later. Again, it was Cavendish who discovered that water is produced when hydrogen burns. However, it took Lavoisier to interpret this information as indicative of the compound nature of water. In modern terms, the burning of hydrogen is represented by the equation

$$2\ H_2(g)\ +\ O_2(g)\ \longrightarrow\ 2\ H_2O(l)$$

This demonstration illustrates the flammability of hydrogen. Its explosive potential is highlighted in Demonstration 6.7 and in Demonstration 1.42 of Volume 1 of this series.

REFERENCES

1. R. C. Weast, Ed., *CRC Handbook of Chemistry and Physics*, 59th ed., CRC Press: Boca Raton, Florida (1978).
2. A. J. Ihde, *The Development of Modern Chemistry*, Dover Publications: New York (1984).

6.7

Explosiveness of Hydrogen

A modified plastic soft-drink bottle,† its top opening having been fitted with a glass tube through a rubber stopper, is filled with hydrogen gas by the downward displacement of air. The hydrogen is ignited where it escapes through the glass tube at the top. As the hydrogen burns, the flame diminishes in size until it suddenly fires back into the bottle. The loud detonation is accompanied by a bright yellow fireball inside the bottle [1, 2]. A special apparatus emits a whistle that decreases in pitch as the hydrogen it contains burns. Eventually, an explosion occurs [3, 4].

MATERIALS FOR PROCEDURE A

cylinder of hydrogen, with valve

> *or*

> 50 g mossy zinc

> 75 mL distilled water

> 100 mL 6M hydrochloric acid, HCl (To prepare 1.0 liter of solution, add 500 mL of concentrated [12M] HCl to 300 mL of distilled water and dilute the resulting solution to 1.0 liter.)

> right-angle glass bend, with outside diameter of 7 mm and length of each arm ca. 5 cm

> 2-holed rubber stopper to fit flask

> long-stemmed glass funnel (or thistle tube)

> 250-mL Erlenmeyer flask

ring stand

15-cm iron ring

clamp with holder

2-liter colorless plastic bottle (such as a carbonated beverage bottle) from which the bottom has been removed

> *or*

> bell jar, 14 cm × 23 cm high, having an opening at the top

8-cm length glass tubing, with outside diameter of 7 mm

1-holed rubber stopper to fit neck opening of plastic bottle

90-cm length plastic or rubber tubing to fit 7-mm glass tubing

matches

---

† We wish to thank Professor J. Arthur Campbell of Harvey Mudd College for suggesting the use of a plastic soft-drink bottle in this demonstration.

MATERIALS FOR PROCEDURE B

cylinder of hydrogen, with valve

    or the materials for a hydrogen generator listed in Materials for Procedure A

special whistle (See description in Procedure B)

ring stand

2 iron rings

ca. 40-cm length plastic tubing to fit valve

wooden matches

rubber stopper to fit larger opening of special whistle

PROCEDURE A

Preparation

Attach the clamp and iron ring to the ring stand so that the clamp holds the neck of the plastic bottle while the open base of the bottle rests on the ring, as illustrated in Figure 1.

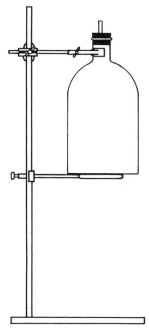

Figure 1.

Insert the 8-cm length of glass tubing through the 1-holed rubber stopper so 2–3 mm protrude beyond the narrow end of the stopper. Seat the stopper in the neck of the bottle.

If a cylinder of hydrogen is not available, fabricate a hydrogen generator in the following manner: Assemble the apparatus as illustrated in Figure 2. Insert one end of

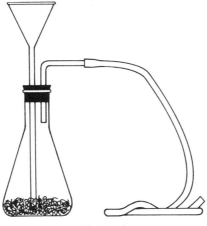

Figure 2.

the right-angle bend through the 2-holed rubber stopper. Attach the 90-cm length of plastic or rubber tubing to the free end of the glass bend. Insert the funnel through the other hole in the stopper so that the tip of its stem is within several millimeters of the bottom of the 250-mL Erlenmeyer flask when the stopper is seated in the mouth of the flask. Place 50 g of mossy zinc in the Erlenmeyer flask and add 75 mL of distilled water. Seal the flask with the stopper. Pour 10 mL of 6M hydrochloric acid through the funnel. The resultant fizzing is evidence of the production of hydrogen gas.

If a cylinder of hydrogen gas is used, attach the plastic or rubber tubing to the valve of the cylinder.

Presentation

Attach the free end of the plastic tubing from the hydrogen generator or cylinder to the glass tube at the top of the bottle. Allow hydrogen gas to flow through the tube into the bottle. If a hydrogen generator is being used, additional 10 mL aliquots of 6M HCl may be needed to continue the production of hydrogen gas. Allow the hydrogen to flow into the bottle at a moderate rate for about 1 minute, until the air in the bottle has been displaced by hydrogen. The hydrogen can be detected in the bottle by inserting a hand up into the bottle; hydrogen gas will feel cooler than air.

When the bottle is full of hydrogen, remove the plastic tubing from the glass tube at the top of the bottle. Using a match, ignite the hydrogen gas escaping through the glass tube. The size of the flame will slowly diminish as the hydrogen is consumed. When an explosive mixture of air and hydrogen has formed within the bottle, the flame will suddenly strike back and detonate the mixture with a loud explosion and a yellow flash.

PROCEDURE B

Preparation

A whistle such as the one illustrated in Figure 3 must be constructed. It can be made by welding together two 13-cm cones of 0.032-inch brass. A hole, 2 cm in diameter, is cut at the apex of the lower cone, and a brass tube, 2 cm in diameter and

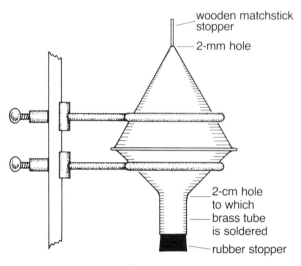

wooden matchstick
stopper

2-mm hole

2-cm hole
to which
brass tube
is soldered

rubber stopper

Figure 3.

5 cm long, is welded to the hole. A 2-mm hole is drilled at the apex of the top cone. Alternatively, a whistle may be fashioned from a metal toilet float by drilling a 2-mm hole at one end, a 2-cm hole at the other, and soldering a metal tube of the dimensions given above to the larger hole.

Mount the whistle on a ring stand between the pair of iron rings, as illustrated in Figure 3.

If a cylinder of hydrogen is unavailable, construct the hydrogen generator as described in Procedure A. Connect the plastic tubing to the valve of the cylinder or to the outlet of the hydrogen generator.

Presentation

Hold the free end of the plastic tubing from the hydrogen cylinder or generator over the hole at the top of the whistle. With a slow flow of hydrogen, flush the air from the whistle. The hydrogen can be detected at the lower end of the whistle by inserting a finger up into the tube; hydrogen gas will feel cooler than air. When the whistle is filled with hydrogen, stopper the top of the whistle with a wooden matchstick, stopper the bottom of the whistle with a rubber stopper, and **close the valve on the cylinder**.

Light a match. Remove the rubber stopper and the matchstick sealing the whistle. Bring the lighted match to the top of the whistle and ignite the escaping hydrogen. Dim the room lights. Observe the flame burning at the top of the whistle. After a short time, the pitch of the whistle will drop into the audible range and continue to fall. When an explosive mixture has formed inside the whistle, a very loud explosion will occur, and flames will shoot out from the base of the whistle.

HAZARDS

Hydrogen gas is very flammable and yields explosive mixtures with air. Therefore, the generated hydrogen should be allowed to sweep the air from the generation apparatus before the collection of gas is started. Do not have any open flames nearby while the plastic bottle or the whistle is being filled.

Hydrochloric acid can cause severe burns. Hydrochloric acid vapors are extremely irritating to the skin, eyes, and respiratory system.

DISPOSAL

The hydrogen-generation flask should be filled with water and the resulting solution flushed down the drain. Any remaining zinc can be collected and dried for reuse.

DISCUSSION

When ignited, hydrogen reacts with oxygen to produce water. The ignition temperature is in the range of 580–590°C [5]. The equation for the reaction is

$$2 \, H_2(g) + O_2(g) \longrightarrow 2 \, H_2O(g)$$

This exothermic reaction yields 232 kJ/mole of water formed [6], and this heat is more than sufficient to keep the temperature of the reactants above the ignition range. (See the Discussion section of Demonstration 1.42 in Volume 1 of this series.)

Hydrogen, which is less dense than air, is buoyant in air and rises out through the jet at the top of the bottle or the hole at the top of the whistle. As the hydrogen escapes from the jet, it burns in the oxygen of the air. As hydrogen is displaced from the interior of the bottle by air, the buoyant force decreases and the hydrogen escapes more slowly from the jet and burns with a smaller flame. In addition, as the hydrogen is displaced from the bottle, the air begins to mix with the hydrogen in the bottle. Eventually, the hydrogen escapes so slowly from the jet that the flame can strike back into the mixture of air and hydrogen in the bottle, detonating an explosion of the mixture.

The temperature at which a mixture of hydrogen and oxygen will explode depends upon the partial pressure of hydrogen, as indicated in Figure 4. At atmospheric pressure (10^5 Pa), the temperature needed to produce an explosion is above the ignition range, so hydrogen at this pressure burns in air. As air mixes with hydrogen, the partial pressure of the hydrogen decreases, thereby lowering the temperature at which the ex-

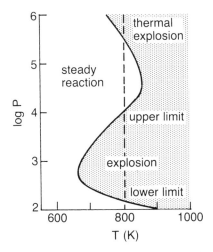

Figure 4. Source: Figure 26.9 [7]

plosion will occur. The temperature within the flame is high enough to detonate the mixture of hydrogen and air in this demonstration.

The hydrogen gas can be detected as the bottle is filled, because it feels cooler than the air it is replacing. The temperature of the hydrogen is very close to that of the air, yet it feels cooler. This is because hydrogen is a better conductor of heat than air is, and it removes heat from the hand more rapidly than air, so it feels cooler. The heat conductivity of a gas is related to its molecular weight. The molecules of hydrogen are less massive than the molecules of air, and therefore move more rapidly than the molecules in air at comparable temperatures. Thus, they are able to remove heat from the hand more quickly.

REFERENCES

1. G. S. Newth, *Chemical Lecture Experiments*, Longmans, Green and Co.: London (1928).
2. H. A. Bent and H. E. Bent, *J. Chem. Educ.* 58:9 (1981).
3. R. D. Eddy, *J. Chem. Educ.* 36:256 (1959).
4. I. Talesnick, *Demonstrations in Chemistry*, Faculty of Education, Queen's University: Kingston, Ontario (undated).
5. E. Thorpe, *Dictionary of Applied Chemistry*, Vol. 2, p. 701, Longmans, Green and Co.: London (1930).
6. "Selected Values of Chemical Thermodynamic Properties," *Natl. Bur. Stand. (U.S.)*, Circ. 500, p. 9 (1952).
7. P. W. Atkins, *Physical Chemistry*, p. 879, W. H. Freeman and Co.: San Francisco (1978).

6.8

Preparation and Properties of Oxygen

A gas is prepared by the reaction of hydrogen peroxide with potassium permanganate and is collected by the downward displacement of water. This gas supports vigorous combustion. The generation of this gas produces a cloud.

MATERIALS FOR PROCEDURE A

100 mL 3% hydrogen peroxide, H_2O_2

60 mL 3M sulfuric acid, H_2SO_4 (To prepare 1.0 liter of solution, dissolve 170 mL of concentrated [18M] H_2SO_4 in 500 mL of water and dilute the resulting solution to 1.0 liter.)

3.0 g potassium permanganate, $KMnO_4$

right-angle glass bend, with outside diameter of 7 mm and length of each arm ca. 5 cm

2-holed #5 rubber stopper

long-stemmed glass funnel (or thistle tube)

250-mL Erlenmeyer flask

90-cm length plastic or rubber tubing to fit 7-mm glass tubing

pan or dish, preferably transparent, with capacity of 4–10 liters

2 500-mL Erlenmeyer flasks

small watch glass or glass plate to cover Erlenmeyer flask

wooden splint

matches

Bunsen burner

tongs

0.5 g steel wool, 000 grade, formed into tight ball

MATERIALS FOR PROCEDURE B

100 mL 3% hydrogen peroxide, H_2O_2

60 mL 3M sulfuric acid, H_2SO_4 (For preparation, see Materials for Procedure A.)

3.0 g potassium permanganate, $KMnO_4$

600-mL beaker (preferably tall form)

wooden splint

matches

MATERIALS FOR PROCEDURE C

4.0 g manganese dioxide, MnO_2

30 mL 30% hydrogen peroxide, H_2O_2

piece of facial tissue, ca. 6 cm square

10-cm length of thread

gloves, plastic or rubber

opaque bottle, ca. 500-mL capacity

2-holed rubber stopper to fit bottle

PROCEDURE A

Preparation

Assemble the apparatus as illustrated in the figure. Insert one end of the right-angle bend through the 2-holed rubber stopper. Insert the stem of the funnel through the other hole in the stopper so that the tip of the stem is within 5 mm of the bottom of the 250-mL Erlenmeyer flask when the stopper is seated in the mouth of the flask. Attach the 90-cm length of plastic or rubber tubing to the free end of the glass bend. Pour 100 mL of 3% hydrogen peroxide into the flask, and seat the stopper assembly in the mouth of the flask.

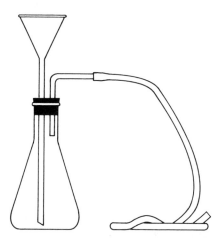

Dissolve 3.0 g of potassium permanganate in 60 mL of 3M sulfuric acid.

Pour water into the pan or dish to a depth of 10 cm. Fill two 500-mL Erlenmeyer flasks to the brim with water. Cover the mouth of each while inverting it in the pan of water.

Presentation

Pour 2–3 mL of the potassium permanganate solution into the funnel. Fizzing will appear where the permanganate mixes with the peroxide. Hold the open end of the plastic or rubber tubing under the water in the pan and observe the bubbles which emerge. Add more permanganate solution to the funnel to keep the bubbles forming at a moderate rate. After the bubbles have been forming for about 30 seconds, place the opening of the tubing under the mouth of one of the inverted Erlenmeyer flasks and fill this flask with gas by the downward displacement of its water. Place the small glass cover over the mouth of the flask and remove it from the bath, setting it upright on the table. Fill the other flask in the same manner.

Light a wooden splint and extinguish the flame so that only glowing embers remain. Plunge the glowing splint into one of the flasks of gas. Observe that the splint bursts into bright flames in the gas.

Light the Bunsen burner. Using tongs, hold the wad of steel wool in the flame of the burner until it glows. Plunge the hot steel wool into the other flask of gas. Observe that the steel wool bursts into flame in the gas.

PROCEDURE B

Preparation

Pour 100 mL 3% hydrogen peroxide into the 600-mL beaker. Dissolve 3.0 g potassium permanganate in 60 mL of 3M sulfuric acid.

Presentation

Pour about 5 mL of the potassium permanganate solution into the hydrogen peroxide in the 600-mL beaker. Observe that the mixture fizzes, indicating the production of a gas, and that the purple color of the permanganate solution disappears.

Light a wooden splint and extinguish the flame leaving a few glowing embers. Pour another 5 mL of the potassium permanganate solution into the beaker. Quickly insert the glowing splint into the gas in the beaker. The glowing embers will burst into flame. This may be repeated until all of the permanganate solution is consumed.

PROCEDURE C

Preparation

Wrap 4.0 g of MnO_2 in a small piece of facial tissue or similar paper. Tie the package with the 10-cm length of thread. Wearing gloves, pour 30 mL of 30% H_2O_2 into the opaque bottle having a capacity of about 500 mL. Carefully suspend the packet of MnO_2 inside the neck of the bottle and stopper the bottle so the stopper holds the packet by its thread.

Presentation

Do not have open flames nearby when presenting this demonstration. Remove the stopper from the bottle and step back. The packet of MnO_2 will drop into the H_2O_2 and cause a sudden decomposition of the peroxide, liberating oxygen gas. The rapid evolution of gas carries small droplets of liquid out of the bottle, producing a cloud over the mouth of the bottle. The bottle will become quite hot.

HAZARDS

Because sulfuric acid is both a strong acid and a powerful dehydrating agent, it must be handled with great care. Spills should be neutralized with an appropriate agent, such as sodium bicarbonate ($NaHCO_3$), and then wiped up. The dilution of concentrated sulfuric acid is a highly exothermic process and releases sufficient heat to cause burns.

Handle potassium permanganate with great care, because explosions can occur if it is brought into contact with organic or other readily oxidizable substances, either in solution or in the solid state.

A 3% solution of hydrogen peroxide is a topical antiseptic and cleaning agent. It decomposes rapidly with evolution of oxygen upon exposure to light, heat, or foreign materials.

Steel wool should be cut with scissors, not pulled apart, because pulling it with the fingers can cause skin lacerations.

Care must be taken when inserting burning materials into pure oxygen; the flames will increase dramatically. Therefore, the burning object should be grasped at a distance from the burning portion.

Because 30% hydrogen peroxide is a strong oxidizing agent, contact with skin and eyes must be avoided. In case of contact, immediately flush with water for at least 15 minutes; get immediate medical attention if the eyes are affected.

Avoid contact between 30% hydrogen peroxide and combustible materials. Avoid contamination from any source, because any contaminant, including dust, will cause rapid decomposition and the generation of large quantities of oxygen gas. Store 30% hydrogen peroxide in its original closed container, making sure that the container vent works properly.

Be careful not to tip the prepared bottle in Procedure C. The reaction can propel the stopper from the bottle with considerable force.

DISPOSAL

The oxygen-generation flask or bottle should be filled with water and the resulting solution flushed down the drain.

DISCUSSION

Oxygen is a colorless, odorless gas at room temperature and atmospheric pressure. It condenses to a pale blue liquid at $-182.962°C$ and 1 atm [1]. The preparation and properties of this liquid are investigated in the subsequent demonstration. Oxygen has a freezing point of $-218.4°C$ [1]. The density of the gas at $0°C$ and 1 atm is 1.429 g/liter; the liquid at $-183°C$ is 1.149 g/mL, and the solid at $-252°C$ is 1.426 g/cm^3 [1].

The discovery of oxygen is generally attributed to Joseph Priestley. In 1774, Priestley produced a gas by the action of sunlight focussed by lens on a sample of mercury calx (HgO) prepared by heating mercury in air. He found that this gas had remarkable properties. It made a candle burn extremely brightly, and a mouse could survive in it about twice as long as in normal air [2].

Although Priestley realized that the gas he had produced differed from any other gas he had prepared, he was unable to recognize it as a new substance. He named it "dephlogisticated air" and considered it to be common air from which most of the phlogiston had been removed. Thus it was able to accept the phlogiston escaping from a burning candle more readily than normal air could. It was Lavoisier who realized that this gas was a unique substance which was a component of air. Lavoisier elucidated the role of this gas, which he named oxygen, in the formation of metal calces and in combustion.

The reaction which these early chemists used to produce oxygen gas is represented by the modern chemical equation.

$$2\ HgO(s) \longrightarrow O_2(g) + 2\ Hg(l)$$

This method is not used in this demonstration because of the toxicity of mercury and its compounds. Instead, oxygen is prepared from the decomposition of an acidic aqueous hydrogen peroxide solution by permanganate ion.

$$5\ H_2O_2(aq) + 2\ MnO_4^-(aq) + 6\ H^+(aq) \longrightarrow$$
$$5\ O_2(g) + 2\ Mn^{2+}(aq) + 8\ H_2O(l)$$

The production of oxygen by this method was developed in the early 19th century by Louis Thenard [3]. The equation indicates that the hydrogen peroxide is oxidized to oxygen by the purple permanganate ion, which is reduced to the nearly colorless manganese(II) ion.

REFERENCES

1. R. C. Weast, Ed., *CRC Handbook of Chemistry and Physics*, 59th ed., CRC Press: Boca Raton, Florida (1979).
2. A. F. Scott, *Sci. Am.* 250:126 (1984).
3. E. Farber, *The Evolution of Chemistry*, Ronald Press Co.: New York (1952).

6.9

Reaction of Oxygen with Dextrose: "The Blue-Bottle Experiment"

A stoppered flask, which is partially filled with a colorless solution, is shaken, and the solution becomes blue. Upon standing, the solution returns to colorless. Further shaking regenerates the blue color. This cycle of color changes can be observed by repeating the shaking/standing procedure. Other color changes can be produced using different indicators [1].

MATERIALS

300 mL distilled water

8 g potassium hydroxide, KOH

10 g dextrose, $C_6H_{12}O_6$

1 or more of the following indicators (for each sequence of color changes demonstrated):

 6–8 drops methylene blue indicator solution (To prepare a stock solution, dissolve 0.2 g of methylene blue in 100 mL of distilled water.)

 6–8 drops resazurin indicator solution (To prepare a stock solution, dissolve 0.1 g of resazurin in 100 mL of distilled water.)

 20 drops indigo carmine indicator solution (To prepare a stock solution, dissolve 0.25 g indigo carmine in 25 mL of distilled water. This solution must be prepared within a few hours of use.)

500-mL Florence (or Erlenmeyer) flask

rubber stopper to fit flask

PROCEDURE

Preparation

Prepare the materials for this demonstration within 15 minutes of its presentation. The solution is not stable for longer periods of time.

Pour 300 mL of distilled water into the 500-mL flask and add 8 g of KOH. Swirl the flask to dissolve the KOH. When the KOH has dissolved, add 10 g of dextrose to the flask and allow the sugar to dissolve completely.

Add 6–8 drops of methylene blue indicator solution to the flask and swirl it. Allow the flask to rest until the solution becomes colorless.

Presentation

Give the flask one or two quick shakes. The blue color will reappear and then slowly fade. The time required for the color to fade depends on how much the flask is shaken. The regeneration and fading of the blue color may be repeated a number of times by shaking the flask and allowing it to rest.

If resazurin indicator solution (6–8 drops) is used in place of methylene blue indicator solution, the alkaline dextrose solution will initially be fluorescent red-blue [1]. After about 2 minutes (depending on how much oxygen is dissolved in the water), the color changes to fluorescent red and gradually fades to colorless. If the flask is given a gentle shake, the fluorescent red reappears and then fades. Shaking the flask vigorously restores the color to the original fluorescent red-blue, which fades to fluorescent red and then to colorless. The demonstration may be repeated several times.

If indigo carmine indicator solution (20 drops) is used, the solution is initially yellow-green [1]. In 3–5 minutes the color changes to orange and finally to yellow. A gentle shake restores the orange color, and, upon more vigorous shaking, the green reappears. Because this indicator solution is not very stable, this mixture, as well as all mixtures containing indigo carmine solution, cannot be used repeatedly.

Two color changes may be observed by using both resazurin indicator and methylene blue indicator solutions [1]. Add 6–8 drops of resazurin indicator solution to the alkaline dextrose solution. When the solution has become colorless, add 6–8 drops of methylene blue indicator solution. After the solution has again become colorless, a gentle shake will restore the fluorescent red of the resazurin. A second, more vigorous shake produces a blue solution. Upon resting, the solution then changes from blue, to violet, to red, to colorless, and the demonstration may be repeated.

If 6–8 drops of methylene blue indicator solution are added to a flask containing 20 drops of indigo carmine indicator solution and the alkaline dextrose solution, the color change will be from green, to orange, to yellow [1].

If 6–8 drops of resazurin indicator solution are added to a flask containing 20 drops of indigo carmine indicator solution and the alkaline dextrose solution, the color change will be from red to yellow [1].

Finally, combining all three indicator solutions with the alkaline dextrose solution produces color changes from violet, to red, to yellow [1]. The shaking procedures may be repeated only a few times, because the indigo carmine indicator solution is not very stable.

Upon standing, all solutions turn yellow-brown and cannot be reused. Even a solution which contains only dextrose and potassium hydroxide develops this color within 10–15 minutes.

HAZARDS

Potassium hydroxide is very caustic and causes severe burns. Eye or skin contact with the solid or solutions must be prevented.

DISPOSAL

The solutions should be flushed down the drain with water.

DISCUSSION

Campbell describes in detail the reaction of oxygen with dextrose in the presence of methylene blue indicator [2]. The origin of the demonstration is not known, but Campbell found it in use in 1954 at the University of Wisconsin. It is useful in stimulating student interest and in initiating discussions about experimentation. Students could be asked to propose simple experiments to determine the nature of the chemical and physical processes responsible for the color changes (caution should be exercised in handling strongly alkaline solutions). For example the role of air in the process can be determined by completely filling the flask with solution. In this case, shaking does not restore the color. Bubbling nitrogen or natural gas through the solution to remove other gases and leave an inert atmosphere above the solution results in no color change upon shaking, except after prolonged standing of the alkaline dextrose solution (see below). Readmitting oxygen gas into the flask reinitiates the cycle. Many other such experiments could also be devised [2].

This demonstration/experiment was proposed by Campbell as an excellent way to introduce students to kinetics and the study of reaction mechanisms [2]. The reaction rate is first-order in OH^-, in methylene blue, and in dextrose; the rate is independent (zero-order) of the concentration of gaseous O_2. This rate law can be obtained easily by finding the dependence of the de-bluing rate on solution concentration. Furthermore, the energy of activation can be obtained by plotting the logarithm of the de-bluing time, log t, for a constant amount of dissolved gas as a function of the reciprocal of the absolute temperature, 1/T. As Campbell points out, this is possible because the rate of the slow step is independent of the concentration of O_2, and thus the time, t, which is required for the total O_2 to disappear is directly related to the rate constant, k. A straight line is obtained from the plot of log t versus 1/T.

This reaction system provides a simple and straightforward way of visually demonstrating the role of a catalyst. Methylene blue acts as a catalyst for the oxidation of dextrose by gaseous oxygen. After repeated cycles, when the reaction mixture turns yellow, it should be discarded and not used to illustrate catalysis or the other kinetics concepts mentioned above. The yellow color results from transformations and degradations of dextrose which occur in alkaline solutions (see below).

The other indicator solutions suggested by Chen function in a manner similar to methylene blue. All are redox indicators having different colors in the oxidized and reduced forms. Indigo carmine solutions are sensitive to light, and their color fades on standing.

Under alkaline conditions, dextrose is irreversibly oxidized by oxygen gas to a variety of products including gluconic acid, glucuronic acid, and δ-gluconolactone [3].

When oxygen gas is bubbled through an alkaline solution of dextrose, a yellow-brown color develops in 10–15 minutes. The exact product, which has a yellow-brown color, is not known.

Indicator Changes

The color changes observed in this demonstration are due to reversible (and some irreversible) oxidation and reduction of the indicator dyes.

Methylene blue indicator is blue in the oxidized form and colorless in the reduced form. The reduction is reversible and may be written as:

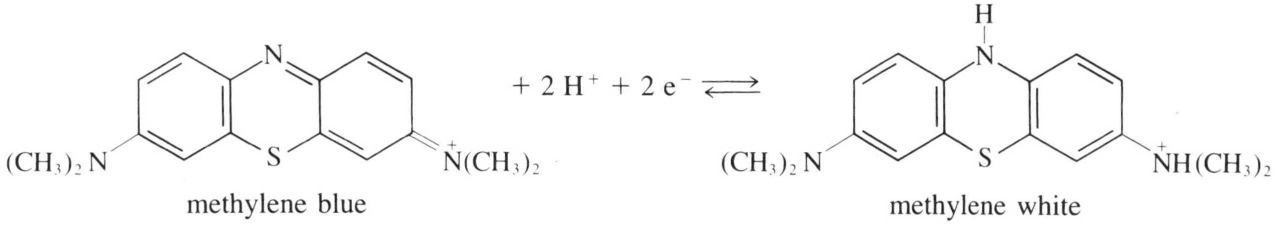

methylene blue
methylene white

Methylene blue presumably acts as a catalyst in this reaction. Its role is not very well understood, but it may be reduced to the colorless methylene white by dextrose. Shaking the flask dissolves oxygen in the solution and oxidizes the dye back to the blue form. If the solution is allowed to stand, the methylene blue is reduced by dextrose. The time required for the color to fade depends on the extent of shaking, because varying amounts of oxygen may be dissolved. The demonstration can be repeated until all of the dextrose has been oxidized or all of the oxygen in the flask has been consumed. The final reaction-product mixture is yellow-brown, and the blue cannot be restored by shaking or by bubbling oxygen gas through the mixture.

Resazurin first undergoes an *irreversible* reduction by dextrose to resorufin.

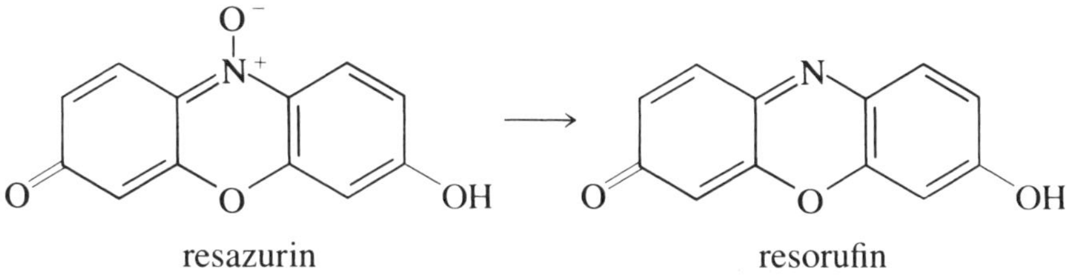

resazurin
resorufin

The resorufin is fluorescent red, and it undergoes a *reversible* reduction to a colorless compound, dihydroresorufin.

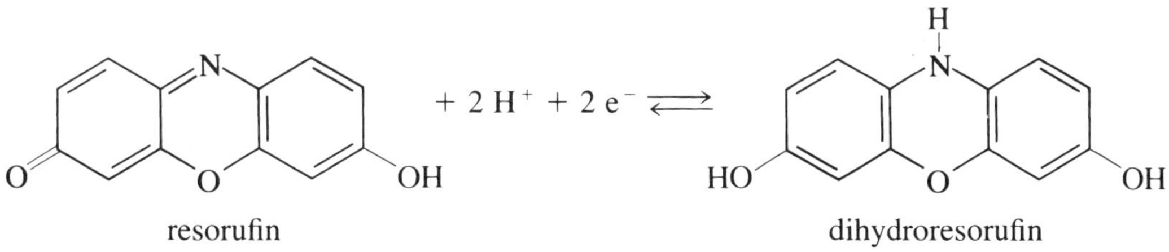

resorufin
dihydroresorufin

This colorless form is readily oxidized, and gentle aeration is sufficient to restore the fluorescent red color.

Indigo carmine also undergoes a reversible reduction.

indigo carmine
reduced species

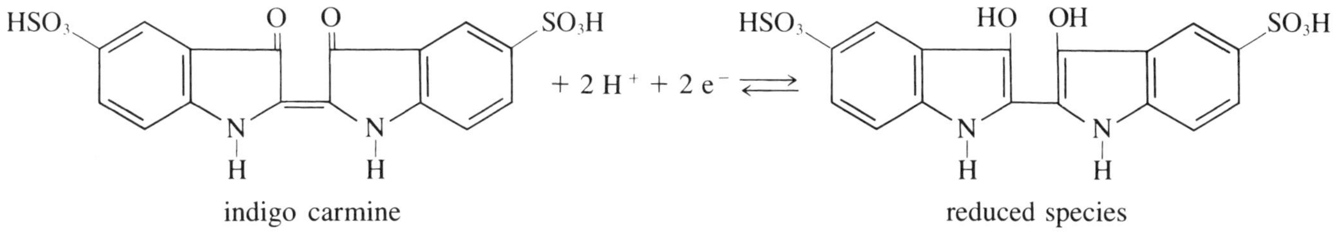

The oxidized form appears blue or green in solution, and the reduced form is pale yellow. The intermediate orange color is possibly due to formation of a red semi-quinone intermediate [4].

The reduced species is easily oxidized and, again, gentle aeration is sufficient to reverse the color changes.

The color changes produced by combining indicators should be related to the relative ease of oxidation or reduction in each case. The color changes that occur when the flask is shaken suggest that the reduced form of indigo carmine is oxidized first and methylene white last, indicating that methylene blue is the easiest to reduce and indigo carmine is hardest to reduce. However, the reduction potentials for the three indicators, listed in the table, do not agree with this deduction.

Reduction Potentials of Indicators [5]

Indicator	E (in volts) at 30°C and pH = 9
methylene blue	−0.050
resorufin	−0.217
indigo carmine	−0.199

The reduction potential of oxygen in alkaline solution is +0.401 volt [6].

$$O_2 + 2\,H_2O + 4\,e^- \longrightarrow 4\,OH^- \qquad E = +0.401 \text{ volt}$$

The reduction potential for the gluconolactone/glucose couple in alkaline solution is not known, but the value in acid solution is +0.09 volts [7].

Despite the need to reconcile the E values with the color changes observed, all the demonstrations described are quite effective in developing the student's ability to relate observations to plausible mechanisms and explanations.

REFERENCES

1. P. S. Chen, *Entertaining and Educational Chemical Demonstrations*, Chemical Elements Publishing Co.: Camarillo, California (1974).
2. J. A. Campbell, *J. Chem. Educ.* 40:578 (1963).
3. A. White, P. Handler, and E. L. Smith, *Principles of Biochemistry*, 5th ed., McGraw-Hill: New York (1973).
4. E. Bishop, Ed., *Indicators*, Pergamon Press: New York (1972).
5. O. Tomicek, *Chemical Indicators*, Butterworths: London (1951). The value for resorufin is calculated from data on p. 501.
6. R. C. Weast, Ed., *CRC Handbook of Chemistry and Physics*, 59th ed., CRC Press: Boca Raton, Florida (1978).
7. W. M. Clark, *Oxidation Reduction Potentials of Organic Systems*, Williams and Wilkens: Baltimore (1960).

6.10

Preparation and Properties
of Liquid Oxygen

When liquid oxygen is poured between the poles of a magnet, some of the liquid is held in the gap until the liquid evaporates. The characteristic blue color of liquid oxygen may also be displayed, as well as the effect of a high concentration of oxygen on the rate of combustion [1].

MATERIALS FOR PROCEDURE A

5 liters liquid nitrogen

cylinder of oxygen, with valve

2-m length copper or aluminum tubing, with outside diameter of 6 mm

60-cm length rubber tubing, with outside diameter of 7–8 mm

1-liter Dewar flask

test tube, 25 mm × 200 mm

stand, with clamp for test tube

250-mL *unsilvered* Dewar flask (optional)

permanent magnet, with pole gap of ca. 1 cm and field strength between the poles of ca. 9 kilogauss†

insulated gloves

overhead projector (optional)

clear acrylic sheet, 1/4 inch thick, large enough to cover surface of overhead projector (optional)

250-mL beaker

cigarette

matches

MATERIALS FOR PROCEDURE B

5 liters liquid nitrogen

cylinder of oxygen, with valve

†Several types of magnets are suitable for this demonstration. The permanent magnet used at the University of Wisconsin–Madison is a "horn gap" magnet from 1940s-vintage radar with an original field strength of 1.8 kilogauss. Conical pole pieces machined from iron were added to the pole faces to reduce the pole gap from 4 cm to about 1 cm. This increased the magnetic field to 9 kilogauss.

1-liter Dewar flask

test tube, 25 mm × 200 mm

stand, with clamp for test tube

30-cm length rubber tubing, with outside diameter of 7–8 mm

30-cm length glass tubing, with outside diameter of 7 mm

250-mL *unsilvered* Dewar flask (optional)

permanent magnet, with pole gap of ca. 1 cm and field strength between the poles of ca. 9 kilogauss†

insulated gloves

overhead projector (optional)

clear acrylic sheet, 1/4 inch thick, large enough to cover surface of overhead projector (optional)

PROCEDURE A

Preparation

Fashion a coil like the one in Figure 1 from about 2 m of copper or aluminum tubing. The coil has 10–12 turns with a separation of about 2 cm between turns. The diameter of the coil is about 6 cm and its height is 22 cm. The ends of the coil extend about 9 cm above the last turn, as shown in Figure 1. Attach a short piece of rubber tubing to each open end of the coil. Attach one of the pieces of tubing to the valve on the oxygen cylinder.

Presentation

Fill the Dewar flask with liquid nitrogen. Clamp the 200-mm test tube on the stand and insert into it the remaining free end of the rubber tubing attached to the coil. Immerse the coil in liquid nitrogen in the Dewar flask. Open the valve on the oxygen cylinder and adjust it to produce a fairly high flow rate, high enough to push the liquid oxygen out of the coil. Replenish the liquid nitrogen in the Dewar flask as necessary. It is possible to prepare about 50 mL of liquid oxygen in less than 1 minute. When the test tube is about half full, close the cylinder valve, disconnect the rubber tubing from the valve, and remove the coil from the liquid nitrogen.

Often moisture condenses quickly on the outside of the test tube and obscures the pale blue color of the liquid oxygen. The color can be seen more readily if the liquid oxygen is transferred to a small, unsilvered Dewar flask.

Demonstrate the strength of the magnet by placing some iron object in the pole gap and having a student pull it out.

Pour some liquid nitrogen over the poles of the magnet. The liquid will run through the gap. This step, as well as showing the diamagnetism of liquid nitrogen, cools the gap prior to the use of liquid oxygen. Next, wearing a glove to handle the tube

---

† See note on page 147.

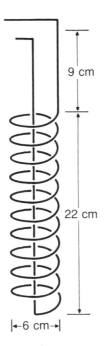

Figure 1. Copper coil for preparing liquid oxygen.

of oxygen, slowly pour several milliliters of liquid oxygen between the poles of the magnet. A small amount of the liquid will be held in the gap for several seconds until the liquid has completely evaporated. This demonstrates the paramagnetism of liquid oxygen. Repeat the procedure several times.

To show the magnetic properties of liquid oxygen to a large group, the magnet may be laid on its side on an overhead projector, as shown in Figure 2. The glass stage

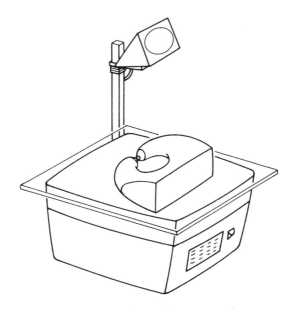

Figure 2.

of the projector should be protected from the extreme cold of the liquids by covering it with the clear acrylic sheet before situating the magnet.

To show the effect of the high concentration of oxygen on the rate of combustion, pour several milliliters of liquid oxygen into the 250-mL beaker, set the beaker on the bench, and quickly toss a lighted cigarette or similar smoldering object into the beaker. The cigarette will be consumed in an immediate flare.

PROCEDURE B

Preparation

Fill the 1-liter Dewar flask with liquid nitrogen. Clamp the 200-mm test tube so it is immersed in the liquid nitrogen. Connect one end of the rubber tubing to the valve of the oxygen cylinder and the other end to the 30-cm length of glass tubing. Place the free end of the glass tube in the test tube.

Presentation

Open the valve of the oxygen cylinder and adjust it to produce a gentle flow of gas through the glass tube. Liquid oxygen will condense in the test tube if the flow rate of the gas is sufficiently low. Replenish the liquid nitrogen in the Dewar flask as necessary. It will take 5–8 minutes to condense about 50 mL of liquid oxygen. When the test tube is approximately half full, close the valve on the cylinder.

Follow the instructions beginning with the second paragraph of the Presentation section of Procedure A.

HAZARDS

The high concentration of oxygen in the liquid phase is capable of supporting violent combustion when mixed with substances which do not burn under ordinary conditions [2]. Open flames are particularly dangerous if liquid oxygen has been spilled on clothing. Clothing and other fabrics will absorb liquid oxygen and retain oxygen for an extended period of time, resulting in its enhanced susceptibility to enflame.

The liquified gases used in this demonstration are extremely cold and can cause severe frostbite. Care should be taken to avoid skin contact with these liquids or with objects which have been in contact with them long enough to become chilled themselves.

Dewar flasks are susceptible to implosion with sudden changes in temperature or mechanical shock. To prevent injuries caused by flying glass, the Dewar flasks should be wrapped with tape or netting. (The unsilvered flask may be wrapped with transparent library tape.)

DISPOSAL

Liquid oxygen should be poured onto a noncombustible surface and allowed to evaporate.

DISCUSSION

Both procedures in this demonstration rely on oxygen's boiling point ($-183°C$) being higher than that of nitrogen ($-196°C$) [3]. This difference in boiling points allows liquid oxygen to be prepared by condensing oxygen gas in liquid nitrogen.

As the liquid oxygen evaporates, it produces a region of high oxygen gas concentration in its container. The effect of this high concentration on the rate of combustion is demonstrated by tossing a smoldering combustible into a beaker containing a small amount of liquid oxygen. The gas in the beaker is nearly pure oxygen, and its high concentration causes the combustible to be consumed almost instantaneously.

The blue color of liquid oxygen is the result of its absorption of light in the red portion of the spectrum. This absorption may be attributed to the simultaneous transition of *two* ground state ($^3\Sigma_g^-$) oxygen molecules to their first excited state ($^1\Delta_g$) in a two-molecule–one-photon process [4].

$$2\,O_2(^3\Sigma_g^-) + h\nu(\sim630nm) \longrightarrow 2\,O_2(^1\Delta_g)$$

The attraction of liquid oxygen by a magnet indicates that oxygen is paramagnetic and suggests that the oxygen molecule contains unpaired electrons. The absence of such an interaction between liquid nitrogen and the magnet indicates that nitrogen is diamagnetic and that all of the electrons in nitrogen molecules are paired.

Valence bond theory predicts that both N_2 and O_2 molecules are diamagnetic. However, as this demonstration shows, oxygen is paramagnetic. Whereas a fairly simple valence-bond description of N_2 adequately accounts for the behavior of liquid nitrogen observed in this demonstration, it is insufficient to explain the behavior of liquid oxygen. To account for the magnetic properties of oxygen, a molecular orbital description of the bonding in O_2 can be used. A simplified molecular orbital diagram for O_2 is given in Figure 3. The unpaired electrons in the π_{2p}^* molecular orbitals produce a triplet state ($^3\Sigma_g^-$).

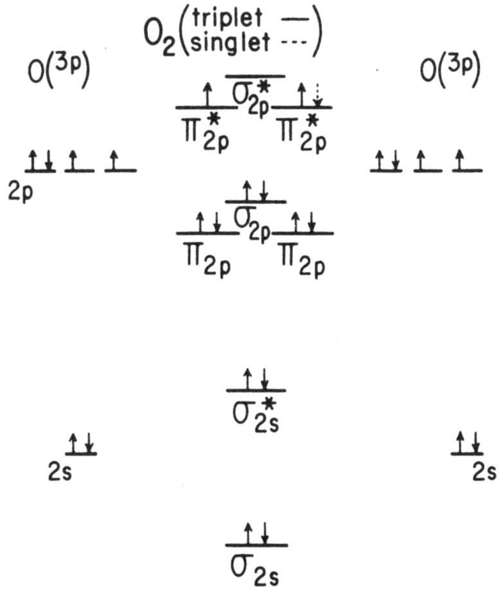

Figure 3. Molecular orbital diagrams of atomic and molecular oxygen.

This demonstration of the paramagnetism and color of liquid oxygen can be used along with a demonstration of the red chemiluminescence of the singlet state of O_2. This is Demonstration 2.1 in Volume 1 of this series. The red chemiluminescence is due to a process which is the reverse of the absorption responsible for the blue color of liquid oxygen. Both the color and the paramagnetism of oxygen are intrinsic properties of its molecules and not bulk properties of the liquid phase [4].

REFERENCES

1. B. Z. Shakhashiri, G. E. Dirreen, and L. G. Williams, *J. Chem. Educ.* 57:373 (1980).
2. I. J. Wilk, *J. Chem. Educ.* 45:A547 (1968).
3. R. C. Weast, Ed., *CRC Handbook of Chemistry and Physics*, 62d ed., CRC Press: Boca Raton, Florida (1982).
4. E. A. Ogryzlo, *J. Chem. Educ.* 42:647 (1965).

6.11

Explosive Reaction
of Hydrogen and Oxygen

When four balloons filled with different gases are ignited, the balloons emit differing intensities of sound. The loudest occurs when a hydrogen-oxygen mixture is ignited. Hydrogen-oxygen mixtures can also be ignited in a glass soft-drink bottle, soap bubbles, or a special cannon. This is described in Demonstration 1.42 in Volume 1 of this series.

6.12

Preparation and Properties of Nitrogen

A gas is prepared by the reaction of ammonium chloride with sodium nitrite in aqueous solution and is collected by the downward displacement of water. This gas is unable to support combustion and is unreactive toward limewater.

MATERIALS

10 g ammonium chloride, NH_4Cl

60 mL distilled water

15 g sodium nitrite, $NaNO_2$

500 mL cold tap water

50 mL saturated aqueous calcium hydroxide, $Ca(OH)_2$ (To prepare 1.0 liter of solution, boil 1.0 liter of distilled water, cover, and allow to cool overnight. Add 1.75 g of solid $Ca(OH)_2$. Shake well and allow to settle. If cloudy at time of use, filter.)

piece of lithium metal, ca. 1 g (optional)

right-angle glass bend, with outside diameter of 7 mm and length of each arm ca. 5 cm

1-holed #5 rubber stopper

90-cm length plastic or rubber tubing to fit 7-mm glass tubing

250-mL Erlenmeyer flask

ring stand, with clamp to hold 250-mL Erlenmeyer flask

pan or dish, preferably transparent, with capacity of 4–10 liters

3 500-mL Erlenmeyer flasks

1-liter beaker

Bunsen burner

matches

2 small watch glasses or glass plates to cover Erlenmeyer flask

wooden splint

combustion spoon (optional)

PROCEDURE

Preparation

Assemble the apparatus as illustrated in the figure. Insert one end of the right-angle bend through the 1-holed rubber stopper. Attach the 90-cm length of plastic or rubber tubing to the free end of the glass bend. Dissolve 10 g of ammonium chloride in 60 mL of distilled water in the 250-mL Erlenmeyer flask. Add 15 g of sodium nitrite to the solution in the flask, and seal the flask with the stopper containing the bend. Mount the flask on the ring stand so that the Bunsen burner will fit under it.

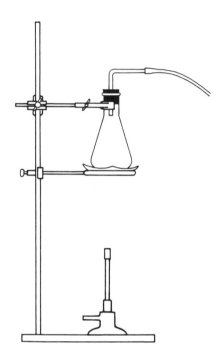

Pour water into a pan or dish to a depth of 10 cm. Fill three 500-mL Erlenmeyer flasks to the brim with water. Cover the mouth of each with the palm of a hand while inverting it in the pan of water.

Pour 500 mL of cold tap water into a 1-liter beaker.

Presentation

Gently heat the contents of the 250-mL Erlenmeyer flask with the Bunsen burner until bubbles of gas begin to form in the flask. When the reaction begins, it may continue without further heating. If the effervescence should become too vigorous, the reaction may be slowed by immersing the flask in the cold water in the 1-liter beaker.

Hold the open end of the plastic or rubber tubing under the water in the pan and observe the bubbles which emerge. After the bubbles have been forming for about 30 seconds, place the opening of the tubing under the mouth of one of the inverted Erlenmeyer flasks and fill the flask with gas by the downward displacement of its water. Place the small watch glass or glass plate over the mouth of the flask and remove it from the bath, setting it upright on the table. Fill the other flasks in the same manner.

Light a wooden splint and plunge the burning splint into one of the flasks of gas. Note that the flame is extinguished.

Pour 50 mL of saturated aqueous calcium hydroxide (limewater) into the second flask of gas. Swirl the solution around within the flask. No apparent reaction occurs between the gas and the solution.

As an optional procedure, cut a fresh piece of lithium metal. Place it in a combustion spoon so that its shiny, freshly cut surface is showing, and suspend it in the third flask of gas. After several minutes the shiny surface of the lithium will have become black because of the formation of lithium nitride (Li_3N).

HAZARDS

Solid sodium nitrite should not contact combustible materials, for this may result in a fire. A mixture of solid sodium nitrite with solid ammonium salts is explosive upon heating. Therefore, the solution used in this demonstration must not be allowed to evaporate to dryness.

Lithium metal should not be handled with the fingers, because it will react with moisture on the skin, forming caustic lithium hydroxide.

Saturated calcium hydroxide solution (limewater) is a base with a pH = 12.4. It can irritate the skin and eyes.

DISPOSAL

The nitrogen-generation flask should be filled with water and the resulting solution flushed down the drain. The contents of each of the 500-mL flasks should be flushed down the drain with water.

DISCUSSION

Nitrogen is a colorless, odorless gas at 1 atm and 25°C. It condenses to a liquid at −195.79°C and freezes to a solid at −210.01°C. It is only slightly soluble in water, 1.6 volumes of gas dissolving in 100 volumes of water at 20°C [1].

Nitrogen gas was first investigated by Daniel Rutherford, a student of Joseph Black, in 1772 [2]. A residual gas remained after carbonaceous material was burned in air. The distinguishing properties of this gas were almost exclusively negative; that is, it failed to react in any way indicative of another known gas. Neither did this gas support combustion, nor was it absorbed by solutions of alkali. Therefore, it was distinct from "fixed air" (carbon dioxide). However, Rutherford did not recognize it as a distinct chemical species, but rather considered it to be common air saturated with phlogiston.

The method used to produce nitrogen in this demonstration involves an actual synthesis rather than simply an isolation as employed by Rutherford. The reaction is represented by the equation

$$NH_4Cl(aq) + NaNO_2(aq) \longrightarrow N_2(g) + NaCl(aq) + 2 H_2O(l)$$

In this reaction, ammonium ion is oxidized to nitrogen by the nitrite ion.

The evidence used in identifying nitrogen, as presented in this demonstration and as used by the early chemists, is mainly negative. Like carbon dioxide, nitrogen does not support combustion. However, unlike carbon dioxide, it does not react with lime-water. In fact, under ordinary conditions, it reacts with very few substances. One such substance is lithium metal, as the procedure demonstrates. It also reacts with hot magnesium, as described in Demonstration 1.15 in Volume 1 of this series. Generally, nitrogen is quite inert. For this reason, it is often used as the gas in inert atmosphere devices, such as glove boxes. Nitrogen does react with other substances under the proper conditions. Under conditions of high temperature (about 450°C) and pressure (200 atm) and with the assistance of a catalyst (a finely divided mixture of iron and aluminum oxide), nitrogen reacts directly with hydrogen in the Haber process for the synthesis of ammonia [3].

$$N_2(g) + 3 H_2(g) \longrightarrow 2 NH_3(g)$$

Under these conditions the yield of the reaction is about 6%; the reactants are recycled to increase the yield.

REFERENCES

1. M. Windholz, Ed., *The Merck Index*, 9th ed., Merck and Co.: Rahway, New Jersey (1976).
2. E. Farber, *The Evolution of Chemistry*, Ronald Press Co.: New York (1952).
3. A. Godman, *Barnes and Noble Thesaurus of Chemistry*, Barnes and Noble Books: New York (1982).

6.13

Combustion of a Candle in Air

The gaseous products of a combustion reaction are collected to show they contain carbon dioxide. The same combustion reaction consumes a portion of the volume of air [1, 2]. The combustion of a candle requires gravity, as demonstrated by dropping a flask containing a burning candle [3].

MATERIALS FOR PROCEDURE A

50 mL saturated aqueous calcium hydroxide, $Ca(OH)_2$ (To prepare 1.0 liter of solution, boil 1.0 liter of distilled water, cover, and allow to cool overnight. Add 1.75 g of solid $Ca(OH)_2$. Shake well and allow to settle. If solution is cloudy at time of use, filter.)

candle, 20 cm tall

matches

15-cm crystallizing dish

500-mL Erlenmeyer flask

1-liter graduated cylinder

MATERIALS FOR PROCEDURE B

wax candle, ca. 10 cm tall

3-liter round-bottomed flask

Bunsen burner

transparent tape

fireplace matches or wooden splint

rubber stopper to fit round-bottomed flask

PROCEDURE A

Preparation

Light the 20-cm candle and drip some of its melted wax into the center of the 15-cm crystallizing dish. Quickly set the candle in the soft wax to fix it to the dish. Fill

the dish two-thirds full with water.

Presentation

Light the candle. Hold the 500-mL Erlenmeyer flask inverted over the flame to catch some of the gases rising from the candle. After about 15 seconds, turn the flask right-side up, and pour 50 mL of saturated aqueous calcium hydroxide $(Ca(OH)_2)$ solution into the flask. Swirl the solution around in the flask. The appearance of the white precipitate of calcium carbonate $(CaCO_3)$ in the solution indicates that a product of the candle's combustion is CO_2.

Quickly invert the graduated cylinder over the candle and rest its mouth on the bottom of the crystallizing dish. As the candle burns, the level of liquid in the graduated cylinder will rise slowly. Eventually the candle flame will go out. At this point the water will rise more rapidly into the cylinder.

PROCEDURE B†

Preparation

Mount the candle inside on the bottom of the 3-liter round-bottomed flask by warming the bottom of the flask with a Bunsen burner and pressing the candle firmly against the glass, so the softened wax adheres to the glass. Allow the flask to cool. Wrap the flask in a grid of transparent tape to prevent scattering of the glass should the flask break.

With an assistant, practice dropping the flask from a height of at least 5 feet into the hands of the assistant.

Presentation

Tilt the flask on its side and light the candle with a fireplace match or a burning splint. Hold the flask at least 5 feet from the floor. Darken the room. Drop the flask into the hands of a waiting assistant. As the flask falls, the candle goes out. Repeat this procedure several times to convince the audience that the candle goes out during the fall and before being caught. (The candle will go out on impact. If it does not go out before being caught, increase the distance of its fall until it does.)

Relight the candle. Stopper the flask and immediately drop it again. The candle will again go out. Relight the candle and stopper the flask. The candle will burn longer in the resting flask than it did when the flask was dropped.

HAZARDS

Saturated calcium hydroxide solution (limewater) is a base with a pH = 12.4. It can irritate the skin and eyes.

---

† For calling this procedure to our attention, we wish to thank Professor Henry A. Bent, of North Carolina State University, who obtained it from his father, Henry E. Bent, Emeritus Professor of Chemistry at the University of Missouri.

DISPOSAL

The solutions used in this demonstration should be flushed down the drain with water.

DISCUSSION

The nature of combustion of carbonaceous material is investigated in this demonstration. The gaseous products of a combustion are collected by holding an inverted container over a flame. The hot, low-density products displace the air from the container. When these products are mixed with limewater, a precipitate forms in the solution, indicating that the products contain carbon dioxide.

$$Ca(OH)_2(aq) + \quad CO_2(g) \quad \longrightarrow CaCO_3(s) + H_2O(l)$$
$$\text{limewater} \quad \text{carbon dioxide} \quad \text{precipitate}$$

The carbon dioxide is produced by the combustion of wax, a mixture of hydrocarbons, symbolized by C_xH_y in the equation

$$C_xH_y(s) + (x + \tfrac{y}{4})\,O_2(g) \longrightarrow x\,CO_2(g) + \tfrac{y}{2}H_2O(l)$$

The equation indicates that more moles of gas are consumed, $x + \frac{y}{4}$ moles of O_2, than are produced, x moles of CO_2. This is illustrated by the second part of the demonstration in which the candle is burned in a confined amount of air in the inverted cylinder. The volume of the air decreases as a result of combustion, as evinced by the rise of the water into the inverted cylinder.

The results of this demonstration cannot be easily interpreted to indicate the value of the fraction of oxygen in the air. There are a number of factors which interfere with a quantitative interpretation. When the cylinder is inverted over the candle, the flame of the candle heats the gases in the cylinder. Therefore, the air initially trapped within the cylinder is warm air. When the candle goes out, the gases cool, contracting to a volume smaller than that of the air initially trapped in the cylinder. Another complicating factor is that the combustion reaction requires a certain minimum concentration of oxygen in order to continue. As the combustion occurs, the amount of oxygen in the cylinder decreases. When the concentration falls below the required minimum, the candle flame is extinguished. This can happen when a large fraction of the oxygen initially present in the air still remains.

Procedure A illustrates only a few of the many processes that occur as a candle burns. In 1860, Michael Faraday presented a series of lectures to a group of young adults on the topic of a burning candle. The copious notes which Professor Faraday made for these lectures have been gathered relatively recently into a book [4].

Procedure B illustrates that gravity is necessary for the combustion of a candle. When a candle burns in air, gaseous products are produced, mainly carbon dioxide and water vapor. These products are hot and, therefore, less dense than air. These low-density products rise as they are displaced by the denser air. The displacement is an effect of gravity acting on the gases involved in the reaction. Gravity pulls the denser gas (air) down, displacing the less dense gas (reaction products). That this is so is illustrated by what happens when the burning candle is in free fall. According to Galileo's law, everything falls at the same rate, regardless of its mass or density. When the flask falls, its contents all fall at the same rate—candle, flame, air, and product gases do not change their relative positions; gaseous convection ceases. Because there is no convection to remove the product gases and supply oxygen to the flame, the candle goes out. A comparison of the situations of a stationary candle and a free-falling candle is represented in the figure, which contains plots of the relative positions of the flame, oxygen molecules, and carbon dioxide molecules versus time. On the left is the stationary candle. Hot carbon dioxide rises, being displaced by cool oxygen (air), which in turn descends to the flame and keeps it burning. On the right is the free-falling candle. Here the hot carbon dioxide is not displaced by the cool oxygen, because both fall at the same rate. Therefore, there is no oxygen to feed the flame, and it goes out.

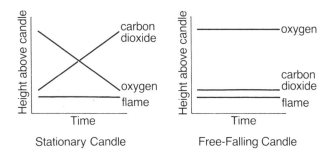

Presenting the demonstration with the sealed flask shows that the turbulence of the air at the open mouth of the flask is not responsible for extinguishing the candle. There is enough oxygen in the 3-liter flask for the candle to burn for more than 30 seconds after the flask has been sealed. This is long enough to show that the candle in the falling, sealed flask is extinguished as a result of the fall, not because the oxygen has been depleted.

In Procedure B, the size of the flask is of some consequence. In a larger flask, the convection currents established as the candle burns involve a greater mass of gas than in a smaller flask. Therefore, the time required for these currents to dissipate during free fall is longer, and the distance the flask must fall before the candle goes out is greater. A smaller flask contains less air than a larger flask, and a candle will go out sooner in a small stoppered flask than in a large stoppered flask. The flask should be large enough that the candle goes out as a result of free-fall, not of having exhausted the oxygen in the sealed flask. A 3-liter flask is suitable for this procedure, being large enough that the candle can burn for some time in the sealed flask before exhausting its oxygen supply and small enough that the convection currents are quickly damped in free fall. However, the demonstrator may use flasks of differing volumes to investigate the effects of the size of the flask on the outcome of the demonstration.

REFERENCES

1. G. S. Newth, *Chemical Lecture Experiments*, Longmans, Green and Co.: London (1928).
2. L. E. Malm, *Laboratory Manual for Chemistry an Experimental Science*, W. H. Freeman and Co.: San Francisco, California (1963).
3. H. A. Bent, *J. Chem. Educ.* 57:395 (1980).
4. M. Faraday, *The Chemical History of a Candle*, The Viking Press: New York (1960).

6.14

Combustion of Magnesium in Air

When a piece of magnesium is ignited in air, light and heat are produced. This is described as Demonstration 1.15 in Volume 1 of this series.

6.15

Preparation and Properties
of Nitrogen(II) Oxide

A colorless gas is generated by the reaction of nitric acid with copper and is collected by the downward displacement of water. When the stopper is removed from a flask containing both this gas and some water, a reddish brown gas forms in the flask. With the stopper replaced, the flask is shaken, and the reddish brown gas disappears. The process is repeated several times, and each time the reddish brown gas fills the flask more quickly than the previous time.

MATERIALS

35 g light copper turnings

250 mL 8.0M nitric acid, HNO_3 (To prepare 1.0 liter of stock solution, slowly pour 500 mL concentrated [16M] HNO_3 into 400 mL of distilled water, and dilute the resulting solution to 1.0 liter with distilled water.)

500-mL Erlenmeyer flask

2-holed rubber stopper to fit Erlenmeyer flask

long-stemmed funnel (or thistle tube)

2 right-angle glass bends, with outside diameter of 7 mm and length of each arm ca. 5 mm

90-cm length rubber tubing to fit 7-mm glass bends

pan or dish, preferably transparent, with capacity of 4–10 liters

2-liter round-bottomed flask

solid rubber stopper to fit round-bottomed flask

gloves, plastic or rubber

cork ring to support flask

white backdrop for round-bottomed flask

PROCEDURE

Preparation

Assemble the apparatus as illustrated in the figure. Place 35 g of copper turnings into the 500-mL Erlenmeyer flask. Insert the long-stemmed funnel through the 2-holed rubber stopper so that its stem extends to within 5 mm of the bottom of the Erlenmeyer

flask when the stopper is seated in the mouth of the flask. Insert one of the glass bends into the other hole of the stopper and seat the stopper assembly in the mouth of the flask. Connect the 90-cm length of rubber tubing to the free end of the glass bend, and insert the remaining glass bend into the free end of the rubber tubing, forming a delivery tube.

Pour water into the pan to a depth of about 10 cm. Fill the round-bottomed flask to the brim with water. Stopper the flask so that it contains no air. In a hood, invert the flask, immerse its neck in the pan of water, and remove the stopper. Support the flask so that its mouth remains below the surface of the water.

Presentation

Wear gloves while preparing the nitrogen(II) oxide in a hood. Through the funnel, slowly add about 100 mL of 8.0M nitric acid to the copper. The gas in the flask will turn reddish brown, because nitrogen dioxide forms when the NO being produced combines with the oxygen trapped in the flask. After about 1 minute, the nitrogen dioxide and oxygen will have been flushed from the apparatus by colorless NO. At this time, place the delivery tube under the mouth of the round-bottomed flask filled with water so that the colorless NO displaces the water from the flask. Agitate the generation flask periodically so that fresh copper contacts the nitric acid. Add more nitric acid as needed to maintain the production of NO. When the round-bottomed collection flask is nearly full of NO and still contains about 100 mL of water, stopper the flask under water. Quench the reaction in the Erlenmeyer flask by filling it with water. Flush the liquid down the drain. Rinse the copper several times with water and store it for future use.

Place the round-bottomed flask on its cork ring in front of a white background. Remove the stopper from the flask for about 5 seconds. Reddish brown nitrogen dioxide (NO_2) will form in the neck of the flask. After replacing the stopper, shake the flask until the gas in the flask returns to a colorless state as the NO_2 dissolves in the water.

Repeat the opening, resealing, and shaking of the flask several times. These repetitions become successively more striking, because the partial vacuum created in the flask by the dissolution of the NO_2 quickly draws air into the flask when the stopper is removed.

HAZARDS

Nitrogen(II) oxide (NO) is rapidly oxidized in air to nitrogen dioxide (NO_2), an extremely toxic gas. Nitrogen dioxide is irritating to the respiratory system; inhaling it may result in severe pulmonary irritation which is not apparent until several hours after exposure. A concentration of 100 ppm is dangerous for even a short period of time, and exposure to concentrations of 200 ppm or more may be fatal.

Concentrated nitric acid is both a strong acid and a powerful oxidizing agent. Contact with combustible materials may cause fires. Contact with the skin may result in severe burns. The vapor irritates the respiratory system, the eyes, and other mucous membranes. Spills should be neutralized with sodium bicarbonate ($NaHCO_3$) before being mopped up.

DISPOSAL

The round-bottomed flask, containing nitrogen(II) oxide, may be stored and used time and time again, until the NO has been consumed. The flask should be clearly labelled, so that it is not opened by mistake. The nitrogen dioxide produced by the reaction is toxic.

The flask can be cleaned by allowing the NO to react with air in a hood. The resulting nitrogen dioxide is dissolved in water by shaking the sealed flask and flushing the dilute nitric acid formed down the drain. This process can be repeated to consume all of the NO.

The generation flask should be rinsed with water and the resulting solution flushed down the drain. The copper turnings can be recovered and dried for repeated use.

DISCUSSION

Nitrogen(II) oxide, or nitric oxide as it is commonly called, is a colorless gas at room temperature and atmospheric pressure. Its normal boiling point is $-151.8°C$, and its melting point is $-163.6°C$ at 1 atm [1].

Nitrogen(II) oxide played an important role in the early development of modern chemistry. Priestley was the first to isolate a colorless, water-insoluble gas, "nitrous air," from the reaction of metals with nitric acid [2]. In studying the properties of nitrous air, Priestley discovered that in air it formed reddish brown fumes which were soluble in water. When the reddish brown fumes had dissolved, the volume of the air had been reduced. In the course of his studies, he noted that the extent of the decrease in the volume of air was related to its fitness for respiration, its "goodness." Priestley used this reaction for a quantitative investigation of the goodness of air, as illustrated in Demonstration 6.16.

Colorless nitrogen(II) oxide reacts rapidly with oxygen gas to form the reddish brown gas, nitrogen dioxide.

$$2 \, NO(g) + O_2(g) \longrightarrow 2 \, NO_2(g) \tag{1}$$

The nitrogen dioxide thus produced dissolves in water and forms a mixture of nitric and nitrous acids through a disproportionation reaction.

$$2\,NO_2(g) + H_2O(l) \longrightarrow H^+(aq) + NO_3^-(aq) + HNO_2(aq) \qquad (2)$$

The nitrous acid itself disproportionates, however, to form nitric acid and nitrogen(II) oxide.

$$3\,HNO_2(aq) \longrightarrow H^+(aq) + NO_3^-(aq) + 2\,NO(g) + H_2O(l) \qquad (3)$$

Therefore, the net reaction of nitrogen dioxide with water is

$$3\,NO_2(g) + H_2O(l) \longrightarrow 2\,H^+(aq) + 2\,NO_3^-(aq) + NO(g) \qquad (4)$$

Each time the flask is opened to the air and then shaken, the amount of gas in the flask decreases. This accounts for the partial vacuum which develops in the flask. However, the NO is not consumed as rapidly as indicated by equation 1, because one-third of the NO consumed each time the flask is opened to the air is regenerated when the product of the reaction, nitrogen dioxide, is dissolved in water by shaking the flask.

The reaction used in this demonstration to generate NO is the method used by Priestley—the reaction of dilute nitric acid with a metal, in this case, copper. The equation for the reaction is

$$3\,Cu(s) + 2\,NO_3^-(aq) + 8\,H^+(aq) \longrightarrow 3\,Cu^{2+}(aq) + 4\,H_2O(l) + 2\,NO(g)$$

This mixture of dilute nitric acid with copper also produces small amounts of NO_2, N_2O, and N_2 [3]. The NO_2 is removed when the generated gas is bubbled through water, because it is soluble in water. The N_2 and N_2O, which constitute about 2% of the generated gas, are inert impurities in the NO used in this demonstration. A reaction that produces more nearly pure nitrogen(II) oxide is that between iron(II) sulfate and sodium nitrite [3].

The reaction of nitric acid with copper to produce nitrogen dioxide has had a significant effect on the career of at least one chemist. Ira Remsen has given us an amusing anecdote regarding his experiences with this reaction. His tale is quoted in "Exocharmic Reactions" by Richard W. Ramette in Volume 1 of this series.

REFERENCES

1. R. C. Weast, Ed., *CRC Handbook of Chemistry and Physics*, 59th ed., CRC Press: Boca Raton, Florida (1978).
2. A. J. Ihde, *The Development of Modern Chemistry*, Dover Publications: New York (1984).
3. P. J. Durant and B. Durant, *Introduction to Advanced Inorganic Chemistry*, 2d ed., Interscience Publishers, John Wiley and Sons: New York (1970).

6.16

Reaction Between Nitrogen(II) Oxide and Oxygen: Combining Volumes of Gases

Measured volumes of gaseous nitrogen(II) oxide and oxygen are mixed and allowed to react over water. The reddish brown fumes of nitrogen dioxide dissolve in the water, and the volume of the residual gas decreases dramatically. The volume of this unreacted gas is measured. The results can be used to demonstrate the stoichiometry of the reaction, the concept of limiting reagent, and the law of combining volumes. A procedure for determining the percent of oxygen gas in the atmosphere is also described.

MATERIALS FOR PROCEDURE A

cylinder of nitrogen(II) oxide, NO, with valve

    or

    35 g light copper turnings

    250 mL 8.0M nitric acid, HNO_3 (To prepare 1.0 liter of stock solution, slowly pour 500 mL concentrated [16M] HNO_3 into 400 mL of distilled water, and dilute the resulting solution to 1.0 liter with distilled water.)

    500-mL Erlenmeyer flask

    2-holed rubber stopper to fit mouth of flask

    long-stemmed funnel (or thistle tube)

    right-angle glass bend, with outside diameter of 7 mm and length of each arm ca. 5 cm

    gloves, plastic or rubber

cylinder of oxygen, with valve

5–10 drops food coloring or other water-soluble dye

2 5-liter round-bottomed flasks

2 solid rubber stoppers to fit round-bottomed flasks

pan, with capacity of at least 10 liters

1-holed rubber stopper to fit round-bottomed flask

3 right-angle glass bends, with outside diameter of 7 mm and length of each arm ca. 5 cm

15-cm length rubber tubing to fit 7-mm glass bend

2 pinch clamps

2-holed rubber stopper to fit round-bottomed flask

3 90-cm lengths rubber tubing to fit 7-mm glass tubing

2 35-cm lengths glass tubing, with outside diameter of 7 mm

glass-working torch

2 cork rings to support round-bottomed flasks

2 5-liter Erlenmeyer flasks, or other glass vessels of similar volume

MATERIALS FOR PROCEDURE B

cylinder of nitrogen(II) oxide, NO, with valve

>   *or* the materials for a NO generator listed in Materials for Procedure A

cylinder of oxygen, with valve

8 250-mL graduated cylinders†

4 100-mL graduated cylinders

3.6-m length copper wire, 16 gauge

waterproof marker, such as a wax pencil or permanent felt-tip marker

transparent tank at least 35 cm deep and having a surface area of at least 900 cm²
>   (A suitable tank may be a glass water bath, battery jar, or aquarium, or may
>   be constructed from 6-mm [¼-inch] acrylic sheets.)

2 35-cm lengths glass tubing, with outside diameter of 7 mm

glass-working torch

2 1-m lengths rubber tubing to fit 7-mm glass tubing

8 cork or solid rubber stoppers to fit the 250-mL graduated cylinders

wooden splint

matches

pH indicator paper, pH 4–8 range

MATERIALS FOR PROCEDURE C

cylinder of oxygen, with valve

cylinder of nitrogen(II) oxide, NO, with valve

>   *or* the materials for a NO generator listed in Materials for Procedure A

6 250-mL graduated cylinders†

1.5-m length copper wire, 16 gauge

transparent tank (See Materials for Procedure B.)

2 35-cm lengths glass tubing, with outside diameter of 7 mm

---

† If the large number of graduated cylinders called for in this procedure is unavailable, the demonstration can be presented using only two 250-mL graduated cylinders—one for nitrogen(II) oxide and one for oxygen. However, the refilling of the cylinders after each trial is time consuming and diminishes the impact of the demonstration. For presentations to small groups, smaller cylinders can be used.

glass-working torch

2 1-m lengths rubber tubing to fit 7-mm glass tubing

MATERIALS FOR PROCEDURE D

cylinder of nitrogen(II) oxide, NO, with valve

    or the materials for a NO generator listed in Materials for Procedure A

4 250-mL graduated cylinders

1.2-m length copper wire, 16 gauge

transparent tank (See Materials for Procedure B.)

2 35-cm lengths glass tubing, with outside diameter of 7 mm

glass-working torch

2 1-m lengths rubber tubing to fit 7-mm glass tubing

PROCEDURE A†

Preparation

Fill the two 5-liter round-bottomed flasks with tap water and seal them with rubber stoppers. Half fill the pan with water and invert the filled flasks in the pan.

Assemble the stopper assemblies as illustrated in Figure 1. Insert one of the right-angle bends into the 1-holed rubber stopper. Connect the free end of the glass bend to the 15-cm length of rubber tubing. Tighten one of the pinch clamps at the center of the rubber tubing. Insert the two remaining right-angle bends into the 2-holed stopper. Connect one of the 90-cm lengths of rubber tubing to the free end of one of these right-angle bends. Tighten the remaining pinch clamp on the rubber tubing near the glass bend.

From the two 35-cm lengths of glass tubing, construct two gas-delivery tubes as

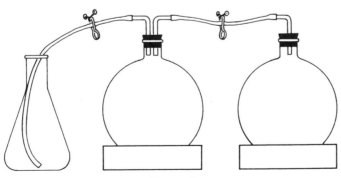

Figure 1.

†We wish to thank Ronald I. Perkins of Greenwich High School, Connecticut, for suggesting this procedure while a Fellow of the Institute for Chemical Education at the University of Wisconsin–Madison.

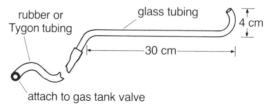

Figure 2.

shown in Figure 2. Use 90-cm lengths of rubber tubing to connect one of these delivery tubes to each of the two cylinders of compressed gas.

If you do not have a cylinder of nitrogen(II) oxide gas, the gas can be prepared as follows: Assemble the apparatus as illustrated in Figure 3. Place 35 g of copper turnings in the 500-mL Erlenmeyer flask. Insert the long-stemmed funnel through the 2-holed rubber stopper so that its stem extends to within 5 mm of the bottom of the Erlenmeyer flask when the stopper is seated in the mouth of the flask. Insert the glass bend into the other hole of the stopper and seat the stopper assembly in the mouth of the flask. Connect the free end of one of the 90-cm lengths of rubber tubing of the gas-delivery-tube assembly to the free end of the glass bend.

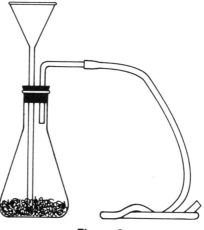

Figure 3.

Wear gloves while preparing the nitrogen(II) oxide in a hood. Through the funnel, slowly add about 100 mL of 8.0M nitric acid to the copper. The gas in the flask will turn reddish brown, because nitrogen dioxide forms when the nitrogen(II) oxide being produced combines with the oxygen trapped in the flask. After about 1 minute, the nitrogen dioxide and oxygen will have been flushed from the apparatus by colorless nitrogen(II) oxide. At this time, the nitrogen(II) oxide may be collected as it escapes from the end of the delivery tube. Agitate the generation flask periodically so that fresh copper contacts the nitric acid. Add more nitric acid as needed to maintain the production of nitrogen(II) oxide.

With the mouth of one of the round-bottomed flasks below the surface of the water in the pan, remove the stopper from the flask. Use the gas-delivery tube to fill the flask with nitrogen(II) oxide by the downward displacement of water. When the flask is filled with nitrogen(II) oxide and while its mouth is still below the surface of the water, seal the flask with the 1-holed stopper with the clamped rubber tubing attached, and set the flask on a cork ring. If you are generating the nitrogen(II) oxide, quench the reaction by filling the Erlenmeyer flask with water. Flush the liquid down the drain. Rinse the copper several times with water and store it for future use.

Use a gas-delivery tube to fill the other round-bottomed flask with oxygen gas by the downward displacement of water. When the flask is filled with oxygen gas, seal it with the 2-holed stopper with the clamped rubber tubing attached, and set the flask on the other cork ring.

Assemble the apparatus as shown in Figure 1. Connect the free end of the rubber tubing from the flask of nitrogen(II) oxide to the open glass bend of the oxygen-filled flask. Insert the free end of the long rubber tubing connected to the oxygen flask into one of the 5-liter Erlenmeyer flasks. Place a few drops of food coloring in the Erlenmeyer flask and fill it with tap water. Fill the other 5-liter Erlenmeyer with tap water.

Presentation

Remove the pinch clamp from the tubing between the two round-bottomed flasks. A reddish brown gas will begin to form inside the flasks. Remove the second pinch clamp from the other piece of rubber tubing. The water in the Erlenmeyer flask will begin to flow into the first round-bottomed flask. As the water drains from the Erlenmeyer flask, replenish it from the other Erlenmeyer flask. The water will eventually fill the first flask completely and half fill the second flask. When no more water is being drawn into the round-bottomed flasks, light a wooden splint and extinguish the flames, leaving only burning embers. Remove the stopper from the flask which contains gas. Plunge the glowing splint into the gas in the flask. The splint will burst into flames, indicating that the gas in the flask is oxygen.

PROCEDURE B†

Preparation

Wrap one end of the piece of copper wire around the base of each of the 12 graduated cylinders and fashion the free end into a hook, so that each of the cylinders can be suspended upside down from the edge of the water tank. Using a waterproof marker, number the bottom of each of the 250-mL graduated cylinders *1* through *8*, and similarly number the 100-mL graduated cylinders *1* through *4*. Fill the water tank to within 8 cm of the top with tap water. Completely fill the graduated cylinders by submerging them in the bath, and suspend them from the edge of the tank.

From the two 35-cm lengths of glass tubing, construct two gas-delivery tubes as shown in Figure 2. Use rubber tubing to connect one of these delivery tubes to each of the two cylinders of compressed gas.

If you do not have a cylinder of nitrogen(II) oxide, use the nitrogen(II) oxide generator described in Procedure A.

Each cylinder is to contain the kind and amount of gas as follows:

---

† This procedure was originally developed by Professor A. Truman Schwartz of Macalester College while on sabbatical in 1980 at the University of Wisconsin–Madison. It is based upon a demonstration used by Professor Aaron Ihde of the University of Wisconsin–Madison.

Cylinder	Gas	Volume
250-mL #1	NO	200 mL
250-mL #2	NO	200 mL
250-mL #3	NO	200 mL
250-mL #4	NO	200 mL
250-mL #5	NO	200 mL
250-mL #6	NO	200 mL
100-mL #1	O_2	25 mL
100-mL #2	O_2	50 mL
100-mL #3	O_2	75 mL
100-mL #4	O_2	100 mL
250-mL #7	O_2	150 mL
250-mL #8	O_2	200 mL

Use the gas-delivery tube to transfer 200 mL of nitrogen (II) oxide to each of the 250-mL graduated cylinders numbered *1* through *6*. In order to obtain accurate volume measurements of the gas, the hydrostatic pressure should be minimized by aligning the meniscus of water inside the cylinder with the level of water in the tank.

In a similar fashion, transfer 25 mL of oxygen into the 100-mL cylinder numbered *1*, 50 mL into the one numbered *2*, 75 mL into that numbered *3*, and 100 mL into the last one. Also transfer 150 mL of oxygen into the 250-mL graduated cylinder numbered *7*, and 200 mL into the one numbered *8*.

Presentation

Unhook one of the graduated cylinders containing nitrogen(II) oxide and the graduated cylinder containing 25 mL of oxygen from the side of the tank, being careful to keep the openings of the cylinders below the surface of the water. Carefully "pour" the oxygen gas upward into the cylinder of nitrogen(II) oxide, as shown in Figure 4. Transfer *all* of the gas, being careful not to lose any gas bubbles in the process.

As the NO and O_2 react, a reddish brown cloud of NO_2 gas forms. To dissolve the NO_2 gas, agitate the cylinder with an up-and-down motion, always keeping its mouth below the surface of the water, until the reddish brown color has disappeared and the volume of gas in the cylinder stabilizes. Align the water level in the cylinder with that in the tank to equalize the pressure of the gas in the cylinder with atmospheric pressure, and record the volume of the residual gas.

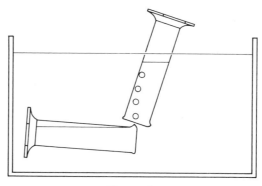

Figure 4.

Table 1. Typical Results from Procedure B (volumes in mL)

Volume of NO	Volume of O_2	Volume of residual gas [a]			Theoretical volume
		1	2	3	
200	25	132	155	147	150
200	50	75	85	84	100
200	75	25	34	42	50
200	100	3	16	12	0
200	150	52	50	54	50
200	200	105	108	104	100

[a] Results were obtained from three separate presentations of this demonstration.

Stopper or cork the reaction cylinder and remove it from the water bath. Set the cylinder upright on the benchtop. Light a wooden splint with a match and blow out the flame, leaving only glowing embers at the tip. Open the cylinder to the air. If the excess gas in the cylinder is nitrogen(II) oxide, the reddish brown color of NO_2 will appear instantly. If the excess gas is oxygen, the glowing splint inserted into the mouth of the cylinder will ignite.

Repeat this procedure five times, combining each of the remaining cylinders of nitrogen(II) oxide with one of the cylinders of oxygen. Typical results of the volume measurements are listed in Table 1. A plot of the volume of the residual gas versus the volume of O_2 used is shown in Figure 5.

Use the pH paper to show that the amount of NO_2 dissolved in the water is insufficient to lower the pH of the water by a significant amount.

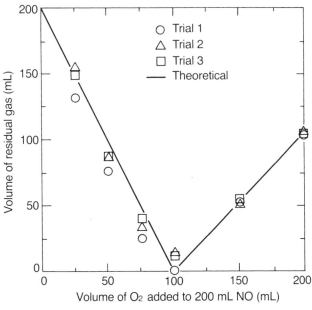

Figure 5.

PROCEDURE C†

Preparation

Wrap one end of the piece of copper wire around the base of each of the graduated cylinders and fashion the free end into a hook, so that each of the cylinders can be suspended upside down from the edge of the water tank. Fill the water tank to within 8 cm of the top with tap water. Completely fill the graduated cylinders by submerging them in the bath, and suspend them from the edge of the tank.

From the two 35-cm lengths of glass tubing, construct two gas-delivery tubes as shown in Figure 2. Use rubber tubing to connect one of these delivery tubes to each of the two cylinders of compressed gas.

Each cylinder is to contain the kind and amount of gas as follows:

Cylinder	Gas	Volume
1st	O_2	200 mL
2d	NO	100 mL
3d	NO	100 mL
4th	NO	100 mL
5th	NO	100 mL
6th	NO	100 mL

Using the gas-delivery tube, transfer 200 mL of oxygen from its cylinder into one of the graduated cylinders. In order to obtain accurate volume measurements of the gas, the hydrostatic pressure should be minimized by aligning the meniscus of water inside the cylinder with the level of water in the tank.

If you are using a cylinder of nitrogen(II) oxide, use the other gas-delivery tube to transfer 100 mL of nitrogen(II) oxide into each of the five remaining graduated cylinders. If you do not have a cylinder of nitrogen(II) oxide, use the nitrogen(II) oxide generator described in Procedure A to introduce 100 mL of NO gas into each of the five remaining graduated cylinders.

Presentation

Unhook the cylinder containing oxygen and one of the cylinders of nitrogen(II) oxide from the side of the tank, being careful to keep the openings of the cylinders below the surface of the water. Carefully "pour" the nitrogen(II) oxide upward into the cylinder of oxygen, as shown in Figure 4. Transfer *all* of the gas, being careful not to lose any gas bubbles in the process.

As the NO and O_2 react, a reddish brown cloud of NO_2 gas forms. To dissolve the NO_2 gas, agitate the cylinder with an up-and-down motion, always keeping its mouth below the surface of the water, until the reddish brown color has disappeared and the volume of gas in the cylinder stabilizes. Record the volume of residual gas in the cylinder.

Repeat this procedure four more times, combining the nitrogen(II) oxide in each

---

† This procedure was originally developed by Professor A. Truman Schwartz of Macalester College while on sabbatical in 1980 at the University of Wisconsin–Madison. It is based upon a demonstration used by Professor Aaron Ihde of the University of Wisconsin–Madison.

Table 2. Typical Results from Procedure C (volumes in mL)

Total volume of NO added to 200 mL O_2	Volume of residual gas	Theoretical volume
100	154	150
200	101	100
300	48	50
400	6	0
500	92	100

of the other cylinders with the residual gas in the oxygen cylinder and noting the volume of the resulting gas. Typical results of the volume measurements are listed in Table 2. A plot of the volume of the residual gas versus the volume of NO used is shown in Figure 6.

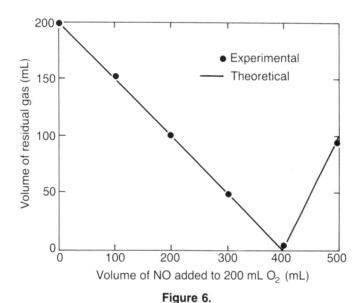

Figure 6.

PROCEDURE D†

Preparation

Wrap one end of the piece of copper wire around the base of each of the graduated cylinders and fashion the free end into a hook, so that each of the cylinders can be suspended upside down from the edge of the water tank. Fill the water tank to within 8 cm of the top with tap water. Completely fill the graduated cylinders by submerging them in the bath, and suspend them from the edge of the tank.

† This procedure was originally developed by Professor A. Truman Schwartz of Macalester College while on sabbatical in 1980 at the University of Wisconsin–Madison. It is based upon a demonstration used by Professor Aaron Ihde of the University of Wisconsin–Madison.

From the two 35-cm lengths of glass tubing, construct two gas-delivery tubes as shown in Figure 2. Use rubber tubing to connect one of these delivery tubes to the nitrogen(II) oxide cylinder.

If you have a cylinder of nitrogen(II) oxide, use the gas-delivery tube to transfer 100 mL of nitrogen(II) oxide into each of two of the 250-mL graduated cylinders. In order to obtain accurate volume measurements of the gas, the hydrostatic pressure should be minimized by aligning the meniscus of water inside the cylinder with the level of water in the tank.

If you do not have a cylinder of nitrogen(II) oxide, use the nitrogen(II) oxide generator described in Procedure A to introduce 100 mL of NO gas into each of two of the 250-mL graduated cylinders.

Presentation

Introduce 200 mL of air into one of the unused 250-mL graduated cylinders by lifting the filled, inverted cylinder out of the water to near its mouth and tipping the cylinder so one edge of its mouth clears the water's surface and bubbles of air enter the cylinder. Using the second gas-delivery tube, introduce 200 mL of exhaled breath into the remaining graduated cylinder.

Unhook one cylinder containing nitrogen(II) oxide and the cylinder containing air, being careful to keep the openings of the cylinders below the surface of the water. Carefully "pour" the air upward into the cylinder of nitrogen(II) oxide, as shown in Figure 4. Transfer *all* of the air, being careful not to lose any bubbles in the process.

As the NO and the O_2 of the air react, a reddish brown cloud of NO_2 gas forms. To dissolve the NO_2 gas, agitate the cylinder with an up-and-down motion, always keeping its mouth below the surface of the water, until the reddish brown color has disappeared and the volume of gas in the cylinder stabilizes. Record the volume of residual gas in the cylinder.

Repeat this process using the remaining nitrogen(II) oxide cylinder and the cylinder of breath.

The volume of the residual gas can be used to determine the volume percent of oxygen in air and in exhaled breath, as illustrated in the Discussion section.

HAZARDS

Nitrogen(II) oxide (NO) is rapidly oxidized in air to nitrogen dioxide (NO_2), an extremely toxic gas. Nitrogen dioxide is irritating to the respiratory system; inhaling it may result in severe pulmonary irritation which is not apparent until several hours after exposure. A concentration of 100 ppm is dangerous for even a short period of time, and exposure to concentrations of 200 ppm or more may be fatal.

DISPOSAL

The residual nitrogen(II) oxide in the cylinders used in this demonstration should be oxidized to nitrogen dioxide by admitting air into the cylinders. The nitrogen dioxide should be dissolved in the water in the tank. The NO_2 which dissolves is insufficient to lower the pH of the water appreciably.

DISCUSSION

Joseph Priestley is credited with the discovery of nitrogen(II) oxide (NO) [*1*]. Stephen Hales (1677–1761) had observed that a reddish brown gas was formed in the action of spirit of niter (nitric acid) on pyrites. Henry Cavendish later suggested to Priestley that metals may be substituted for pyrites in the production of this reddish brown gas. In the course of Priestley's investigations of the action of nitric acid on metals, he suceeded in isolating a colorless, water-insoluble gas. He observed that this gas reacted with common air to produce a reddish brown gas (nitrogen dioxide) which was soluble in water. He called this colorless gas "nitrous air."

In quantitative studies of the reaction of nitrous air with common air, he observed that the volume of gas was reduced when the reaction was carried out over water. By varying the volume ratios of nitrous air to common air, he found that a ratio of 1 volume of nitrous air to 2 volumes of common air produced the largest diminution in the volume of the air, a decrease of one-fifth. The extent of the diminution of volume was apparently related to the fitness of the air for respiration. This reaction provided Priestley with a better test for fitness than was provided by measuring the length of time in which a mouse could survive in the air.

Priestley also used this test with the gas he had prepared by the decomposition of red mercury calx (mercury oxide). With the 1:2 ratio of combined volumes, he again found a volume decrease of approximately one-fifth. Therefore, he initially concluded that the gas he had prepared was good common air. However, sometime later he was surprised to find that, when he added more nitrous air, it consumed almost the entire volume of this gas. On the basis of this additional evidence, Priestley estimated that the new gas was "five or six times as good as common air" [*2*]. That meant, for example, that it was far more effective than air in supporting combustion. A committed phlogistonist, Priestley believed that burning involved the transfer of phlogiston from the combustible material to air. The behavior of the new gas suggested that it had a much greater capacity for phlogiston than did ordinary air. Consequently, Priestley called it "dephlogisticated air." To his dying day, he refused to accept Lavoisier's name, "oxygene," and the Frenchman's interpretation of combustion which revolutionized chemistry.

In modern terms, the reactions which Priestley observed are

$$2\,NO(g)\ +\ O_2(g)\ \longrightarrow\ 2\,NO_2(g)$$
nitrous air   brown gas

$$2\,NO_2(g)\ +\ H_2O(l)\ \longrightarrow\ H^+(aq)\ +\ NO_3^-(aq)\ +\ HNO_2(aq)$$

From this, it can be seen that a mixture of 2 volumes of nitrogen(II) oxide with 1 volume of oxygen should result in a complete consumption of the gases. If a ratio of NO to O_2 smaller than 2:1 is used, the remaining gas will be oxygen. If the ratio is larger than 2:1, the excess is nitrogen(II) oxide. Because air is approximately 20% oxygen by volume, five times as much air as oxygen must be used to consume a given amount of nitrogen(II) oxide. Thus, a volume ratio of 2 parts nitrogen(II) oxide to 5 parts air would provide a stoichiometric ratio of nitrogen(II) oxide to oxygen. The 1:2 volume ratio that Priestley selected for his test of the goodness of air used a slight excess of nitrogen(II) oxide, which is reasonable for a quantitative analysis for oxygen. However, when this volume ratio of 1:2 is used with nitrogen(II) oxide and pure oxygen, the oxygen is in excess and 75% of it will remain unreacted, which is approximately the same percent of air that would remain, that is, 80%.

The law of combining volumes of gases may be illustrated by the results of Procedure A. Procedure A demonstrates that 5 liters of nitrogen(II) oxide react with 2.5 liters of oxygen gas. Therefore, these two gases react in the volume ratio of 2 volumes of nitrogen(II) oxide to 1 volume of oxygen, which is indeed a ratio of small whole numbers.

A sample of the results which may be obtained in Procedure B is presented in Table 1. These results are plotted in Figure 5, along with the theoretical results as predicted by the stoichiometry of the reaction. Procedure C is a variation of Procedure B that uses a single sample of oxygen gas to which are added increments of nitrogen(II) oxide. Typical results of this procedure are listed in Table 2 and plotted in Figure 6.

Procedure D uses Priestley's method for assessing the "goodness" of air. Based on the results and the stoichiometry of the reaction, the percent of oxygen in air can be calculated. For example, in a typical demonstration using air, 200 mL of air are combined with 100 mL of NO, and the final volume of the mixture is 180 mL. If x is the volume of O_2 reacted, then 2x is the volume of NO reacted, and

$$(200 \text{ mL} - x) + (100 \text{ mL} - 2x) = 180 \text{ mL}$$

Solving for x yields 40 mL as the volume of oxygen in 200 mL of air. Thus, the volume percent of oxygen in air is

$$100 \times \frac{40 \text{ mL}}{200 \text{ mL}} = 20\%$$

As can be seen in the tables, the experimentally observed volumes do not conform exactly to the values predicted from simple stoichiometric theory. The major cause of this deviation is not clear. Priestley's own writings indicate that he observed gradual decreases in the volume of residual gas over periods of several hours, suggesting that reactions in addition to the principal one may also occur.

One of these possible reactions is the combination of nitrogen(II) oxide with nitrogen dioxide to form dinitrogen trioxide, which is soluble in water.

$$NO(g) + NO_2(g) \rightleftharpoons N_2O_3(g)$$

At 25°C and 1 atm, this equilibrium contains 10% N_2O_3 [3]. When NO is in excess in a mixture of NO and O_2, some of the residual NO may react with the product NO_2, leading to anomalously small volumes of residual gas. Furthermore, when O_2 is slowly added to the NO, the NO is initially in excess and can react with the product. When NO is added to oxygen, the NO is in excess only after all of the O_2 has been consumed. Thus, the volume of residual gas may depend on whether the O_2 is poured into the NO or vice versa. In any case, the relative values of the rate of the reaction between NO and NO_2 and the rate of the dissolution of NO_2 would influence the outcome.

Results seem to indicate that an anomalously large volume of residual gas is more common. One explanation is the disproportionation of the nitrous acid to produce nitric acid and nitrogen(II) oxide.

$$3 \text{ HNO}_2(aq) \longrightarrow H^+(aq) + NO_3^-(aq) + H_2O(l) + 2 \text{ NO}(g)$$

Thus, the overall reaction between NO_2 and water is

$$3 \text{ NO}_2(g) + H_2O(l) \longrightarrow 2 \text{ H}^+(aq) + 2 \text{ NO}_3^-(aq) + NO(g)$$

If the original reactant mixture contains an excess of NO, the residual gas could have a volume larger than predicted because of the excess NO produced by the disproportionation of nitrous acid. On the other hand, if the original reactant mixture contains an excess of O_2, the residual gas may have a smaller volume than predicted, because some of the excess O_2 may react with the NO produced by the disproportionation reaction. However, because the graduated cylinders are agitated to dissolve the NO_2 in water and the water is in an open container, the NO produced by disproportionation of nitric acid does not necessarily collect in the graduated cylinders.

Solubility phenomena, as well, may be responsible for departures from theoretical behavior in this demonstration. The solubility of NO at 25°C and a partial pressure of 1 atm is 0.043 liter NO/liter H_2O [4]. Under comparable conditions, O_2 is less soluble: 0.028 liter O_2/liter H_2O [4]. The partial solubility of these gases might influence the results to a small degree. In addition, gases initially dissolved in the water of the bath—most likely air—could be released into the reaction mixture.

Impurities in the test gases would, of course, introduce error. A likely impurity is N_2 if the NO is generated by the action of nitric acid on copper [3]. Variations in the temperature and pressure at which the gas volumes are measured would also lead to errors. In short, although the procedure is adequate for demonstration purposes, it is not sufficiently refined to yield quantitative results of high accuracy and precision.

REFERENCES

1. A. J. Ihde, *The Development of Modern Chemistry*, Dover Publications: New York (1984).
2. J. T. Rutt, *Life and Correspondence of Joseph Priestley*, Vol. 1, London (1832).
3. P. J. Durant and B. Durant, *Introduction to Advanced Inorganic Chemistry*, 2d ed., Interscience Publishers, John Wiley and Sons: New York (1970).
4. R. C. Weast, Ed., *Handbook of Chemistry and Physics*, 59th ed., CRC Press: Boca Raton, Florida (1978).

6.17

Equilibrium Between Nitrogen Dioxide and Dinitrogen Tetroxide

The intensity of the reddish brown color in a sealed tube of nitrogen(IV) oxide decreases when the tube is immersed in a cold bath and increases when it is immersed in a hot bath [1].

MATERIALS FOR PROCEDURE A

4 sealed glass containers of nitrogen dioxide, NO_2†

    or

    cylinder of nitrogen dioxide, with valve

    4 500-mL flat-bottomed boiling flasks (Florence flasks)

    glass-working torch

    90-cm length rubber or plastic tubing to fit gas valve

1000 g ice

100 g sodium chloride, NaCl

3 2-liter beakers

hot plate

white backdrop

MATERIALS FOR PROCEDURE B

cylinder of nitrogen dioxide, NO_2, with valve

5-cm length glass or plastic tubing, with outside diameter of 20 mm

rubber septum to fit tube

rubber balloon

10-cm length copper wire, 18–22 gauge

syringe needle

50-cm length rubber tubing to fit gas tank valve

---

† Suitable tubes are commercially available from a number of laboratory-equipment supply companies, such as Sargent-Welch (#4426).

2 30-mL glass syringes, with needles

2 plastic syringe caps

PROCEDURE A

Preparation

If prefilled sealed glass containers of nitrogen dioxide are not available, they can be made by the method described in the following two paragraphs:

Using the glass-working torch, constrict the necks of the boiling flasks so that the rubber or plastic tubing will just fit into the flasks. Attach the tubing to the valve on the cylinder of nitrogen dioxide.

The next steps must be performed in a well-ventilated fume hood! Insert the tubing to the bottom of one of the flasks. Slowly fill the flask with nitrogen dioxide by the upward displacement of air. The reddish brown color of the gas will indicate the level of the gas in the flask. When the flask is filled, remove the tubing and seal the flask at the constriction, using the torch. Fill the remaining three flasks in the same manner. The flasks must then be checked for leaks. To do this, fill one of the 2-liter beakers with water and immerse each of the flasks, in turn, upside down in the water. Heat the water to boiling and then let it cool with the flask immersed. Water drawn into the flask will signal a leak. If the flask leaks, it cannot be used and should be opened in the hood to vent the NO_2.

Fill one of the 2-liter beakers with water and warm it to 70–80°C on the hot plate. Place 500 g of ice in a second beaker and fill it with cold water. Place 500 g of ice in the third beaker, fill it with cold water, and mix in 100 g of sodium chloride.

Presentation

Place one of the flasks containing NO_2 in each of the beakers and leave one flask at room temperature. Allow several minutes for the flasks to reach thermal equilibrium, remove them from the baths, and place them before a white backdrop. Compare the intensity of the colors. The warm flask will be more intensely colored than the one at room temperature, which, in turn, will be more intensely colored than the cold flasks.

PROCEDURE B

Preparation

Insert one end of the 5-cm length of 20-mm glass or plastic tubing into a rubber septum. Insert the other end of the tube into the mouth of the rubber balloon, and tighten the balloon to the tube by wrapping a wire around the neck of the balloon. Insert the base of the single syringe needle into the opening at one end of the 50-cm length of rubber tubing and tighten the tubing to the needle with wire. Attach the free end of the rubber tubing to the valve of the NO_2 cylinder.

The next steps must be performed in a well-ventilated fume hood! Fill the balloon with NO_2 gas by inserting the needle through the septum and opening the valve on the NO_2 cylinder. When the balloon is about half inflated, close the valve, and remove the needle from the septum. Insert the needle of one of the syringe assemblies through the septum and draw 30 mL of gas into the syringe. Remove the needle from the syringe and seal the syringe with a plastic syringe cap. Press the plunger into the syringe to check that the syringe does not leak. Fill the second syringe in the same manner.

Presentation

Place both NO_2-filled syringes on an overhead projector. Grasp one syringe firmly by the barrel and rapidly press in the plunger to compress the gas as much as possible. The gas in this syringe will first become darker and gradually become lighter than the gas in the other, "control" syringe. Release the plunger, allow the gas to return to its initial color, and then repeat the process.

HAZARDS

Nitrogen dioxide (NO_2) is an extremely toxic gas. It is irritating to the respiratory system; inhaling it may result in severe pulmonary irritation which is not apparent until several hours after exposure. A concentration of 100 ppm is dangerous for even a short period of time, and exposure to concentrations in excess of 200 ppm may be fatal.

DISPOSAL

The flasks of nitrogen(IV) oxide, once sealed, can be stored indefinitely for repeated use.

Discharge the nitrogen(IV) oxide from the balloon and syringes in a fume hood.

DISCUSSION

Nitrogen(IV) oxide has a melting point of $-11.2°C$ and a boiling point of $21.2°C$ [2]. The solid state is composed exclusively of N_2O_4, while the liquid, like the gas, is composed of a mixture of NO_2 and N_2O_4 [3]. Solid N_2O_4 is colorless when pure, although impurities can impart a blue-green color to it, most likely through the formation of blue N_2O_3, as represented in the following equation [3, 4]:

$$2\,N_2O_4(s) + H_2O(l) \longrightarrow N_2O_3(s) + 2\,HNO_3(l)$$

The liquid is yellow because of the presence of small amounts of NO_2.

Nitrogen(IV) oxide gas exists as an equilibrium mixture of a monomer (NO_2) and a dimer (N_2O_4).

$$2\,NO_2(g) \rightleftharpoons N_2O_4(g)$$

Nitrogen dioxide gas has a reddish brown color, while dinitrogen tetroxide is colorless. Therefore, the more intense the color of a mixture of these substances, the greater the concentration of the colored NO_2. Because N_2O_4 condenses at $21.2°C$, liquid may appear in the container that is cooled in the ice bath.

Table 1. Standard State Thermodynamic Data for NO_2 and N_2O_4 [5]

Gas	ΔH_f° (kJ/mol)	ΔG_f° (kJ/mol)	S° (J/mol−K)
NO_2	33.18	51.31	240.06
N_2O_4	9.16	97.89	304.29

Table 1 contains standard state thermodynamic data for NO_2 and N_2O_4 [5]. The data in Table 1 indicate that the dimerization reaction is exothermic, with $\Delta H^\circ = -57.20$ kJ/mol N_2O_4. However, the dimerization results in a decrease in entropy, with $\Delta S^\circ = -175.83$ J/K-mol N_2O_4. Because of this decrease in entropy upon dimerization, the free energy change for the reaction $\Delta G = \Delta H - T\Delta S$ increases as the temperature increases. This effect is illustrated by the data in Table 2, which were calculated from the data in Table 1. Table 2 also presents the effect of the temperature dependence of ΔG on the value of the equilibrium constant, $\Delta G = -RT ln K_c$. This indicates that as the temperature of the system increases, the equilibrium shifts toward NO_2, resulting in a deepening of the color of the gas mixture.

Table 2. Temperature Dependence of the NO_2 Dimerization[a]

Temperature	ΔG (kJ/mol N_2O_4)	K_c
23°C (296K)	−5.13	8.03
70°C (343K)	3.14	0.334
100°C (373K)	8.41	0.0665

[a] Calculated from the data of Table 1.

The syringes used in Procedure B contain an equilibrium mixture of NO_2 and N_2O_4. When the plunger is depressed, the concentrations of the gases inside increase because of the decrease in the volume. The increased concentration of NO_2 accounts for the deepening of the color; now the system is no longer at equilibrium. A decrease in the volume of the mixture should lead, according to LeChatelier's principle, to a smaller number of moles of gas. This is accomplished through the dimerization of NO_2 into N_2O_4. The dimerization reduces the amount and concentration of NO_2 in the syringe, leading to a less intense color.

REFERENCES

1. A. W. Sangster, *J. Chem. Educ.* 36:A159 (1959).
2. R. C. Weast, Ed., *CRC Handbook of Chemistry and Physics*, 59th ed., CRC Press: Boca Raton, Florida (1978).
3. K. Jones, "Nitrogen," in *Comprehensive Inorganic Chemistry*, Vol. 2, ed. J. C. Bailar, Jr., H. J. Emeleus, R. Nyholm, and A. F. Trotman-Dickenson, Pergamon Press: Oxford (1973).
4. C. A. Jacobson, Ed., *Encyclopedia of Chemical Reactions*, Vol. 5, Reinhold Publishing Co.: New York (1953).
5. "The NBS Tables of Chemical Thermodynamic Properties," *J. Phys. Chem. Ref. Data* 11, Supp. 2 (1982).

6.18

Preparation and Properties of Sulfur Dioxide

Sulfur is burned in air to form sulfur dioxide, which forms an acidic solution in water. Two colorless gases are prepared by the reaction of hydrochloric acid with sodium sulfite and with iron(II) sulfide. The gases are collected by the upward displacement of air. When the two gases are allowed to mix, a yellow solid forms. When a purple solution is poured into a cylinder containing the first gas, the solution becomes colorless. When an orange solution is poured into the same cylinder, the solution becomes green [1, 2].

MATERIALS FOR PROCEDURE A

400 mL tap water

2 mL bromocresol green indicator solution (To prepare 50 mL of stock solution, dissolve 0.02 g of bromocresol green in 3 mL of 0.01M NaOH and dilute the resulting solution to 50 mL with distilled water.)

5M aqueous ammonia, NH_3 (Amount needed depends on pH of tap water. To prepare 50 mL of stock solution, pour 17 mL of concentrated [15M] NH_3 into 25 mL of distilled water, and dilute the resulting mixture to 50 mL with distilled water.)

2 g powdered sulfur, S_8

cork to fit 5-liter round-bottomed flask

combustion spoon

5-liter round-bottomed flask

Bunsen burner

MATERIALS FOR PROCEDURE B

50 g sodium sulfite, Na_2SO_3

16 g iron(II) sulfide, FeS

40 mL 1M aqueous sulfuric acid, H_2SO_4 (To prepare 1.0 liter of solution, pour 56 mL of concentrated [18M] H_2SO_4 into 500 mL of distilled water and dilute the resulting solution to 1.0 liter.)

5 mL 0.001M aqueous potassium permanganate, $KMnO_4$ (To prepare 1.0 liter of solution, dissolve 0.16 g of $KMnO_4$ in 500 mL of distilled water and dilute the resulting solution to 1.0 liter.)

5 mL 0.008M aqueous potassium dichromate, $K_2Cr_2O_7$ (To prepare 1.0 liter of solution, dissolve 2.4 g of $K_2Cr_2O_7$ in 500 mL of distilled water and dilute the resulting solution to 1.0 liter.)

1300 mL distilled water

310 mL 6M hydrochloric acid, HCl (To prepare 1.0 liter of stock solution, carefully pour 500 mL of concentrated [12M] HCl into 400 mL of distilled water and dilute the resulting solution to 1.0 liter with distilled water.)

500-mL Erlenmeyer flask

1-liter Erlenmeyer flask

2 long-stemmed funnels (or thistle tubes)

2 right-angle glass bends, with outside diameter of 7 mm and length of each arm ca. 5 cm

3 2-liter glass cylinders

stopcock grease

2 90-cm lengths plastic or rubber tubing to fit 7-mm glass bends

3 glass plates to cover 2-liter cylinders

2 1-liter glass cylinders

blue litmus paper

lead acetate paper

PROCEDURE A

Preparation

Pierce the cork and insert the handle of the combustion spoon through the hole, so that when the cork is seated in the mouth of the flask the bowl of the spoon will be slightly above the flask's equator, as illustrated in Figure 1.

Pour 400 mL of tap water into the flask. Add 2 mL of bromocresol green indicator solution. If the solution is not blue, add just enough drops of 5M NH_3 to turn it blue.

Figure 1.

Presentation

Fill the combustion spoon with powdered sulfur and ignite the sulfur in the flame of a Bunsen burner. While the sulfur is burning, insert the spoon into the flask, loosely seating the cork in the mouth of the flask. The sulfur will continue to burn for about a minute, producing smoke and fog in the flask. When the flame has gone out, seat the stopper firmly and shake the flask vigorously. The color of the solution will change to yellow.

PROCEDURE B

Preparation

Assemble two gas generators as illustrated in Figure 2, one with the 500-mL Erlenmeyer flask and the other with the 1-liter Erlenmeyer flask. Insert a long-stemmed funnel and a right-angle bend through each stopper. Adjust each funnel so its tip is within 1 cm of the bottom of the flask when the stopper assembly is seated in the mouth of the flask. Attach a 90-cm length of rubber tubing to the free end of each bend.

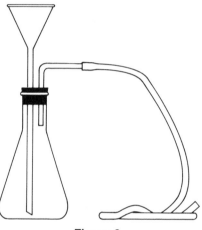

Figure 2.

Grease the rim of each of the three 2-liter cylinders.

Place 50 g of Na_2SO_3 in the 1-liter Erlenmeyer flask, and add water to a depth of about 2 cm in the flask. Seat the stopper in the mouth of the flask. Insert the free end of the rubber tubing to the bottom of one of the 2-liter cylinders. Cover the opening of the cylinder with a glass plate.

Place 16 g of FeS in the 500-mL Erlenmeyer flask and add water to a depth of about 2 cm in the flask. Seat the stopper in the mouth of the flask. Insert the free end of the rubber tubing to the bottom of another 2-liter cylinder. Cover the opening of the cylinder with a glass plate.

Pour 20 mL of 1M H_2SO_4 into each of the 1-liter cylinders. Add 5 mL of 0.001M $KMnO_4$ to one of the cylinders and 5 mL of 0.008M $K_2Cr_2O_7$ to the other. Pour 650 mL of distilled water into each of these two cylinders.

Presentation

This demonstration should be presented only in a well-ventilated fume hood.
Slowly pour about 50 mL of 6M HCl into the funnel of the 1-liter flask containing Na_2SO_3, adding it at a rate which produces a gentle bubbling in the mixture. As the sulfur dioxide (SO_2) gas is produced, it will displace the air from the cylinder. Keep

adding 6M HCl to maintain the gas formation until a piece of dampened blue litmus paper held at the mouth of the cylinder becomes pink, indicating that the cylinder is filled with SO_2. This will require no more than 125 mL of 6M HCl. Remove the rubber tubing from the cylinder and immediately insert it to the bottom of another cylinder. Fill the second cylinder in the same way. Cover both cylinders with glass plates. Flush the contents of the flask down the drain with plenty of water.

Slowly pour 6M HCl through the funnel into the 500-mL Erlenmeyer flask containing FeS, adding it at a rate which produces gentle bubbling in the mixture. As the hydrogen sulfide (H_2S) gas is produced, it will displace the air from the cylinder. Keep adding 6M HCl to maintain the gas formation until a piece of dampened lead acetate paper held at the mouth of the cylinder becomes black, indicating that the cylinder is filled with H_2S. This will require no more than 60 mL of 6M HCl. Remove the rubber tubing from the cylinder and cover the cylinder with a glass plate. Fill the flask with water to stop the reaction, flush the solution down the drain with plenty of water, and recover any unreacted FeS for reuse.

Invert one of the covered cylinders of SO_2 and rest it on top of the covered cylinder of H_2S. Remove both glass plates from between the two cylinders. As the gases in the cylinders mix, they will react, depositing yellow sulfur dust on the sides of the cylinders.

Place the second cylinder of SO_2 before a white backdrop. Remove the cover plate and quickly pour the $KMnO_4$ solution into the cylinder. The purple solution will become colorless. If the color has not completely disappeared, replace the cover plate and agitate the solution until it does. Next, remove the cover and pour the $K_2Cr_2O_7$ solution into the cylinder of SO_2. The orange solution will become green. If it does not become green, replace the cover and agitate the solution.

HAZARDS

This demonstration should be presented only in a location having adequate ventilation. Sulfur dioxide gas is an irritating and toxic gas. At concentrations of 3 ppm, the odor of sulfur dioxide is easily detectable. At concentrations over 8 ppm, SO_2 irritates the eyes and throat and induces coughing. Even brief exposure to concentrations over 400 ppm can be fatal.

Hydrogen sulfide is very toxic by inhalation. High concentrations can cause immediate unconsciousness followed by respiratory paralysis. At high concentrations, the offensive odor of hydrogen sulfide may not give sufficient warning because of fatique of the sense of smell. Low concentrations can cause irritation of the eyes and mucous membranes and can produce headaches, nausea, dizziness, and general weakness.

Potassium permanganate is a powerful oxidizing agent, and the solid must be kept away from combustible materials. Its combination with reducing agents may result in an explosion.

Potassium dichromate is a very strong oxidizing agent, and the solid must be kept away from combustible materials. Dust from the solid irritates the respiratory tract. Long-term exposure of skin to the dust may cause ulceration. Absorption of the salt into the body's system may result in liver and kidney disease and cancer.

Because sulfuric acid is a strong acid and a powerful dehydrating agent, it can cause burns. Spills should be neutralized with an appropriate agent, such as sodium bicarbonate, and then rinsed clean.

Hydrochloric acid may cause severe burns. Hydrochloric acid vapors are extremely irritating to the skin, eyes, and respiratory system.

DISPOSAL

The solution in the 5-liter flask should be flushed down the drain with water.

The 2-liter glass cylinders should be placed in a hood and the cover plate removed until all of the gases have dissipated. Then the solution should be flushed down the drain with water.

DISCUSSION

Sulfur dioxide is a colorless, nonflammable gas with a suffocating odor. It condenses to a liquid at $-10°C$ under atmospheric pressure, and freezes at $-72°C$. It is quite soluble in water—a saturated solution at $25°C$ contains 8.5% w/w of SO_2. Hydrogen sulfide is a colorless, flammable, and poisonous gas with the odor of rotten eggs, detectable at a level of only 0.002 mg/liter of air. Its normal boiling point is $-60.3°C$, and its freezing point is $-85.5°C$. One gram of H_2S will dissolve in 242 mL of water at $20°C$ [3].

In Procedure A, sulfur dioxide is produced by the combustion of sulfur in air, as represented by the following equation:

$$S_8(s) + 8\ O_2(g) \longrightarrow 8\ SO_2(g)$$

Because sulfur dioxide is quite soluble in water, some of the product dissolves in the water in the flask. Sulfur dioxide reacts with water to form sulfurous acid.

$$SO_2(g) + H_2O(l) \longrightarrow H_2SO_3(aq)$$

Sulfurous acid is a weak acid, and some of it dissociates, producing hydrogen ions, which make the solution acidic.

$$H_2SO_3(aq) \longrightarrow H^+(aq) + HSO_3^-(aq)$$

The acidity of the solution is demonstrated by the change in the color of the pH indicator, bromocresol green. Bromocresol green undergoes its blue-to-yellow color change in the pH range of 5.4–3.8. Any other pH indicator that changes color in a similar range can be used in place of bromocresol green.

In Procedure B, sulfur dioxide is prepared by the action of hydrochloric acid on sodium sulfite. The reaction is represented by the following equation:

$$2\ HCl(aq) + Na_2SO_3(aq) \longrightarrow 2\ NaCl(aq) + H_2O(l) + SO_2(g)$$

Hydrogen sulfide is prepared in a similar fashion from hydrochloric acid and iron(II) sulfide.

$$2\ HCl(aq) + FeS(s) \longrightarrow FeCl_2(aq) + H_2S(g)$$

Sulfur dioxide can undergo either oxidation or reduction. Oxidation takes sulfur from the $+4$ to the $+6$ oxidation state, as in, for example, SO_3, SO_4^{2-}, and HSO_4^-. Reduction can lead to a number of different oxidation states ranging from $+2$ to -2—for example, $+2$ in $S_2O_3^{2-}$, 0 in S_8, and -2 in H_2S. Experimentally, SO_2 has been found more likely to undergo oxidation in slightly acidic media and reduction in strongly acidic media [4].

In Procedure B, sulfur dioxide is both oxidized and reduced. It is oxidized to sulfate ions, reducing the deep purple permanganate ion (MnO_4^-) to the nearly colorless Mn^{2+} ion.

$$2\ MnO_4^-(aq) + 5\ SO_2(g) + H^+(aq) + 2\ H_2O(l) \longrightarrow$$
$$2\ Mn^{2+}(aq) + 5\ HSO_4^-(aq)$$

In the next reaction of the demonstration, sulfur dioxide reduces the orange dichromate ion ($Cr_2O_7^{2-}$) to the green chromium(III) ion (Cr^{3+}).

$$Cr_2O_7^{2-}(aq) + 3\ SO_2(g) + 5\ H^+(aq) \longrightarrow$$
$$2\ Cr^{3+}(aq) + 3\ HSO_4^-(aq) + H_2O(l)$$

Sulfur dioxide is reduced to sulfur in the reaction with hydrogen sulfide, in which the H_2S is oxidized to sulfur as well. The equation for the reaction is

$$8\ SO_2(g) + 16\ H_2S(g) \longrightarrow 16\ H_2O(l) + 3\ S_8(s)$$

REFERENCES

1. H. N. Alyea and F. B. Dutton, Eds., *Tested Demonstrations in Chemistry*, 6th ed., Journal of Chemical Education: Easton, Pennsylvania (1965).
2. T. J. Greenbowe, *Proceedings of the Third Annual Symposium on Chemical Demonstrations*, Western Illinois University (1980).
3. M. Windholz, Ed., *The Merck Index*, 9th ed., Merck and Co.: Rahway, New Jersey (1976).
4. J. W. Mellor, *A Comprehensive Treatise on Inorganic and Theoretical Chemistry*, Vol. 10, Longmans, Green and Co.: London (1930).

6.19

Combining Volume of Oxygen with Sulfur

The pressure of the gas inside a flask containing burning sulfur is monitored. When the system reaches thermal equilibrium, there is no significant change in the pressure inside the flask, demonstrating that the product of the combustion is sulfur dioxide rather than sulfur trioxide [1].

MATERIALS

cylinder of oxygen, with valve

10 g flowers of sulfur

50 mL tap water

500-mL filter flask

stand, with clamp to hold flask

mercury-filled, U-tube manometer, open at both ends, having arms at least 80 cm long

rubber tubing to connect filter flask to manometer

2 solid rubber stoppers to fit flask

combustion spoon

rubber tubing to fit valve of oxygen cylinder

Bunsen burner

meter stick

PROCEDURE

Preparation

Connect the filter flask to the manometer with rubber tubing, as shown in the figure. In one of the rubber stoppers, bore a hole small enough to hold the handle of the combustion spoon. Insert the handle of the spoon through the stopper and adjust it so the spoon is at the center of the flask when the stopper is seated in the mouth of the flask.

Connect rubber tubing to the oxygen cylinder and insert the open end of the tubing into the flask so that it rests on the bottom of the flask. Open the valve of the oxygen cylinder slightly to produce a moderate flow of gas into the flask. Allow the gas to flow for a minute or two to fill the flask with oxygen by the upward displacement of air. Remove the tubing and seal the flask with the remaining stopper.

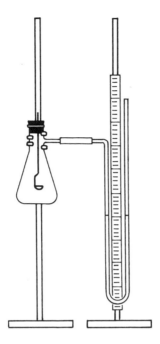

Presentation

Fill the combustion spoon with sulfur, and ignite the sulfur over a Bunsen burner. Remove the stopper from the flask and note that the levels of mercury are the same in the two arms of the manometer. Insert the spoonful of burning sulfur into the filter flask and seal the flask. The sulfur will burn for as long as several minutes, until the oxygen in the flask has been consumed.

As soon as the reaction stops, measure the difference in the levels of mercury in the arms of the manometer. Allow the flask and its contents to cool to room temperature. Again, measure the difference in the manometer's mercury levels. Open the flask, pour 50 mL of water into it, and reseal it with the solid stopper. Note what happens to the levels of the mercury in the manometer.

HAZARDS

This demonstration should be presented only in a location having adequate ventilation. The sulfur dioxide produced when the sulfur is ignited is an irritating and toxic gas. At concentrations of 3 ppm, the odor of sulfur dioxide is easily detectable. At concentrations over 8 ppm, SO_2 irritates the throat and induces coughing. Even brief exposure to concentrations over 400 ppm can be fatal.

DISCUSSION

A number of observations can be made during this demonstration including:
(a) the pressure in the flask increases during the course of the reaction;
(b) the pressure decreases as the flask cools to room temperature;
(c) when the system returns to room temperature, there is no significant difference in

the level of mercury in the two arms of the manometer, indicating that the pressure in the flask is the same after the reaction as before; and

(d) when water is added to the apparatus, the pressure inside the flask slowly decreases. The increase in pressure during the course of the reaction may be attributed to the increase in the temperature of the gases in the flask as the exothermic combustion of sulfur takes place. Similarly, the decrease in pressure as the flask cools is due to the decrease in temperature.

The third observation may be interpreted in terms of the stoichiometry of the reaction between sulfur and oxygen. There are only two stable oxides of sulfur, SO_2 and SO_3 [2, 3]. Simple thermodynamic calculations suggest that sulfur should burn in oxygen to form sulfur trioxide rather than sulfur dioxide, because the former reaction is more highly exothermic [4].

$$2\,S(s) + 3\,O_2(g) \longrightarrow 2\,SO_3(g) \qquad \Delta H° = -395.72 \text{ kJ/mol S}$$

$$S(s) + O_2(g) \longrightarrow SO_2(g) \qquad \Delta H° = -296.83 \text{ kJ/mol S}$$

If sulfur trioxide is indeed produced in this reaction, 2 moles of SO_3 are produced for every 3 moles of O_2 consumed. This decrease in the number of moles of gas in the flask will be accompanied by a proportional decrease in the total pressure: from atmospheric pressure to two-thirds of that value, or a decrease of 250 mm Hg. If SO_2 is the product of the reaction, however, there will be no change in the number of moles of gas in the system, and, therefore, no change in the pressure. Because no change in the pressure is observed, the product of the reaction must be SO_2.

Students may misinterpret the fourth observation. They may conclude that SO_2 reacts with O_2 in the presence of water to form SO_3, or that SO_2 reacts with the water itself to form SO_3. However, when the mixture comes to equilibrium, the difference between the levels of mercury in the two arms of the manometer is greater than the 250 mm expected if SO_3 were produced. In fact, the reaction of SO_2 with O_2 is quite slow and requires high temperatures and a catalyst to occur. This reaction is carried out in the contact process for the production of sulfuric acid, in which the catalyst is finely divided platinum or vanadium(V) oxide, and the temperatures employed are about 400°C [5]. Sulfur trioxide reacts violently with water to form sulfuric acid. The fourth observation, then, must be explained through the dissolution of SO_2 in the water. The solubility of SO_2 in water is 79.8 liters of SO_2 per liter of water at 0°C and 1 atm partial pressure [6].

REFERENCES

1. H. N. Alyea and F. B. Dutton, Eds., *Tested Demonstrations in Chemistry*, 6th ed., Journal of Chemical Education: Easton, Pennsylvania (1965).

2. J. W. Mellor, *A Comprehensive Treatise on Inorganic and Theoretical Chemistry*, Vol. 10, Longmans, Green and Co.: London (1930).

3. M. Schmidt and W. Siebert, *Comprehensive Inorganic Chemistry*, Vol. 2, Pergamon Press: Oxford (1973).

4. "The NBS Tables of Chemical Thermodynamic Properties," *J. Phys. Chem. Ref. Data* 11, Supp. 2 (1982).

5. W. W. Duecker and J. R. West, Eds., *Manufacture of Sulfuric Acid*, Reinhold Book Corp.: New York (1959).

6. J. A. Dean, Ed., *Lange's Handbook of Chemistry*, 12th ed., McGraw-Hill: New York (1979).

6.20

Preparation and Properties
of Methane

A gas is prepared by the reaction of sodium acetate with sodium hydroxide and is collected by the downward displacement of water. This gas is flammable [1].

MATERIALS

4.0 g anhydrous sodium acetate, $NaC_2H_3O_2$

12 g sodium hydroxide, NaOH

right-angle glass bend, with outside diameter of 7 mm and length of each arm ca. 5 cm

1-holed rubber stopper to fit test tube

90-cm length plastic or rubber tubing to fit 7-mm glass tubing

test tube, 25 mm × 200 mm

ring stand, with clamp to hold test tube

pan or dish, with capacity of 4–10 liters

250-mL Erlenmeyer flask

matches

Bunsen burner

solid rubber stopper to fit Erlenmeyer flask

wooden splint

PROCEDURE

Preparation

Insert one end of the right-angle bend through the 1-holed rubber stopper. Attach the 90-cm length of plastic or rubber tubing to the free end of the glass bend. Place 4 g of anhydrous sodium acetate and 12 g of sodium hydroxide in the test tube. Seat the stopper assembly in the mouth of the test tube, and mount the test tube at a 45° angle on the ring stand.

Pour water into the pan or dish to a depth of 10 cm. Fill a 250-mL Erlenmeyer flask to the brim with water. Cover the mouth of the flask with the palm of the hand while inverting it in the pan of water.

Presentation

Gently heat the contents of the test tube with the Bunsen burner. Place the open end of the tubing under the surface of the water in the pan and heat the test tube more strongly. Note the bubbles of gas escaping from the submerged tubing. Allow the bubbles to form for about 30 seconds to flush the air from the apparatus, and then place the opening of the tubing under the inverted flask to catch the bubbles. Fill the flask completely with the gas. Keep the bubbles forming at a moderate rate to prevent water from being drawn back into the hot test tube and breaking it. When the flask is filled, seal it with the solid rubber stopper and remove it from the pan. Remove the delivery tube from the water, and stop heating the test tube.

Light the wooden splint and, holding the flask inverted, remove the stopper. Plunge the burning splint into the flask. The flame on the splint will be extinguished, but the gas will be ignited at the mouth of the flask. Set the flask upright. The flame will gradually sink into the flask. Because the flame may not be very luminous, it can be better seen in a darkened room.

HAZARDS

Mixtures of air and methane can be explosive. Therefore, care must be taken to avoid mixing any air with the methane in the Erlenmeyer flask before the burning splint is plunged into it.

If the delivery tube is not removed from the pan of water before the reaction mixture is allowed to cool, water may be drawn back into the tube, causing it to crack.

Sodium hydroxide can cause severe burns to the skin and eyes. Dust from solid sodium hydroxide is very caustic.

The rim of the Erlenmeyer flask becomes hot as the methane burns, so it should be allowed to cool before the flask is handled.

DISPOSAL

The methane-generation tube should be allowed to cool and then filled with water. The resulting solution should be flushed down the drain.

DISCUSSION

Methane is an odorless, colorless, flammable gas at room temperature and atmospheric pressure. Its normal boiling point is $-164°C$, and its freezing point is $-182.5°C$ [2]. It is only slightly soluble in any liquid at atmospheric pressure and, therefore, may be collected by the displacement of water, as in this demonstration.

Methane was recognized as a distinct chemical substance by the early 19th century. Dalton was familiar with it as "carburetted hydrogen." It was known to be a compound of hydrogen and carbon. Also known was a similar compound of hydrogen and carbon called "olefiant gas" (ethane). The elemental composition of these gases was determined: carburetted hydrogen is 75% carbon and 25% hydrogen; olefiant gas is

86% carbon and 14% hydrogen. Dalton saw that these compositions could be expressed as 6:2 and 6:1, respectively. These compounds served as an example in Dalton's exposition of his law of multiple proportions, which states that, when two elements form more than one compound, the ratio of the weights of the elements in the compounds may be expressed in small whole numbers [3].

Methane is also referred to as marsh gas, because of its formation from decaying organic materials at the bottom of swamps. It was collected from that source in the early days of chemistry. Methane is sometimes formed from the decay of organic matter in septic tanks and landfill areas, where it can lead to explosions and asphyxiations. Today, most methane is obtained from natural gas, which is found in large underground deposits, frequently in connection with petroleum. The methane from this source is also attributed to organic matter.

The method of synthesis in this demonstration can be viewed as producing methane from organic matter, as well. Sodium acetate is a salt of acetic acid, which is the product of oxidation of ethyl alcohol, the result of the fermentation of sugars. The actual reaction is represented by the equation

$$NaC_2H_3O_2(s) + NaOH(s) \longrightarrow Na_2CO_3(s) + CH_4(g)$$

Methane is flammable, as illustrated in this demonstration, and burns to form carbon dioxide and water.

$$CH_4(g) + 2\ O_2(g) \longrightarrow CO_2(g) + 2\ H_2O(l)$$

The reaction is highly exothermic; the heat of combustion of methane is -802.3 kJ/mol CH_4 [4]. It is the highly exothermic nature of this combustion that makes methane an important fuel.

Additional properties of methane are illustrated in Demonstration 6.21 in this volume.

REFERENCES

1. G. S. Newth, *Chemical Lecture Experiments*, Longmans, Green and Co.: London (1928).
2. R. C. Weast, Ed., *CRC Handbook of Chemistry and Physics*, 59th ed., CRC Press: Boca Raton, Florida (1978).
3. A. J. Ihde, *The Development of Modern Chemistry*, Dover Publications: New York (1984).
4. "Selected Values of Chemical Thermodynamic Properties," *Natl. Bur. Stand. (U.S.)*, Circ. 500 (1952).

6.21

Combustion of Methane

Mixtures of methane and oxygen gases are ignited either under continuous gas flow or static conditions. Controlled gas flow in a burner results in a smooth continuous flame. Combustion of methane in a balloon or in soap bubbles results in an explosion. This is described in Demonstration 1.43 in Volume 1 of this series. Described here is a method for filling a balloon with methane from a natural gas outlet (Procedure A). Also described is an alternate procedure for demonstrating the combustion of methane (Procedure B).

MATERIALS FOR PROCEDURE A

natural gas outlet

pan or dish, with capacity of 4–10 liters

5-liter round-bottomed flask

2 right-angle glass bends, with outside diameter of 7 mm and length of each arm ca. 5 cm

2-holed rubber stopper to fit flask

60-cm length plastic or rubber tubing to fit 7-mm glass tubing

rubber balloon, minimum inflated diameter ca. 25 cm

MATERIALS FOR PROCEDURE B

natural gas outlet

drill, with 1-cm (3/8-inch) and 6-mm (1/4-inch) bits

1-gallon empty paint can, with lid

plastic or rubber tubing to fit natural gas outlet

fireplace matches, or wooden splint

PROCEDURE A†

Preparation and Presentation

Fill the pan with water to a depth of about 10 cm. Fill the 5-liter round-bottomed flask with tap water. Cover the mouth of the flask with the palm of the hand while in-

196   † We wish to thank Ronald I. Perkins, Greenwich High School, Connecticut, for suggesting this procedure.

verting the flask in the pan of water. Insert the two right-angle bends through the two holes in the stopper.

Connect the plastic or rubber tubing to the laboratory gas outlet and place the open end under the mouth of the flask. Open the gas outlet and fill the flask with gas by the displacement of water. Close the outlet, seat the stopper in the mouth of the flask, and set the flask upright. Connect one end of the tubing to a water tap and the other end to one of the glass bends in the stopper. Hold the mouth of the balloon securely over the free end of the other glass bend in the stopper. Open the water tap and allow the water to displace the gas from the flask into the balloon.

PROCEDURE B

Preparation

Drill a hole 1-cm in diameter in the side and near the bottom of the empty paint can. Drill a 6-mm hole in the center of its lid. Press the lid firmly onto the can, but do not seal it tightly.

Presentation

Connect the tubing to a natural gas outlet, and insert the free end of the tubing into the hole in the side of the can. Open the outlet for 15–20 seconds. Remove the tubing. Place the can in an area away from overhead light fixtures or other obstructions. Light a match and bring it to the hole in the lid of the can. A bright yellow flame about 5 cm high will form at the opening. After a few minutes, the size of the flame will diminish, becoming less and less visible. Suddenly, the flame will strike back into the can, causing a sudden explosion which forces the lid of the can to pop off and fly several feet into the air.

HAZARDS

Because methane is extremely flammable, the gas source must be isolated from unintentional ignition.

DISCUSSION

The density of methane (0.714 g/liter at STP) is lower than that of air (1.29 g/liter at STP); therefore, methane is buoyant in air, and it rises out of the can through the hole in the lid. As the methane leaves the can through the upper hole, air replaces it through the lower hole. The air and methane in the can mix, eventually producing an explosive mixture which is ignited by the flame.

For a more detailed description of the combustion of methane, see the Discussion section of Demonstration 1.43 in Volume 1 of this series.

6.22

Preparation and Properties of Hydrogen Chloride

A gas is prepared by the reaction of sodium chloride with sulfuric acid and is collected by the upward displacement of air. This gas is soluble in water, producing an acid as it dissolves, and is reactive toward ammonia gas [1].

MATERIALS

25 g sodium chloride, NaCl

100 mL 12M sulfuric acid, H_2SO_4 (To prepare 1.0 liter of 12M H_2SO_4, *slowly* add 670 mL of concentrated [18M] H_2SO_4 to 200 mL of distilled water. The H_2SO_4 should be added in increments of about 100 mL and the resulting solution allowed to cool before more is added. Once all of the H_2SO_4 has been added, the solution should be diluted to 1.0 liter with water. Allow the solution to cool before using it.)

820 mL cold tap water

10 mL concentrated (15M) ammonia solution, NH_3

right-angle glass bend, with outside diameter of 7 mm and length of each arm ca. 5 cm

1-holed #5 rubber stopper

90-cm length plastic or rubber tubing to fit 7-mm glass tubing

250-mL Erlenmeyer flask

ring stand, with clamp and ring to hold 250-mL Erlenmeyer flask

wire gauze to fit on ring

Bunsen burner

17-cm crystallizing dish

1-liter beaker

2-liter round-bottomed flask

thistle tube, with mouth small enough to fit through neck of round-bottomed flask

1-holed rubber stopper to fit round-bottomed flask

rubber balloon, with minimum inflated diameter of 25 cm

3 pieces blue litmus paper, 46 mm × 6 mm

2 500-mL Erlenmeyer flasks

2 solid rubber stoppers to fit 500-mL Erlenmeyer flasks

solid rubber stopper to fit round-bottomed flask

PROCEDURE

Preparation

Assemble the apparatus as illustrated in Figure 1. Insert one end of the right-angle bend through the 1-holed #5 stopper. Attach the 90-cm length of plastic or rubber tubing to the free end of the glass bend. In a hood, place 25 g of sodium chloride and 100 mL of 12M sulfuric acid in the 250-mL Erlenmeyer flask. Seal the flask with the stopper containing the bend. Mount the flask on the ring stand with the wire gauze beneath it and clamp it to the stand, as illustrated in Figure 1. Place the Bunsen burner under the flask.

Pour 800 mL of cold tap water into a 17-cm crystallizing dish. Pour 10 mL of concentrated ammonia solution into a 1-liter beaker.

Insert the thistle tube through the 1-holed stopper that fits the round-bottomed

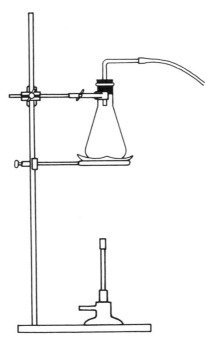

Figure 1.

flask so that the mouth of the thistle tube is inside the flask when the stopper is seated in the neck of the flask, as shown in Figure 2. Stretch the neck of the balloon over the mouth of the thistle tube.

Presentation

In a hood, gently heat the contents of the 250-mL Erlenmeyer flask with the Bunsen burner. Test the effluent of the attached tubing with dampened blue litmus paper. When the litmus turns red, insert the open end of the attached tubing all the way to the bottom of one of the 500-mL Erlenmeyer flasks. Hold a piece of dampened blue litmus paper at the mouth of this receptacle flask. When the flask is full of hydrogen

Figure 2.

chloride gas, as indicated by the color change in the litmus paper, remove the tubing and seal the flask with a solid rubber stopper. Fill the second 500-mL Erlenmeyer flask and the 2-liter round-bottomed flask in the same manner, then stopper them both with solid rubber stoppers. Once these flasks have been filled, stop heating the mixture in the generation flask.

Hold the mouth of one of the sealed Erlenmeyer flasks below the surface of the water in the crystallizing dish. Quickly remove the stopper. The water will rise into the flask. Stopper the flask under water, then remove it from the water. Test the solution in the flask with blue litmus paper. The paper will turn red, indicating that the solution is acidic.

Hold the mouth of the other sealed Erlenmeyer flask over the beaker containing concentrated ammonia solution. Remove the stopper. White smoke will form at the mouth of the flask and trail down into the beaker.

Remove the solid stopper from the round-bottomed flask containing hydrogen chloride gas and replace it with the balloon–thistle tube–stopper assembly, as shown in Figure 2. Open the flask, pour 20 mL of water into it, immediately reseal it, then swirl the flask. As the hydrogen chloride dissolves in the water, the balloon will expand to fill the flask.

HAZARDS

Hydrogen chloride gas is severely irritating to the eyes and respiratory system and can cause burns to the skin. Inhalation must be prevented.

Because concentrated sulfuric acid is both a strong acid and a powerful oxidizing agent, it must be handled with great care. Spills should be neutralized with an appropriate agent, such as sodium bicarbonate ($NaHCO_3$), and then wiped up.

Concentrated aqueous ammonia solution causes burns and is irritating to the skin, eyes, and respiratory system.

DISPOSAL

After it has cooled, the hydrogen chloride–generation flask should be filled with water and the resulting solution flushed down the drain. The contents of the 500-mL flasks, the round-bottomed flask, and the 1-liter beaker should be flushed down the drain with water.

DISCUSSION

Hydrogen chloride is a colorless, corrosive, nonflammable gas with a characteristic pungent odor. Its normal boiling point is −84.9°C, and its melting point is −114.8°C [2]. The gas has a high affinity for water, which is responsible for its fuming in damp air, and which makes it impossible to collect by the displacement of water. Hydrochloric acid is an aqueous solution of hydrogen chloride.

Gaseous hydrogen chloride was first isolated by Joseph Priestley by heating "spirit of salt" (hydrochloric acid solution) [3]. It was Priestley's innovation—the substitution of mercury for water in the pneumatic trough—that allowed him to collect this highly water-soluble gas (500 liters of HCl at 1 atm partial pressure dissolve in 1 liter of water at 0°C) [4]. Later, he discovered that the gas was better prepared by the action of sulfuric acid on common salt, the method used in this demonstration.

$$H_2SO_4(aq) + NaCl(s) \longrightarrow HCl(g) + Na^+(aq) + HSO_4^-(aq)$$

Priestley also noted the acidic properties of this gas when it was dissolved in water: its solutions dissolved iron with the evolution of "flammable air" (hydrogen). Because of its origin in common sea salt and its acidic properties, Priestley dubbed this gas "marine acid air."

Because his hydrogen chloride gas formed acidic solutions in water, and his samples of ammonia gas formed basic solutions, Priestley wondered if he could form a neutral soluble gas by the combination of these two gases. To his surprise, when he mixed these two gases, they formed a smoke, which turned out to be sal ammoniac (ammonium chloride), as illustrated in Demonstration 6.25.

$$HCl(g) + NH_3(g) \longrightarrow NH_4Cl(s)$$

Because hydrogen chloride gas is very soluble in water, it cannot be collected by the displacement of water. Priestley used the displacement of mercury as his method of collection. However, because of the toxicity of mercury, his method is not used in this demonstration. Although collecting the gas by the displacement of air does not allow one to see the gas as it is collecting, this method is convenient. Other liquids, such as hydrocarbons, in which hydrogen chloride is not soluble, could be substituted for mercury. Those which are not flammable, such as mineral oil, are quite messy and difficult to clean from the apparatus. However, those which evaporate quickly are highly flammable and are not recommended.

REFERENCES

1. G. S. Newth, *Chemical Lecture Experiments*, Longmans, Green and Co.: London (1928).
2. R. C. Weast, Ed., *CRC Handbook of Chemistry and Physics*, 59th ed., CRC Press: Boca Raton, Florida (1978).
3. A. J. Ihde, *The Development of Modern Chemistry*, Dover Publications: New York (1984).
4. P. Arthur, *Lecture Demonstrations in General Chemistry*, McGraw-Hill: New York (1939).

6.23

Preparation and Properties of Ammonia

A gas is prepared by the reaction of ammonium chloride with calcium hydroxide and is collected by the downward displacement of air. This gas is soluble in water and is reactive with hydrogen chloride gas [1].

MATERIALS

10 g ammonium chloride, NH_4Cl

10 g calcium hydroxide, $Ca(OH)_2$

10 mL distilled water

800 mL cold tap water

10 mL concentrated (12M) hydrochloric acid, HCl

right-angle glass bend, with outside diameter of 7 mm and length of each arm ca. 5 cm

1-holed #5 rubber stopper

90-cm length plastic or rubber tubing to fit 7-mm glass tubing

250-mL Erlenmeyer flask

ring stand, with clamp and ring to hold 250-mL Erlenmeyer flask

wire gauze to fit ring

Bunsen burner

2 500-mL Erlenmeyer flasks

ring stand, with two clamps to hold 500-mL Erlenmeyer flasks

17-cm crystallizing dish

1-liter beaker

red litmus paper

2 solid rubber stoppers to fit 500-mL Erlenmeyer flasks

PROCEDURE

Preparation

Assemble the apparatus as illustrated in the figure. Insert one end of the right-angle bend through the 1-holed rubber stopper. Attach the 90-cm length of plastic or rubber tubing to the free end of the glass bend. Place 10 g of ammonium chloride and

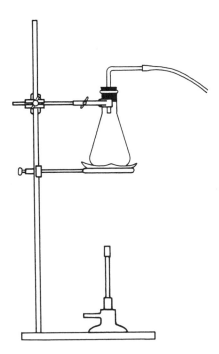

10 g of calcium hydroxide in the 250-mL Erlenmeyer flask. Add about 10 mL of distilled water and seal the flask with the stopper assembly. Mount the flask with the wire gauze beneath it on the ring stand, and clamp it to the stand, as illustrated in the figure. Place the Bunsen burner under the flask.

Mount the two 500-mL Erlenmeyer flasks inverted on the second ring stand. Pour 800 mL of cold tap water into a 17-cm crystallizing dish. Pour 10 mL of concentrated hydrochloric acid into a 1-liter beaker.

Presentation

In a hood, gently heat the contents of the 250-mL Erlenmeyer flask with the Bunsen burner until bubbles of gas begin to form in the flask. Test the effluent of the attached tubing with dampened red litmus paper. When the litmus turns blue, insert the open end of the tubing all the way to the bottom of one of the inverted flasks. Hold a piece of dampened red litmus paper at the mouth of the flask. When the flask is full of ammonia, as indicated by the color change in the litmus paper, remove the tubing and seal the flask with a solid rubber stopper. Fill the second flask in the same manner.

Hold the mouth of one of the sealed flasks below the surface of the water in the 17-cm crystallizing dish. Remove the stopper. The water will rush into the flask.

Hold the mouth of the other sealed flask over the beaker containing concentrated hydrochloric acid. Remove the stopper. White smoke will form at the mouth of the flask.

HAZARDS

Concentrated hydrochloric acid may cause severe burns. The vapors are extremely irritating to the skin, eyes, and respiratory system.

Ammonia gas irritates all parts of the respiratory system and is severely irritating to the eyes. Ammonia gas also causes burns to the skin and is toxic by inhalation. Mixtures of ammonia gas and air can be explosive and should be kept away from sparks or open flame.

DISPOSAL

The ammonia-generation flask should be filled with water and the resulting solution flushed down the drain. The contents of the 500-mL flasks and the HCl in the 1-liter beaker should be flushed down the drain with water.

DISCUSSION

Ammonia is a colorless gas with a very pungent odor. Its normal boiling point is $-33.35°C$, and its melting point is $-77.7°C$ [2]. It is extremely soluble in water—a saturated solution at 25°C contains 34% by weight of ammonia [1].

Gaseous ammonia was first isolated by Joseph Priestley by heating "volatile alkali" (aqueous ammonia solution) [3]. It was Priestley's innovation—the substitution of mercury for water in the pneumatic trough—that allowed him to collect this highly water-soluble gas (1130 liters of NH_3 at 1 atm partial pressure dissolve in 1 liter of water at 0°C) [4]. Later, he improved upon the method of preparation by treating the aqueous ammonia with quicklime (CaO). The method used in this demonstration is even more efficient than Priestley's "improved" method.

$$2 \, NH_4Cl(aq) + Ca(OH)_2(aq) \longrightarrow 2 \, NH_3(g) + CaCl_2(aq) + 2 \, H_2O(l)$$

Priestley also noted the basic properties of this gas when it is dissolved in water. Because of its basic properties, Priestley dubbed this gas "alkaline air."

Ammonia can be made directly from its elements, nitrogen and hydrogen. However, this requires high temperatures, high pressures, and a catalyst, as employed in the Haber process described in the Discussion section of Demonstration 6.12. One of the major industrial uses of ammonia is in the manufacture of nitric acid, a process in which the ammonia is oxidized catalytically. The catalytic oxidation of ammonia is described in Demonstration 6.26.

REFERENCES

1. G. S. Newth, *Chemical Lecture Experiments*, Longmans, Green and Co.: London (1928).
2. M. Windholz, Ed., *The Merck Index*, 9th ed., Merck and Co.: Rahway, New Jersey (1976).
3. A. J. Ihde, *The Development of Modern Chemistry*, Dover Publications: New York (1984).
4. J. A. Dean, Ed., *Lange's Handbook of Chemistry*, 12th ed., McGraw-Hill: New York (1979).

6.24

Gas Solubility: The Fountain Effect

A small amount of water is injected into an inverted round-bottomed flask connected by a glass tube to a reservoir of water below it. Soon after the injection, the water from the reservoir rushes into the flask, turning red as it enters and forming a fountain inside the flask. Other color changes are also described [*1, 2*].

MATERIALS FOR PROCEDURE A

source of dry ammonia gas (cylinder with valve, or see Demonstration 6.23 for preparation)

10 mL phenolphthalein indicator solution (To prepare 100 mL of solution, dissolve 0.05 g of phenolphthalein in 50 mL of 95% ethanol, and dilute the solution to 100 mL with distilled water.)

2-liter round-bottomed flask

2-holed rubber stopper to fit 2-liter flask

100-cm length glass tubing, with outside diameter of 8 mm

3-liter round-bottomed flask

2 ring stands

2 rings to support 2-liter flask

cork ring to support 3-liter flask

dropper

10-cm length rubber tubing to fit over dropper's open end

15-cm length copper wire, 16 gauge

gloves, plastic or rubber

90-cm length plastic or rubber tubing

solid rubber stopper to fit 2-liter flask

pressure bulb

MATERIALS FOR PROCEDURE B

source of dry hydrogen chloride gas (cylinder with valve, or see Demonstration 6.22 for preparation)

10 mL bromothymol blue indicator solution (To prepare 250 mL of solution, first prepare 1.0 liter of 0.01M NaOH solution by dissolving 0.40 g NaOH in 1.0 liter of water. Then dissolve 0.1 g of bromothymol blue in 16 mL of 0.01M NaOH and dilute the mixture to 250 mL with distilled water.)

2-liter round-bottomed flask

2-holed rubber stopper to fit 2-liter flask

100-cm length glass tubing, with outside diameter of 8 mm

3-liter round-bottomed flask

ring stand

ring to support 2-liter flask

cork ring to support 3-liter flask

dropper

10-cm length rubber tubing to fit over dropper's open end

15-cm length copper wire, 16 gauge

gloves, plastic or rubber

90-cm length plastic or rubber tubing

solid rubber stopper to fit 2-liter flask

pressure bulb

MATERIALS FOR PROCEDURE C

2 mL phenolphthalein indicator solution (For preparation, see Materials for Procedure A.)

800 mL tap water

100 mL concentrated (15M) aqueous ammonia, NH_3

1 g sodium hydroxide, NaOH

35-cm length glass tubing, with outside diameter of 7 mm

1-holed rubber stopper to fit 250-mL Erlenmeyer flask

2 10-cm iron rings

ring stand, at least 75 cm in height

Bunsen burner

iron gauze

250-mL Erlenmeyer flask

500-mL round-bottomed flask

50-cm length glass tubing, with outside diameter of 7 mm

2-holed rubber stopper to fit 500-mL round-bottomed flask

dropper, with bulb

PROCEDURE A

Preparation

Assemble the glassware as illustrated in Figure 1. The glass tube should extend to within 10 cm of the bottom of the inverted upper, 2-liter flask and to within 1 cm of the bottom of the lower, 3-liter flask. When the dropper is inserted through the stopper, the constricted end of the dropper should be inside the 2-liter flask. Tighten the rubber tubing to the dropper with wire.

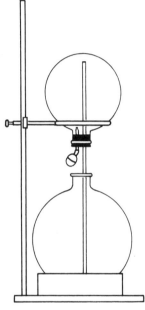

Figure 1.

Unstopper and remove the 2-liter flask from the apparatus. Support it inverted on the second ring stand in a well-ventilated hood. Wearing gloves, attach the 90-cm length of plastic or rubber tubing to the valve of the cylinder of dry ammonia and insert the other end into the inverted flask. Fill the flask with ammonia gas by the downward displacement of air. Stopper the flask with the solid rubber stopper.

Place 6–10 mL of phenolphthalein indicator solution in the 3-liter flask and fill the flask with water. Fill the pressure bulb with water and wire its exhaust valve to the rubber tubing attached to the dropper. Holding the ammonia-filled 2-liter flask inverted over the apparatus, remove its solid stopper and reassemble the apparatus.

Presentation

Squeeze the pressure bulb to deliver several milliliters of water into the ammonia-filled flask. As ammonia dissolves in the water, the level of water in the long glass tube will rise until it overflows into the upper flask. When the water begins to flow from the tube into the flask, more ammonia will dissolve, and the rate at which the water rises in the tube will increase dramatically, producing a fountain effect within the flask. As the ammonia dissolves in the water, it forms a basic solution, causing the phenolphthalein in the water to turn magenta.

PROCEDURE B

Preparation

Assemble the glassware as illustrated in Figure 1. The glass tube should extend to within 10 cm of the bottom of the inverted upper, 2-liter flask and to within 1 cm of the bottom of the lower, 3-liter flask. When the dropper is inserted through the stopper, the constricted end of the dropper should be inside the 2-liter flask. Tighten the rubber tubing to the dropper with wire.

Unstopper and remove the 2-liter flask from the apparatus. Support it upright on the cork ring in a well-ventilated hood. Wearing gloves, attach the 90-cm length of plastic or rubber tubing to the valve of the cylinder of dry hydrogen chloride gas and insert the other end into the flask. Fill the flask with hydrogen chloride gas by the upward displacement of air. Stopper the flask with the 2-holed stopper containing the glass tube and dropper.

Place 6–10 mL of bromothymol blue indicator solution in the 3-liter flask and fill the flask with water. Fill the pressure bulb with water, and wire its exhaust valve to the rubber tubing attached to the dropper. Invert the 2-liter flask of hydrogen chloride together with its stopper assembly and reassemble the apparatus shown in Figure 1.

Presentation

Squeeze the pressure bulb to deliver several milliliters of water into the flask of hydrogen chloride. As HCl gas dissolves in the water, the level of water in the long glass tube will rise until it overflows into the upper flask. When the water begins to flow from the tube into the flask, more HCl will dissolve, and the rate at which the water rises in the tube will increase dramatically, producing a fountain effect within the flask. As the HCl dissolves in the water, it forms an acidic solution, causing the bromothymol blue in the water to turn from green to yellow.

PROCEDURE C

Preparation

Assemble the apparatus as illustrated in Figure 2. Insert the 35-cm length of glass tubing through the 1-holed rubber stopper so that its tip extends about 2 cm beyond the narrow end of the stopper. Position one of the iron rings on the ring stand so that it is about 8 cm above the top of the Bunsen burner, and place the iron gauze on the ring. Set the 250-mL Erlenmeyer flask on the iron gauze and seat the 1-holed stopper in its mouth. Position the second iron ring so that the upper end of the glass tube in the stopper assembly of the Erlenmeyer flask extends to within 3 cm of the bottom of the 500-mL round-bottomed flask when it is inverted on the upper iron ring. Remove the flasks from the ring stand.

Insert the 50-cm length of glass tubing through the 2-holed rubber stopper so that the tip of the tube is about 1 cm from the bottom of the round-bottomed flask when the stopper is seated in the mouth of the flask. Fill the dropper with water and insert it into the other hole of the stopper. Remove the stopper assembly.

Pour 2 mL of phenolphthalein solution into the 1-liter beaker and add 800 mL of

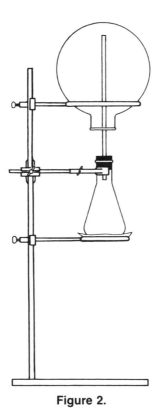

Figure 2.

water. Pour 100 mL of concentrated aqueous ammonia into the 250-mL Erlenmeyer flask and add 1 g of solid sodium hydroxide. Seat the 1-holed rubber-stopper assembly in the mouth of the flask and set the flask on the iron gauze. Invert the round-bottomed flask on the upper ring with the glass tube extending into its mouth.

Presentation

Gently heat the Erlenmeyer flask until its contents begin to simmer. Simmer the solution for about 60–90 seconds to fill the upper flask with ammonia gas. Keeping the round-bottomed flask inverted, quickly remove it from the ring stand and seat the 2-holed stopper assembly in its mouth. Immerse the free end of the glass tube in the water in the 1-liter beaker. Squeeze the bulb of the dropper to inject water into the flask. As the ammonia gas in the flask dissolves in the injected water, the water in the beaker will rise up the glass tube. When the water begins to flow from the tube into the flask, more ammonia will dissolve, and the rate at which the water rises in the tube will increase dramatically, producing a fountain effect within the flask. As the ammonia dissolves in the water, the solution becomes basic, turning the phenolphthalein in the water from colorless to magenta.

HAZARDS

Only round-bottomed flasks should be used as the upper flask in this demonstration. Erlenmeyer and other flat-bottomed flasks may implode under the stress of the vacuum created during this procedure.

Ammonia gas irritates all parts of the respiratory system and is severely irritating to the eyes. Ammonia gas also causes burns to the skin and is toxic by inhalation. Mixtures of ammonia gas and air can be explosive and should be kept away from sparks or open flame.

Hydrogen chloride gas is severely irritating to the eyes and the respiratory system and can cause burns to the skin. Inhalation must be prevented.

Sodium hydroxide can cause severe burns of the skin and eyes. Dust from solid sodium hydroxide is very caustic.

DISPOSAL

The contents of the flasks should be flushed down the drain with water.

DISCUSSION

Ammonia and hydrogen chloride are among the most water-soluble gases known. The volume of ammonia that will dissolve in water at 0°C and a partial pressure of 1 atm is 1130 liters [3], while 506 liters of hydrogen chloride dissolve under the same conditions [4]. These gases are highly soluble in water because they react with water. Hydrogen chloride gas forms hydronium ions and chloride ions when it dissolves in water.

$$HCl(g) + H_2O(l) \longrightarrow H_3O^+(aq) + Cl^-(aq)$$

After ammonia gas has dissolved in water, the solution contains an equilibrium mixture of aquated ammonia molecules, ammonium ions, and hydroxide ions.

$$NH_3(aq) + H_2O(l) \rightleftharpoons NH_4^+(aq) + OH^-(aq)$$

The difference in the solubilities of these two gases can be seen when Procedures A and B are presented together. The ammonia dissolves more rapidly, and its fountain is more vigorous. Because the ammonia fountain is more spectacular, it is recommended that the hydrogen chloride fountain be presented first when the two are presented together.

The pH indicators specified in these procedures are just two of the many that can be used. Quite a variety of color changes are possible with other indicators. Any indicator which changes color in the pH range of 9 through 11 can be used with the ammonia fountain, while one that changes in the range of 2 through 5 can be used with the hydrogen chloride fountain.

REFERENCES

1. G. S. Newth, *Chemical Lecture Experiments*, Longmans, Green and Co.: London (1928).
2. H. N. Alyea and F. B. Dutton, Eds., *Tested Demonstrations in Chemistry*, 6th ed., Journal of Chemical Education: Easton, Pennsylvania (1965).
3. J. A. Dean, Ed., *Lange's Handbook of Chemistry*, 12th ed., McGraw-Hill: New York (1978).
4. P. Arthur, *Lecture Demonstrations in General Chemistry*, McGraw-Hill: New York (1939).

6.25

Reaction Between
Ammonia and Hydrogen Chloride

Two test tubes containing hydrochloric acid and aqueous ammonia are held next to each other. When air is blown across the mouths of the tubes, a white smoke is formed. Smoke rings can also be formed [1].

MATERIALS FOR PROCEDURE A

50 mL concentrated (12M) hydrochloric acid, HCl

50 mL concentrated (15M) aqueous ammonia, NH_3

2 test tubes, 25 mm $\times$ 200 mm

MATERIALS FOR PROCEDURE B

20 mL concentrated (12M) hydrochloric acid, HCl

20 mL concentrated (15M) aqueous ammonia, NH_3

special bellows (See description in Procedure B.)

2 50-mL beakers

PROCEDURE A

Preparation

Pour 50 mL of concentrated hydrochloric acid into one of the test tubes. Pour 50 mL of concentrated aqueous ammonia into the other test tube.

Presentation

Hold one of the test tubes vertically with its mouth about 1 foot from your mouth. Blow over the mouth of the test tube. Repeat this with the other test tube. No smoke will be observed.

While blowing over the mouth of one of the test tubes, bring the mouth of the other tube downwind of the first tube and observe the smoke which forms. Reverse the positions of the tubes and note that the smoke still forms.

PROCEDURE B

Preparation

The special bellows is assembled from 1 × 4–inch planking, 1/8-inch plywood, and an automobile inner tube (see Figure 1). From the planking, construct a frame 16 inches square on the outside and 4 inches deep. Cut a circular hole 13 cm in diameter near the center of a 16-inch square of the 1/8-inch plywood. Nail the plywood to one side of the frame. Cut a 16-inch square from an automobile inner tube and staple it with overlapping staples to the other side of the frame.

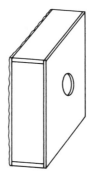

Figure 1.

A similar bellows may be constructed from a cardboard box modified as illustrated in Figure 2. Cut a circular hole 8 cm in diameter in the center of one side of the box. Cut out the opposite side of the box and reattach it with the duct tape along the bottom edge to form a flap. To the top edge of the flap, attach a small tab made of the tape to serve as a handle. Seal any other openings, such as the top of the box, with tape.

Set the bellows so that the circular hole faces outward from the edge of a table or bench. Pour 20 mL of concentrated HCl into one of the 50-mL beakers and 20 mL of concentrated aqueous ammonia into the other. Place the two beakers in the bellows through the round hole. Allow about 1 minute for ammonium chloride (NH_4Cl) smoke to fill the box before presenting the demonstration.

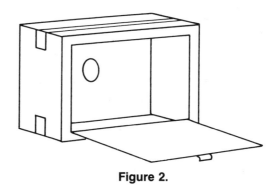

Figure 2.

Presentation

If you are using the wood and rubber bellows, grasp it firmly at the top and strike the rubber membrane opposite the hole. If you are using the cardboard box, grasp the tab and quickly push the flap into the box. A smoke ring will be blown out through the circular hole. A number of trials may be needed to discover the striking force or the flap motion which produces the best rings.

HAZARDS

Hydrochloric acid may cause severe burns. The vapors are extremely irritating to the skin, eyes, and respiratory system.

Concentrated aqueous ammonia solution can cause burns and is irritating to the skin, eyes, and respiratory system.

DISCUSSION

Ammonia reacts with hydrogen chloride in the gas phase to form a white solid of ammonium chloride.

$$NH_3(g) + HCl(g) \longrightarrow NH_4Cl(s)$$

This reaction is responsible for the white film that is found on bottles of hydrochloric acid and aqueous ammonia that are stored together. In this demonstration, the particles of solid NH_4Cl are very small and suspended colloidally in the air as smoke.

Both ammonia and hydrogen chloride gas escape from their concentrated aqueous solutions. According to Henry's law, the solubility of a gas increases as its partial pressure increases. In a sealed bottle, the gas can be in equilibrium with its dissolved form, its partial pressure having reached a constant value. When the bottle is opened, some of the gas escapes, and the partial pressure of the gas decreases. This means that the solubility of the gas decreases proportionately, and gas escapes from the solution. This demonstration relies on the escaping gas as its source of reactants. Concentrated solutions of ammonia and hydrogen chloride are used because they have the highest partial pressures of the gases; therefore, the gases escape rapidly from an open container.

The reaction between ammonia gas and hydrogen chloride gas is used to illustrate gaseous diffusion in Demonstration 5.15 of this volume.

REFERENCE

1. H. N. Alyea and F. B. Dutton, Ed., *Tested Demonstrations in Chemistry*, 6th ed., Journal of Chemical Education: Easton, Pennsylvania (1965).

6.26

Catalytic Oxidation of Ammonia

A coiled platinum wire is heated in a flame until it glows red hot. Then it is placed in the mouth of a beaker containing concentrated aqueous ammonia. When the room lights are dimmed, the metal catalyst can be seen glowing brightly [1–4].

MATERIALS

200 mL concentrated (15M) aqueous ammonia, NH_3

15–20-cm length platinum wire

10-cm glass rod, 7 mm in diameter

500-mL beaker

Bunsen burner

tongs

PROCEDURE

Preparation

Coil the platinum wire around the glass rod. Pour 200 mL of concentrated aqueous ammonia into the 500-mL beaker. Attach one end of the coiled platinum wire to the glass rod so that the other end of the coil hangs 4–5 cm above the surface of the ammonia when the rod is resting across the mouth of the beaker.

Presentation

Heat the coil of platinum in the flame of a Bunsen burner until it glows red, and quickly suspend it over the ammonia solution in the beaker. Dim the room lights. The platinum wire will continue to glow for as long as half an hour.

HAZARDS

Concentrated aqueous ammonia solution can cause burns and is irritating to the skin, eyes, and respiratory system.

Nitrogen dioxide (NO_2) is an extremely toxic gas. It is irritating to the respiratory system; inhaling it may result in severe pulmonary irritation which is not apparent until several hours after exposure. A concentration of 100 ppm is dangerous for even a short period of time, and exposure to concentrations in excess of 200 ppm may be fatal.

DISPOSAL

The catalyst should be removed from the beaker of ammonia and allowed to cool. It can be reused indefinitely.

The concentrated ammonia should be flushed down the drain with water.

DISCUSSION

Platinum catalyzes the oxidation of ammonia to nitrogen(II) oxide.

$$4 \ NH_3(g) + 5 \ O_2(g) \longrightarrow 4 \ NO(g) + 6 \ H_2O(g)$$

The nitrogen oxide reacts immediately with oxygen to form nitrogen(II) dioxide.

$$2 \ NO(g) + O_2(g) \longrightarrow 2 \ NO_2(g)$$

These reactions form the basis of the Ostwald process for the synthesis of nitric acid. Throughout the 19th century, the primary source of nitrates, used in explosives and fertilizers, was Chile saltpeter ($NaNO_3$), mined in South America. However, in the early part of the 20th century, shortly after the development of the Haber process—an inexpensive method for the synthesis of ammonia—Ostwald devised a procedure for making nitric acid from the catalyzed oxidation of ammonia [5]. A 10% mixture of NH_3 with air is passed over a platinum catalyst at temperatures of 900–1000°C. The nitrogen oxide produced then reacts with oxygen to form nitrogen dioxide, as is represented by the second equation above. The nitrogen dioxide dissolves in water in an absorption tower to give a mixture of nitric acid and nitrogen oxide, which is recycled.

$$3 \ NO_2(g) + H_2O(l) \longrightarrow 2 \ H^+(aq) + 2 \ NO_3^-(aq) + NO(g)$$

Copper metal also catalyzes this reaction, although not as effectively as platinum. The demonstration may be presented using a thin copper disk, but the results are variable and, at best, not as visible as when platinum is used.

REFERENCES

1. H. N. Alyea and F. B. Dutton, Eds., *Tested Demonstrations in Chemistry*, 6th ed., Journal of Chemical Education: Easton, Pennsylvania (1965).
2. G. Fowles, *Lecture Experiments in Chemistry*, 5th ed., G. Bell and Sons, Ltd.: London (1959).
3. N. S. Baylis, *J. Chem. Educ.* 20:510 (1943).
4. H. M. State, *J. Chem. Educ.* 34:A375 (1957).
5. A. J. Ihde, *The Development of Modern Chemistry*, Harper and Row: New York (1964).

6.27

Vapor-Phase Oxidations

A small amount of ethanol is placed in a bottle having electrodes through its sides. When a Tesla coil is touched to one of the electrodes, a spark is produced within the bottle and the ethanol explodes, popping the cork from the bottle. A copper disk is heated with a Bunsen burner until red hot and then suspended over a boiling liquid. When the room lights are dimmed, the copper disk can be seen glowing [1–3].

MATERIALS FOR PROCEDURE A

2–3 mL 95% ethanol, C_2H_5OH

2 nails, ca. 7 cm long

250-mL polyethylene bottle

cork to fit bottle

Tesla coil

MATERIALS FOR PROCEDURE B

200 mL methanol, CH_3OH, acetone, CH_3COCH_3, or 2-propanol, $CH_3CHOHCH_3$

500-mL beaker

copper disk, ca. 3 cm in diameter and 0.2 mm thick, or a copper 1-cent coin

drill, with $\frac{1}{16}$–$\frac{1}{4}$-inch bit

20-cm length copper wire, 16–20 gauge

glass rod

hot plate

Bunsen burner

PROCEDURE A

Preparation

Construct the reactor as shown in the figure. Insert one of the nails through the side of the polyethylene bottle until its tip is near the center of the bottle. Insert the other nail through the opposite side of the bottle so its tip is about 0.5 cm from the tip of the first nail. Pour 2–3 mL of 95% ethanol into the bottle. Stopper the bottle with the cork. Set the bottle in an area free of overhead obstruction.

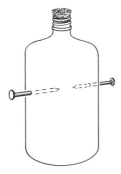

Presentation

Turn on the Tesla coil and bring its tip to one of the nails. A spark will jump to the nail and from one nail to the other. This spark inside the bottle detonates the mixture of air and ethanol vapor, causing the cork to shoot out of the bottle.

PROCEDURE B

Preparation

Pour 200 mL of methanol, acetone, or 2-propanol into the 500-mL beaker. Drill a small hole through the copper disk near its edge. Fashion a hook from the copper wire, and hang the disk on the hook. Wrap the other end of the wire around the glass stirring rod. Adjust the length of the wire so that the bottom edge of the disk is about 0.5 cm above the surface of the liquid when the rod is rested across the mouth of the beaker.

Presentation

This demonstration should be presented in a hood. Place the beaker on the hot plate and adjust the heat so the liquid simmers gently. Heat the copper disk in the flame of the Bunsen burner until it glows red. Quickly suspend the disk above the liquid and extinguish the room lights. The disk will continue to glow until almost all of the liquid has evaporated.

HAZARDS

In Procedure A, care should be taken to avoid damage to light fixtures or other objects, because the cork is expelled with considerable force.

Procedure B of this demonstration should be presented in a hood. A product of the pyrolysis of acetone is ketene, a poisonous gas similar to phosgene in its toxicity. Ketene is also formed when 2-propanol is used as the starting material. When methanol is used, the product is formaldehyde vapor, which is extremely irritating to mucous membranes and has been indicated as a possible carcinogen.

Acetone, methanol, and 2-propanol are highly flammable. The flash point (the lowest temperature at which the vapors of a volatile substance ignite in air in the presence of a flame) of acetone is $-20°$, and that of methanol and 2-propanol is 12°C [4]. Therefore, care must be taken when heating the copper disk to avoid igniting the liquid. A fire extinguisher should be ready when the demonstration is presented.

DISPOSAL

The copper disk should be saved for future use. Any remaining liquids should be poured into a suitable container for waste solvents or flushed down the drain with water.

DISCUSSION

The combustion of the vapors of these organic liquids in air can occur in a controlled manner, such as with the copper catalyst, or in an uncontrolled manner, such as when detonated by a spark. In the former case, energy is released in a continuous fashion, heating the copper to glowing red. In the latter case, the energy is released suddenly, resulting in an explosion.

In Procedure A, a small amount of ethanol liquid is placed in a bottle that contains air. Some of the ethanol vaporizes, producing an explosive mixture of vapor and oxygen. When a spark is created between the electrodes, the mixture detonates, forcing the stopper from the bottle.

$$C_2H_5OH(g) + 3 O_2(g) \longrightarrow 2 CO_2(g) + 3 H_2O(g)$$

The demonstration cannot be repeated without flushing the bottle, presumably because the explosion consumes all of the oxygen in the bottle, replacing it with carbon dioxide.

Although platinum is a more effective catalyst for the dehydrogenation of alcohols, copper can also catalyze this reaction. When methanol is dehydrogenated, the product is formaldehyde.

$$CH_3OH(g) \xrightarrow{\text{Cu}} H_2(g) + H_2CO(g) \qquad \Delta H° = +85.3 \text{ kJ/mole}$$

The reaction is endothermic, and requires an input of energy to occur [5]. This is supplied initially by heating the copper. However, the copper remains hot because of the oxidation of hydrogen at the copper surface, which is highly exothermic [5].

$$H_2(g) + \tfrac{1}{2} O_2(g) \longrightarrow H_2O(l) \qquad \Delta H° = -242 \text{ kJ/mole}$$

This oxidation is the source of the energy which keeps the copper glowing.

The dehydrogenation product of 2-propanol is acetone.

$$CH_3CHOHCH_3(g) \xrightarrow{\text{Cu}} H_2(g) + CH_3COCH_3(g) \qquad \Delta H° = +56.0 \text{ kJ/mol}$$

Again the energy required for this reaction can be supplied by the oxidation of hydrogen. The fact that acetone itself can be used indicates that acetone itself may be subject to dehydrogenation or oxidation at the copper surface. The pyrolysis of acetone at 700–750°C yields ketene, $H_2C{=}C{=}O$, and methane, CH_4.

REFERENCES

1. S. Sharpe, Ed., *The Alchemist's Cookbook*, Instructional Development Centre, McMaster University: Hamilton, Ontario (undated).
2. R. E. Ashmore, *J. Chem. Educ.* 45:243 (1968).
3. E. S. Olson and R. E. Ashmore, *J. Chem. Educ.* 59:1042 (1982).
4. M. Windholz, Ed., *The Merck Index*, 9th ed., Merck and Co.: Rahway, New Jersey (1976).
5. R. C. Weast, Ed., *CRC Handbook of Chemistry and Physics*, 59th ed., CRC Press: Boca Raton, Florida (1978).

6.28

Preparation and Properties
of Chlorine

A yellow-green gas is prepared by the reaction between manganese dioxide and hydrochloric acid. This gas supports the combustion of hydrogen and methane [1]. Additional procedures demonstrate other reactions of this gas. Other demonstrations of the properties of chlorine may be found in Demonstrations 1.25, 1.26, 1.27, 1.29, and 1.45 of Volume 1 in this series.

MATERIALS FOR PROCEDURE A

natural gas outlet, or cylinder of hydrogen, with valve

15 g manganese dioxide, MnO_2

200 mL tap water

8 g sodium hydroxide, NaOH

40 mL concentrated (12M) hydrochloric acid, HCl

30-cm length glass capillary tubing, with 2-mm bore and outside diameter of 8 mm

glass-working torch

1-m length rubber tubing to fit capillary tubing

500-mL Erlenmeyer flask

20-cm length glass tubing, with outside diameter of 7 mm

2-holed rubber stopper to fit 500-mL Erlenmeyer flask

right-angle glass bend, with outside diameter of 7 mm and length of each arm ca. 5 cm

6-cm length rubber tubing to fit 7-mm glass tubing

short-stemmed funnel (or thistle tube)

pinch clamp

wire gauze

ring stand with 2 rings

2 clamps, with clamp holders

30-cm length rubber tubing to fit 7-mm glass tubing

right-angle glass bend, with outside diameter of 7 mm and length of one arm ca. 5 cm, the other ca. 30 cm

1-liter Erlenmeyer flask

glass plate to cover 1-liter Erlenmeyer flask

Bunsen burner

matches

glass plate to cover 500-mL Erlenmeyer flask

blue litmus paper, ca. 2 cm × 10 cm

tongs

MATERIALS FOR PROCEDURE B

100 mL 1M hydrochloric acid, HCl (To prepare 1.0 liter of stock solution, slowly pour 83 mL of concentrated [12M] HCl into 500 mL of distilled water, and dilute the resulting solution to 1.0 liter with distilled water.)

10 mL 5% sodium hypochlorite solution, NaClO (liquid laundry bleach)

0.5 g calcium carbide, CaC_2

1- or 2-liter glass cylinder (or beaker)

glass plate to cover cylinder

MATERIALS FOR PROCEDURE C

40 g iodine

4 g red phosphorus

5 mL distilled water

cylinder of chlorine, with valve

>   *or*

>   15 g manganese dioxide, MnO_2

>   200 mL tap water

>   8 g sodium hydroxide, NaOH

>   40 mL concentrated (12M) hydrochloric acid, HCl

>   500-mL Erlenmeyer flask

>   20-cm length glass tubing, with outside diameter of 7 mm

>   2-holed rubber stopper to fit 500-mL Erlenmeyer flask

>   right-angle glass bend, with outside diameter of 7 mm and length of each arm ca. 5 cm

>   36-cm length rubber tubing to fit 7-mm glass tubing

>   short-stemmed funnel (or thistle tube)

>   pinch clamp

>   wire gauze

>   ring stand, with ring

>   2 clamps, with clamp holders

right-angle glass bend, with outside diameter of 7 mm and length of one arm ca. 5 cm, the other ca. 30 cm

dropping funnel, 100-mL capacity or less

2 right-angle glass bends, with outside diameter of 7 mm and length of each arm ca. 5 cm

2-holed stopper to fit 500-mL Erlenmeyer flask

right-angle glass bend, with outside diameter of 7 mm and length of one arm ca. 5 cm, the other ca. 25 cm

2-holed rubber stopper to fit 1-liter wide-mouth bottle

2 1-liter wide-mouth bottles

2 60-cm lengths rubber tubing to fit 7-mm glass tubing

gloves, plastic or rubber

Bunsen burner

blue litmus paper

stopcock grease

2 glass plates to cover wide-mouth bottles

PROCEDURE A

Preparation

Bend one end of the capillary tubing into a J shape, as illustrated in Figure 1, so that this U-bend can be lowered into a 1-liter Erlenmeyer flask. Use the 1-m length of rubber tubing to connect the long end of the tube to the natural gas outlet or the valve of the cylinder of hydrogen.

Assemble the apparatus as illustrated in Figure 2. Place 15 g of MnO_2 in the 500-mL Erlenmeyer flask. Insert the 20-cm length of glass tubing through the 2-holed rubber stopper so its tip is within 1 cm of the bottom of the Erlenmeyer flask when the stopper is seated in the mouth of the flask. Insert one end of the smaller right-angle bend through the other hole in the stopper. Use the 6-cm length of rubber tubing to

Figure 1.

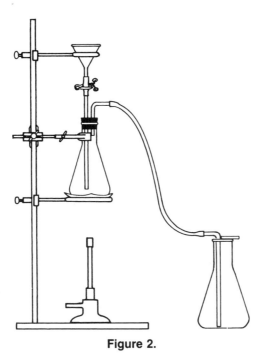

Figure 2.

attach the funnel to the free end of the straight glass tube in the stopper, and seal the rubber tubing with the pinch clamp. Set the flask on a wire gauze atop a ring on the stand, clamp the flask to the stand, and support the funnel with the other ring. Use the 30-cm length of rubber tubing to attach the glass bend in the stopper to the shorter end of the larger glass bend. Insert the free end of the glass bend to the bottom of a 1-liter Erlenmeyer flask and cover as much of the opening as possible with a glass plate.

Pour 200 mL of water into the 500-mL beaker, and dissolve 8 g of NaOH in the water.

Presentation

This demonstration should be presented only in a well-ventilated fume hood. Open the pinch clamp and pour 40 mL of concentrated hydrochloric acid through the funnel into the flask. Close the pinch clamp. Gently warm the flask with the flame of the Bunsen burner until the mixture of HCl and MnO_2 bubbles moderately, and continue the warming to sustain the reaction at this level. As chlorine is generated, it will displace the air from the 1-liter Erlenmeyer flask. When the flask is filled with pale yellow-green chlorine, remove the delivery tube and insert it into the solution of NaOH in the 500-mL beaker. Stop heating the generation flask. Cover the receptacle flask with the glass plate.

Open the natural gas outlet or hydrogen valve and adjust it to produce a gentle flow of gas. Ignite the gas at the end of the U-bend with a match. Remove the cover from the flask of chlorine and lower the U-bend into the flask. The flame will burn gray, and the color of the chlorine gas will disappear. Remove the U-bend from the flask, cover the flask, and turn off the gas.

Remove the cover from the flask and blow across the mouth of the flask. The gas in the flask will produce fumes in the moist exhaled air. Dampen a piece of blue litmus paper and lower it into the flask with tongs; the litmus will turn pink.

PROCEDURE B [2, 3]

Preparation and Presentation

Pour 100 mL of 1M HCl into the glass cylinder. Add 10 mL of 5% NaClO solution to the acid. Cover the cylinder with the glass plate. Within a minute the cylinder will fill with yellow-green chlorine gas.

Remove the cover plate from the cylinder, add a few pellets of CaC_2, and quickly step back. Large flames will erupt from the cylinder, and deposits of black soot will form on its sides. Add a few more pieces of CaC_2 to repeat the reaction.

PROCEDURE C [1]

Preparation

Assemble the apparatus as illustrated in Figure 3. Insert the stem of the dropping funnel and one arm of one of the small right-angle bends through the holes of the flask stopper. Insert the other small bend and the long arm of the large bend through the bottle stopper. Adjust the long bend so that it reaches to within 1 cm of the bottom of the bottle when the stopper is seated in the mouth of the bottle. Use one of the pieces of rubber tubing to connect the long bend in the bottle to the bend in the flask, and connect the other piece to the free arm of the small bend in the bottle.

While wearing gloves, do the following procedure for the production of HI vapor in a well-ventilated hood. Place 40 g of iodine and 4 g of red phosphorus in the Erlenmeyer flask. Seal the flask with the stopper assembly. Pour 5 mL of distilled water into the dropping funnel. Insert the free end of the rubber tubing (from the wide-mouth bottle)

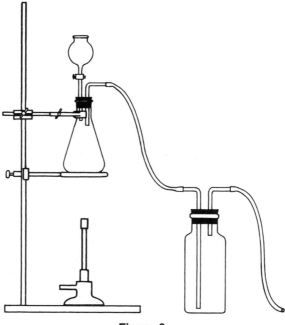

Figure 3.

down a drain and open a water tap to flush any escaping gas down the drain. Add several drops of water from the funnel to the flask. *Gently* warm the contents of the flask with the Bunsen burner until the mixture bubbles slowly. The hydrogen iodide gas will displace the air from the bottle. The bottle is full when the gas escaping from the tubing in the drain turns a piece of blue litmus paper pink. Add more water (no more than 20 drops should be used) if necessary to keep the reaction going. When the bottle is full of HI, remove it from the apparatus, grease its rim, and cover it with a glass plate.

Fill the second bottle with Cl_2 from a gas cylinder or from the chlorine generator described in Procedure A. Grease the rim of this bottle and seal it with a glass plate.

Invert the covered bottle of Cl_2 and rest it on top of the bottle of HI.

Presentation

Remove the glass plates separating the mouths of the two bottles and allow the gases to mix inside the bottles. As the gases mix, violet vapors will form and condense into black crystals on the sides of the bottles. Some orange crystals of I_2Cl_6 may also form after several minutes.

HAZARDS

The procedures in this demonstration involve noxious gases and should be presented only in a well-ventilated hood.

Chlorine gas irritates the eyes and mucous membranes and, if inhaled, can cause severe lung irritation and fatal pulmonary edema. In high concentrations, the gas irritates the skin.

Chlorine is a strong oxidizing agent, and combustible materials will burn in an atmosphere of Cl_2. Chlorine forms explosive mixtures with flammable vapors such as acetylene.

Hydrogen gas is very flammable and yields explosive mixtures with air and oxygen.

Because methane is highly flammable, the gas source must be isolated from possible accidental ignition.

Sodium hypochlorite is an eye and skin irritant. Contact of NaClO with acids liberates chlorine gas.

Hydrochloric acid may cause severe burns. Hydrochloric acid vapors are extremely irritating to the skin, eyes, and respiratory system.

Calcium carbide in contact with water liberates acetylene, a highly flammable gas. Acetylene reacts spontaneously with chlorine (the basis for the demonstration in Procedure B) and also forms explosive mixtures with oxygen over a wide range of concentrations (2–20% by volume).

Hydrogen iodide is corrosive and severely irritating to the eyes, skin, and mucous membranes.

Contact with solid iodine causes burns. The vapor irritates the eyes and respiratory system.

The chlorides of iodine can cause severe burns to the skin and eyes, and their vapors are irritating to mucous membranes.

Excess I_2 should always be present while preparing HI by its reaction with phosphorus and water. Excess I_2 suppresses the formation of explosive phosphines.

Sodium hydroxide can cause severe burns to the eyes and skin. Dust from solid sodium hydroxide is very caustic.

Red phosphorus is flammable and potentially explosive when mixed with oxidizing agents.

DISPOSAL

The chlorine generators should be filled with water and their contents flushed down the drain with water.

The chlorine-containing flask from Procedure A should be rinsed with water and the rinse flushed down the drain.

The cylinder from Procedure B should be filled with water and scrubbed. The rinse should be flushed down the drain with water.

The HI-generation flask and the bottles used in Procedure C should be rinsed with acetone and the rinse discarded in a receptacle for waste solvents.

DISCUSSION

Chlorine is a pale yellow-green gas with a suffocating odor. Its normal boiling point is $-34.0°C$, and its melting point is $-101.0°C$. It is soluble in water to a concentration of 0.092 mole/liter at 25°C [4].

Chlorine is commonly produced from chloride salts, commercially by the electrolysis of fused sodium chloride. Procedure A in this demonstration prepares chlorine through the oxidation of the chloride ions in hydrochloric acid by manganese dioxide. The reaction can be represented by

$$MnO_2(s) + 4\ H^+(aq) + 4\ Cl^-(aq) \longrightarrow MnCl_2(aq) + 2\ H_2O(l) + Cl_2(g)$$

In Procedure B, chlorine is prepared by the oxidation of hydrochloric acid by hypochlorous acid.

$$Cl^-(aq) + HOCl(aq) + H^+(aq) \longrightarrow Cl_2(g) + H_2O(l)$$

The hypochlorous acid is supplied in the form of liquid laundry bleach, nominally a 5% solution of NaOCl in water, but which contains hypochlorous acid from the hydrolysis of the hypochlorite ions.

$$OCl^-(aq) + H_2O(l) \rightleftharpoons HOCl(aq) + OH^-(aq)$$

The reaction between chlorine and methane produces hydrogen chloride and chlorinated methane. The hydrogen chloride can be detected by its effect on moist blue litmus paper—it turns pink. The chlorinated methane is a mixture of CH_3Cl, CH_2Cl_2, $CHCl_3$, and possibly CCl_4. The reactions are represented by the following equations:

$$CH_4(g) + Cl_2(g) \longrightarrow HCl(g) + CH_3Cl(g)$$

$$CH_3Cl(g) + Cl_2(g) \longrightarrow HCl(g) + CH_2Cl_2(g)$$

$$CH_2Cl_2(g) + Cl_2(g) \longrightarrow HCl(g) + CHCl_3(g)$$

$$CHCl_3(g) + Cl_2(g) \longrightarrow HCl(g) + CCl_4(g)$$

The reaction between hydrogen and chlorine to form hydrogen chloride—

$$H_2(g) + Cl_2(g) \longrightarrow 2\ HCl(g)$$

—is extremely slow at room temperature in the absence of light. A mixture of these two gases is photosensitive with a high quantum yield, so that brief exposure to short-wavelength light can lead to an explosion. (This photochemical reaction is described in Demonstration 1.45 of Volume 1 in this series.) The reaction can also be carried out under controlled conditions to produce a flame, as Procedure A of this demonstration illustrates.

Acetylene gas is formed by the reaction between calcium carbide and water. Calcium carbide is one of the salt-like carbides that contains the carbide ion (C_2^{2-}). Calcium carbide can be considered to be a salt of the weak acid acetylene (H_2C_2). This ion reacts with water in a hydrolysis reaction, whose product is the insoluble gas.

$$C_2^{2-}(aq) + 2\ H_2O(l) \longrightarrow H_2C_2(g) + 2\ OH^-(aq)$$

Acetylene contains two triple-bonded carbon atoms, each bonded to a hydrogen atom in a linear fashion.

$$H—C\equiv C—H$$

One might expect chlorine to attack the triple bond and add to it, forming dichloroethene and tetrachloroethane. However, chlorine abstracts the hydrogen atoms from acetylene, forming hydrogen chloride and carbon.

$$C_2H_2(g) + Cl_2(g) \longrightarrow 2\ HCl(g) + 2\ C(s)$$

The carbon product is visible as the soot produced when this reaction occurs in Procedure B of this demonstration.

Hydrogen iodide is produced in Procedure C by the reaction of iodine with the reducing agent phosphorus in the presence of water. In the course of the reaction, iodine picks up hydrogen from the water, and the phosphorus acquires the oxygen [5]. The net reaction is given by the following equation:

$$P_4(s) + 10\ I_2(s) + 16\ H_2O(l) \longrightarrow 4\ H_3PO_4(l) + 20\ HI(g)$$

This reaction was first used by Joseph Louis Gay-Lussac, and it is presumed to give HI via the reaction of phosphorus with iodine, forming phosphorus triiodide (PI_3), which is subsequently hydrolyzed to hydrogen iodide.

The chlorine reacts with hydrogen iodide to form hydrogen chloride and iodine.

$$Cl_2(g) + 2\ HI(g) \longrightarrow 2\ HCl(g) + I_2(g)$$

The iodine produced in this reaction can combine with residual Cl_2 to give a mixture of iodine monochloride (ICl) and "iodine trichloride" (I_2Cl_6). In the presence of large excesses of chlorine, needle-like crystals of I_2Cl_6 are obtained. When Cl_2 is not in excess, the product is the reddish brown liquid ICl, which is easily mistaken for bromine. Because I_2Cl_6 is formed in excess chlorine, its crystals are most likely to form near the top of the chlorine bottle used in Procedure C.

REFERENCES

1. H. N. Alyea and F. B. Dutton, Eds., *Tested Demonstrations in Chemistry*, 6th ed., Journal of Chemical Education: Easton, New Jersey (1965).
2. R. E. Dunbar, *J. Chem. Educ.* 35:A299 (1958).
3. S. Sharpe, Ed., *The Alchemist's Cookbook*, Instructional Development Centre, McMaster University: Hamilton, Ontario (undated).
4. M. Windholz, Ed., *The Merck Index*, 9th ed., Merck and Co.: Rahway, New Jersey (1976).
5. J. W. Mellor, *A Comprehensive Treatise on Inorganic and Theoretical Chemistry*, Vol. 2, Longmans, Green and Co.: London (1922).

6.29

Facilitated Transport of Carbon Dioxide Through a Soap Film †

A soap bubble grows and changes color when it is immersed in a box filled with carbon dioxide gas. When the bubble is removed, it shrinks. A soap bubble floats on carbon dioxide gas and grows as it floats [1, 2].

MATERIALS FOR PROCEDURE A

cylinder of carbon dioxide, CO_2, with valve and rubber tubing

  or

  dry ice, solid CO_2, in block or chunks sufficient to cover bottom
    of box

  gloves suitable for handling dry ice

  or

  100 g sodium bicarbonate, $NaHCO_3$

  200 mL 6M hydrochloric acid, HCl (To prepare 1.0 liter of stock solution,
    carefully pour 500 mL of concentrated [12M] HCl into 400 mL of dis-
    tilled water and dilute the cooled solution to 1.0 liter with distilled water.)

  shallow dish (e.g., crystallizing dish), with a capacity of at least 500 mL, to
    fit inside box

100 mL soap solution (sold in toy stores for blowing bubbles)

wand for blowing soap bubbles

box with transparent walls, such as an aquarium

black backdrop for box (e.g., poster board or cloth)

MATERIALS FOR PROCEDURE B

100 mL soap solution (sold in toy stores for blowing bubbles)

dry ice, solid CO_2, in block or chunks sufficient to cover bottom of box or beaker

gloves suitable for handling dry ice

box with transparent walls, such as an aquarium, or 4-liter beaker

black backdrop for box (e.g., poster board or cloth)

† We wish to thank Dr. Ilan Chabay, Director of The New Curiosity Shop, San Francisco, for calling this demonstration to our attention.

ca. 500-mL glass jar (e.g., mayonnaise jar)

slide projector (optional)

shallow pan (e.g., pie pan or petri dish)

PROCEDURE A

Preparation

Cover the back of the box with the black backdrop.

Presentation

If a cylinder of carbon dioxide gas is to be used, place the open end of the rubber tubing on the bottom of the box and open the valve of the cylinder to produce a gentle flow of carbon dioxide into the box. After a minute or two, close the valve and remove the rubber tubing from the box.

If dry ice is available, cover the bottom of the box with it while wearing gloves. Allow a minute or two for some of the dry ice to sublime into carbon dioxide gas.

If carbon dioxide gas is to be generated, place 100 g of $NaHCO_3$ in the shallow dish and set the dish inside on the bottom of the box. Pour the 6M HCl onto the $NaHCO_3$ slowly enough to avoid any frothing of the mixture over the dish.

Blow one or more soap bubbles and catch one of them on the wand. Note the color pattern in the soap film. Lower the captured bubble into the box of carbon dioxide gas. The bubble will immediately begin to expand, and its color pattern will change as it grows. Remove the bubble from the box and it will shrink and change color again. This immersing and removing of the bubble can be repeated until the bubble breaks.

Gently blow soap bubbles over the box so that some of them settle into the box. The bubbles will float on the carbon dioxide gas in the box. Within several seconds the bubbles will begin to grow and change color. As they grow, they will gradually sink farther into the box. When dry ice is used as the source of carbon dioxide, occasionally a bubble may settle onto it and freeze.

PROCEDURE B

Preparation

Cover the back of the box with the black backdrop. If a slide projector is available, it can be used to enhance the visibility of the demonstration by aligning its beam so that the mouth of the jar is illuminated from the side when the jar is placed in the box. Pour enough bubble soap into the pan to cover the bottom to a depth of about 1 cm.

Presentation

Wearing gloves, cover the bottom of the box with the dry ice. Turn on the slide projector. Dip the mouth of the jar into the bubble soap to form a film across the mouth. Set the jar upright on the dry ice in the box.

Within seconds, a soap bubble will grow on the mouth of the jar until it grows so large that it breaks. The presentation can be repeated several times. If the jar is removed from the box of carbon dioxide before the bubble breaks, the bubble will shrink back to the mouth of the jar.

HAZARDS

Dry ice, solid CO_2, has a temperature of $-78°C$ and can cause frostbite. Thermal protection in the form of gloves should be used when handling dry ice.

DISPOSAL

The dry ice should be allowed to sublime in a hood or other well-ventilated area.

DISCUSSION

When an air-filled soap bubble is immersed in an atmosphere of carbon dioxide, the bubble expands. When the expanded bubble is removed from the carbon dioxide atmosphere, the bubble shrinks. This illustrates that carbon dioxide diffuses more rapidly than air across the film of a soap bubble. Procedure A shows the reversibility of the diffusion of carbon dioxide from a region of high concentration to one of lower concentration. It also shows that the air-filled bubble floats on the CO_2, and, therefore, is less dense than carbon dioxide gas. Procedure B shows essentially the same phenomenon, with the complicating factor that the jar is cooled by the dry ice on which it is resting. The cooling effect could lead to the prediction that the soap film will sink into the jar because of the contraction of the cooled gas, but this contraction proves to be insignificant compared with the diffusion of carbon dioxide into the jar.

The growth of the air-filled bubble when it is immersed in carbon dioxide gas is, for most observers, an unexpected occurrence. This growth is counter to what would be predicted on the basis of a simple application of Graham's law of gaseous diffusion: the heavier carbon dioxide molecules diffuse through the soap film more slowly than the lighter air molecules.

In Procedure B, where dry ice (sublimation point, $-78°C$) is used, the growth of the bubble is also counter to what would be predicted by a straightforward application of Charles's law. Applying Charles's law to the gas in the jar leads to the prediction that, as the gas within the jar is cooled by the dry ice, the gas will contract, causing the soap film to sink into the jar. However, just the opposite is observed—the soap film expands out of the jar, forming a bubble.

The key to explaining the growth of the bubble is that CO_2 dissolves in water to a much greater extent than does air. The solubility of CO_2 in water at 25°C is 0.145 g/100 mL of water, while that of air is 0.0029 g/100 mL of water [3]. On one side of the soap film, where the partial pressure of carbon dioxide is high, the carbon dioxide dissolves in the water. This produces a concentration gradient of dissolved carbon dioxide across the soap film, and the dissolved carbon dioxide diffuses to the side of lower concentration. On the other side, where the partial pressure of carbon dioxide is low, carbon dioxide gas is released. Because CO_2 is more soluble in water than is air, its concentration gradient across the soap film is greater, and its transport of CO_2 across the film is faster. If the partial pressure of CO_2 is higher outside the bubble than inside, carbon dioxide diffuses into the bubble more rapidly than air diffuses out, and the bubble expands.

When carbon dioxide dissolves in water, some of it is slowly hydrated to carbonic acid, and the following equilibrium is established:

$$CO_2(aq) + H_2O(l) \rightleftharpoons H_2CO_3(aq)$$

However, the soap solution is alkaline, so the concentration of carbonic acid is decreased by the following reaction, which produces bicarbonate ions:

$$OH^-(aq) + H_2CO_3(aq) \longrightarrow H_2O(l) + HCO_3^-(aq)$$

This has the effect of shifting the equilibrium between dissolved carbon dioxide and carbonic acid toward carbonic acid, and thereby increasing the solubility of the carbon dioxide. Therefore, the transport of carbon dioxide across the soap film is even faster than it would be through a thin film of water.

The colors observed in the film of a bubble (and in an oil film on water) are produced by the phenomenon of interference between rays of light reflected at the opposite surfaces of the film. The colors depend on the thickness of the film, and as the thickness varies, the color varies also. A discussion of this phenomenon can be found in most introductory physics texts, for example, reference 4. When a ray of light strikes the film of a bubble, a portion of it may be reflected from the outer surface and some of it from the inner surface of the film. The portion of the ray reflected from the inner surface is out of phase with that reflected from the outer surface. The degree of the phase shift depends upon the thickness of the film. Because the film is very thin, the phase shift may be only a portion of a period, and rays reflected from the inner surface can interfere with their counterparts reflected from the outer surface. This means that the light reflected from the bubble will be missing the wavelengths (and colors) which interfere destructively at the point of observation. Only those colors which do not interfere destructively will be seen.

As a bubble grows, the material composing its film spreads out and becomes thinner. As the film becomes thinner, the wavelengths of light whose rays interfere destructively also change, leading to the change in the colors of the bubble.

In this demonstration, soap bubbles are used to demonstrate the transport of carbon dioxide through a thin film. Soap bubbles are used also as carriers for exploding hydrogen and oxygen gases in Demonstration 1.42 of Volume 1 of this series. For descriptions of other experiments with soap bubbles, see C. V. Boys [5].

REFERENCES

1. R. C. Millikan, *J. Chem. Educ.* 55:807 (1978).
2. L. L. Jones and M. K. Ogawa, in *Chemical Demonstrations Proceedings*, Western Illinois University: Macomb, Illinois (1981).
3. W. C. Weast, Ed., *CRC Handbook of Chemistry and Physics*, 59th ed., CRC Press: Boca Raton, Florida (1978).
4. F. W. Sears, M. W. Zemansky, and H. D. Young, *College Physics*, 4th ed., Addison-Wesley Publishing Co.: Reading, Massachusetts (1977).
5. C. V. Boys, *Soap Bubbles and the Forces Which Mould Them*, Educational Services Incorporated, A Doubleday Anchor Book: New York (1959).

7

Oscillating Chemical Reactions †

Earle S. Scott, Rodney Schreiner, Lee R. Sharpe,
Bassam Z. Shakhashiri, and Glen E. Dirreen

Oscillating reactions are among the most fascinating chemical demonstrations. In one type of reaction, a mixture of chemicals goes through a sequence of color changes, and this sequence repeats periodically. In another, the mixture periodically emits a burst of gas, foaming up. To many, oscillating reactions are engaging examples of "chemical magic." To those having some acquaintance with chemistry, these reactions are a mystery and a challenge. To everyone, they are memorable demonstrations of the wonder of chemistry.

Perhaps what makes oscillating reactions so fascinating to chemists is that they seem to contradict common sense. Experience tells us that, under a given set of conditions, chemical reactions go in only one direction. We rarely find a chemical reaction that appears to reverse itself, much less to do so repeatedly. When we do encounter such a reaction, we may be inclined to draw an analogy to a simple physical oscillator such as a pendulum. A pendulum oscillates from side to side through its equilibrium position, and these oscillations can be attributed to the interconversion between the potential and kinetic energy of the pendulum. Analogous to this physical process, the chemical oscillator may seem to swing through its equilibrium composition. However, this is contrary to the second law of thermodynamics, which asserts that once a chemical system reaches equilibrium, it cannot deviate from that condition spontaneously. Therefore, oscillations in chemical reactions cannot be like the oscillations of a pendulum; chemical reactions cannot oscillate through the equilibrium condition.

A more appropriate physical model for an oscillating chemical reaction would be a grandfather's clock. The hands of the clock pass repeatedly (twice daily) through the same position, but the energy stored in the elevation of the weights decreases continuously as the clock runs. In an oscillating chemical reaction, the concentrations of some components of the reaction mixture pass repeatedly through the same value, but the energy-releasing reaction that drives the oscillations proceeds continuously toward completion. Just as the clock is running down while its hands rotate, the chemical oscillator is running down toward equilibrium as it oscillates. In order to exhibit oscillations, a chemical system must be far from its equilibrium composition.

The oscillations in an oscillating chemical reaction are driven by the decrease in free energy of the mixture. This decrease is what drives *all* chemical reactions, but not

† We wish to thank Professor R. M. Noyes of the University of Oregon for making detailed suggestions to improve the clarity of the Introduction and Discussion sections of this chapter.

all chemical reactions exhibit oscillations. There must be some feature peculiar to oscillating reactions that allows them to display this unusual behavior. This feature occurs in the pathways these reactions take as they approach equilibrium, in other words, in their reaction mechanisms. The pathway that a reaction follows determines how the concentrations of its components change as the reaction proceeds. The more complex the pathway of the reaction is, the more complex the changes in concentration of the components can be.

The problem of how oscillating chemical reactions occur has been examined theoretically to establish ways in which oscillations can happen. The reaction mechanisms of all known chemical oscillators have at least three common features. First, while the oscillations occur, the chemical mixture is far from equilibrium, and an energy-releasing reaction occurs whose energy drives the oscillating "sideshow." Second, the energy-releasing reaction can follow at least two different pathways, and the reaction periodically switches from one pathway to another. Third, one of these pathways produces a certain intermediate, while another pathway consumes it, and the concentration of this intermediate functions as a "trigger" for the switching from one pathway to the other. When the concentration of the intermediate is low, the reaction follows the producing pathway, leading to a relatively high concentration of the intermediate. When the intermediate's concentration is high, the reaction switches to the consuming pathway, and the concentration of the intermediate decreases. Eventually the reaction reverts to the producing pathway. The reaction repeatedly switches from one pathway to the other.

REACTION MECHANISMS AND CONCENTRATION CHANGES

In order to see how a reaction pathway affects the concentrations of components of a reaction mixture, we shall examine several hypothetical reaction mechanisms and the pattern of concentration changes that they produce. The first hypothetical reaction mechanism to be considered is a simple one. The overall reaction is the conversion of A to P, and the mechanism involves a single intermediate, X. (In all of the following discussion, reactants are represented by letters from the beginning of the alphabet, products from the middle, and intermediates from the end of the alphabet.) The mechanism involves two steps.

step 1: $\qquad$ A $\longrightarrow$ X

step 2: $\qquad$ X $\longrightarrow$ P

Mechanism 1

Each step in this mechanism proceeds at its characteristic rate. Their relative rates can be divided into three cases: (1) step 1 is much faster than step 2, (2) step 1 occurs at about the same rate as step 2, and (3) step 1 is much slower than step 2. In the first case, A will be converted quickly to X, but X will be only slowly converted to P. Thus, [A] will drop quickly, [X] will rise quickly and fall slowly, and [P] will rise only slowly. This situation is illustrated in Figure 1a. In the second case, the initial increase in [X] occurs at about the same rate as its later decrease. Therefore, [A] decreases at about the same rate as [P] increases, as shown in Figure 1b. In the third case, X will be converted to P almost as fast as it is formed, and its concentration will never become significant. It appears as though A is converted directly to P. This situation is depicted in Figure 1c. In the second case, as shown in Figure 1b, [X] rises and falls in a nearly symmetrical curve. This resembles one cycle of an oscillation, but this mechanism is too simple to produce repeated fluctuations in the concentrations of any of the species involved.

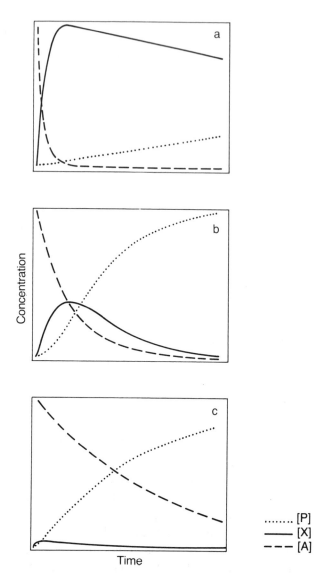

Figure 1. Concentration versus time for the species in Mechanism 1.

If the concentration of X is to repeat its rise and fall, the mechanism must be altered. In the simple two-step mechanism, as A is converted to X in step 1, [X] increases. As [X] increases, the rate of the second step increases, eventually becoming faster than that of the first step, and then [X] falls and [P] rises. In order for [X] to rise again, a situation similar to the initial conditions must be restored; that is, both [X] and [P] must be low. The concentration of P can be lowered if another step is added to the mechanism, one which consumes P, such as P ⟶ Q. Adding this step changes the overall reaction to A ⟶ Q, and P becomes an intermediate, which we shall rename as Y in keeping with the symbolism described earlier. The mechanism now is

step 1:	A ⟶ X	
step 2:	X ⟶ Y	Mechanism 2
step 3:	Y ⟶ Q	

The third step, Y ⟶ Q, causes [Y] to decrease after it has risen, as illustrated in Figure 2. However, the addition of step 3 has not yet caused [X] to rise again, after its

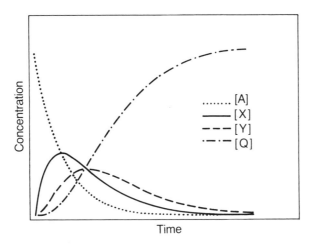

Figure 2. Concentration versus time for the species in Mechanism 2.

initial decline. We must modify the mechanism in such a way that [X] will rise whenever [Y] is low. This rise will occur if the rate of the step that consumes X is slow when [Y] is low. This can be accomplished by making Y a reactant in step 2, changing the step to X + Y ⟶ 2 Y. The mechanism becomes

step 1:   A ⟶ X

step 2:   X + Y ⟶ 2 Y   Mechanism 3

step 3:   Y ⟶ Q

The rate of the second step is proportional to both [X] and [Y]. Therefore, X is consumed only slowly when [Y] is low, and [X] can increase as X is produced in step 1.

Mechanism 3 was first described in 1910 by Alfred Lotka [1]. He analyzed the changes in the concentrations of the intermediates, X and Y, and found that this mechanism did lead to periodic oscillations in their concentrations under certain conditions. In order to simplify the situation, he examined the case where [A] is constant. (In an actual reaction, the constancy of [A] can be produced by adding A at the same rate at which it is consumed; or it can be simulated by using a large excess of A in the case where step 1 is slow, so [A] changes negligibly.) Lotka showed that the mathematical model for this mechanism is a set of differential equations whose solution exhibits damped oscillations for some values of the rates of the steps. Figure 3 shows how the

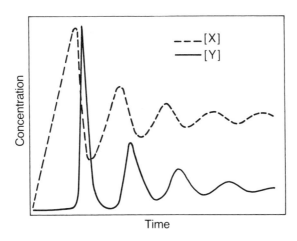

Figure 3. Concentration versus time for the intermediates of Mechanism 3.

concentrations of the intermediates change when the rates of the steps meet Lotka's criterion: oscillations do occur! However, the oscillations are damped and quickly disappear. We shall investigate other mechanisms to see what is needed to lead to sustained oscillations, but we have at least found what is necessary in a mechanism to produce oscillations. What is needed is some sort of "feedback loop" in which the rate of one step can be affected by a later product of the mechanism. This condition exists in step 2 of mechanism 3, because the rate depends on the concentration of the product of the step, Y. In other words, step 2 is autocatalytic.

A SIMPLE MECHANISM LEADS TO SUSTAINED OSCILLATIONS

In 1920, Lotka presented a second mechanism which also led to oscillations in the concentrations of its intermediates, but with this mechanism the oscillations were sustained [2]. The mechanism he described is

step 1:	$A + X \longrightarrow 2X$	
step 2:	$X + Y \longrightarrow 2Y$	Mechanism 4
step 3:	$Y \longrightarrow Q$	

In this mechanism the overall reaction is $A \longrightarrow Q$, and the intermediates are X and Y. Both step 1 and step 2 are autocatalytic—their rates increase as the concentrations of their products increase. Again assuming a constant [A], Lotka examined this mechanism and found that these steps led to sustained oscillations in the concentrations of X and Y. Figure 4 shows the oscillatory behavior of [X] and [Y].

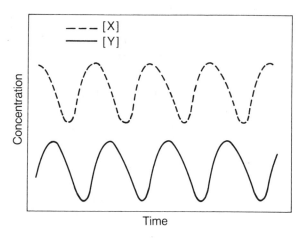

Figure 4. Concentration versus time for the intermediates of Mechanism 4.

THE BELOUSOV-ZHABOTINSKY REACTIONS

Neither of Lotka's two mechanisms corresponds to an actual reaction. In fact, when Lotka described his mechanisms, no homogeneous oscillating reaction was known. The first report of a homogeneous oscillating reaction was made by Bray in 1921 [3]. He was investigating the effect of iodate on the decomposition of hydrogen peroxide, when he accidentally discovered oscillations in the concentrations of iodine and in the evolution of oxygen in a mixture of hydrogen peroxide and potassium iodate in dilute sulfuric acid. It was in some part because of the skepticism of other chemists about the possible existence of homogeneous oscillating reactions that this reaction stood for a long time as the only reported example of such a reaction. However, in 1958 Belousov by chance discovered another example of a closed, homogeneous oscillating reaction [4]. He mixed potassium bromate, cerium(IV) sulfate, and citric acid in dilute sulfuric acid and found that the color of the solution oscillated. Zhabotinsky studied this reaction extensively [5–7]. He found that the oscillations still occurred when citric acid was replaced by any of a number of carboxylic acids with the common structural feature

$$
\begin{array}{ccc}
& O & O \\
& \| & \| \\
R-&C-CH_2-C-OH
\end{array}
$$

The cerium ions could be replaced by manganese ions, and oscillations would still occur. Together, these reactions form a family called the Belousov-Zhabotinsky (BZ) reactions.

The BZ reactions have been studied extensively, both experimentally and theoretically. The agreement between observed oscillations and those generated from theoretical models is not perfect, but it is close enough to generate confidence that the basic kinetic pattern employed in the models is correct, and that improvement in the models will result from fine-tuning and not from a fundamental change. The fact that the models have been successfully applied to other oscillating chemical reactions further increases the confidence in their correctness.

One of these models for the BZ reaction was developed by Richard Field and Richard Noyes at the University of Oregon; it is called the Oregonator [8]. The model mechanism has five steps.

step 1:	$A + Y \longrightarrow X + P$	
step 2:	$X + Y \longrightarrow 2P$	
step 3:	$A + X \longrightarrow 2X + Z$	Mechanism 5
step 4:	$2X \longrightarrow P + A$	
step 5:	$Z \longrightarrow Y$	

The simplest combination of these five steps so that there is no net consumption or destruction of the intermediates, X, Y, and Z, is (step 1) + (step 2) + 2(step 3) + (step 4) + 2(step 5). This leads to the overall net reaction

$$2A \longrightarrow 4P$$

Step 3 of this mechanism provides the necessary feedback; it is autocatalytic in X. Step 4 keeps the autocatalytic step from consuming all of the reactants; it is second order in the catalytic product of the autocatalytic step. Therefore, as the concentration of X increases and speeds up its own production in step 3, it also speeds up its own consumption in step 4, and eventually a pseudo-steady-state condition is reached, in which the concentration of X is virtually constant. Figure 5 shows the oscillatory behavior of several of the intermediates of this model mechanism. An application of this model mechanism to the Belousov-Zhabotinsky reaction will be described later.

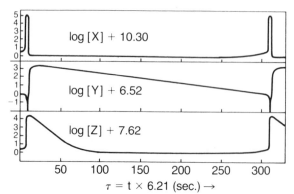

Figure 5. Logarithmic plot of intermediate concentrations versus time for the Oregonator mechanism [8].

EXPERIMENTAL INVESTIGATIONS

A typical BZ reaction involves many components, including an organic substrate, bromate ions, a catalyst, and an acidic medium, as well as numerous intermediates. Because the composition of an oscillating reaction mixture is so complex and because the mixture must be held far from equilibrium, the reaction is often studied in a special reactor, called a continuous-flow stirred tank reactor (CSTR). The CSTR allows the concentrations of the reactants to be carefully controlled, permitting investigation of the effects of reactant concentrations on the state of the system. Figure 6 is a schematic representation of a CSTR. It consists of a chamber which can be maintained at a constant temperature and into which a number of reactant solutions can be admitted at

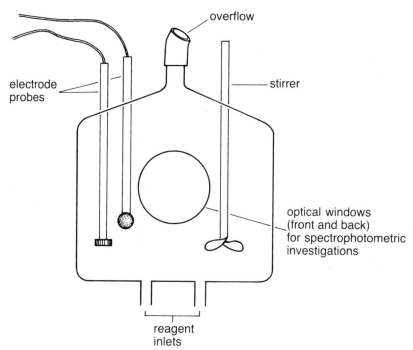

Figure 6. Schema of a continuous-flow stirred tank reactor.

known flow rates. The contents of the chamber are stirred vigorously to create homogeneity, and they leave the chamber through an overflow. Electrodes of various types monitor the concentrations of ions in the solution, and an optical window in the chamber allows the solution to be monitored spectrophotometrically.

The reactants flow into the CSTR at a constant rate. The concentrations of the reactants in the reactor are determined by this flow rate and by the speed of the reaction occurring in the reactor. When the flow of reactants into the CSTR is fast compared with their rate of consumption by the reaction, the mixture in the reactor can be maintained in a pseudo–steady state far from equilibrium. If the reactants flowing into the CSTR are the components of a simple reaction, then the concentrations of the reactants, products, and intermediates in the reactor will reach a steady state determined by the flow rates and the mechanism of the reaction. However, if the reactants flowing into the CSTR are the components of an oscillatory system, then the concentrations of reactants, products, and intermediates can reach either of two steady states, each corresponding to one of the alternative pathways which the reaction can follow. The conditions under which each pathway is followed can be investigated, and the factors responsible for switching between pathways can be determined. In a CSTR the concentrations of the reactants are constant and oscillations can be maintained indefinitely, thereby producing data which are simpler to analyze than those from a batch reactor, where the concentrations are continually changing.

When the flow rate of reactants into the CSTR is constant, the concentrations of those reactants can be maintained at a nearly steady state. With an oscillatory system in a CSTR, the oscillations have a constant period and amplitude, allowing careful measurement and correlation of the oscillating properties for a number of species in the solution. One of the most complete sets of such experimental observations has been reported by Vidal, et al. [9]. In this set, a BZ reaction using malonic acid, bromate ions, and Ce(IV) catalyst was studied in a CSTR. The reaction was monitored spectrophotometrically at two different wavelengths, and electrodes were used to monitor the electrical potential of the solution and the concentrations of dissolved O_2, CO_2, and Br^- ions. A graphical representation of the results is given in Figure 7. (In this figure, the vertical scale differs for each plot, in order to expand each to full scale.) There is a correspondence between the electrical potential of the solution and the con-

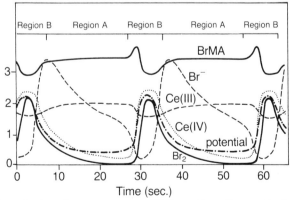

Figure 7. Temporal oscillations in components of BZ reaction. One vertical unit corresponds to the following concentrations (mol/liter): bromomalonic acid (BrMA), 2.5×10^{-3}; bromine (Br_2), 5×10^{-5}; bromide ion (Br^-), 1.2×10^{-4}; Ce (III), 2.5×10^{-4}; Ce (IV), 5×10^{-5}. The scale of the potential is arbitrary. (Adapted from reference 9.)

centration of Ce(IV), although the correspondence is not 1:1, because the potential depends on the ratio of Ce(IV) to Ce(III) concentrations rather than on the concentration of Ce(IV) alone. However, there is not a simple correlation between the concentrations of Br_2 and Br^-. The concentration of bromide ions covers a much greater range than does the concentration of bromine. The surge in the concentration of Br^- follows closely upon the increase in the concentrations of Ce(IV) and Br_2. As we shall see, both of these species undergo reactions which produce bromide ions. The bromomalonic acid concentration remains relatively high with only minor fluctuations. The formation of bromomalonic acid is essential to the onset of oscillations when this system is used as a demonstration in a beaker. The rate of bromination of malonic acid is an important factor in determining the period of the oscillations in this system. For all components of the reaction, each period in the oscillation can be divided into two portions: one in which the concentrations change relatively slowly (Region A), and a shorter portion in which the concentrations change quickly (Region B). It may be helpful to keep these observations in mind as we describe the mechanism which has been proposed for this reaction.

A MECHANISM FOR THE BELOUSOV-ZHABOTINSKY REACTION

The mechanism of the Belousov-Zhabotinsky reaction, as described here, was proposed by Field, Körös, and Noyes in 1972 [10] and later summarized by Tyson [11]. The basic elements of this mechanism have stood the test of time and are successful in explaining the major features of this reaction. During the course of the BZ reaction, malonic acid is brominated by molecular bromine. The production of molecular bromine is accomplished through two competing processes. One of these processes (Process A, described below) involves ions, and its steps are two-electron transfers. The other (Process B) involves radicals and one-electron transfers. In Figure 7, Region A corresponds to Process A, while Region B corresponds to Process B. Which process is dominant at a particular time is determined by the bromide ion concentration. Figure 8 is a representation of the major steps in the mechanism of this reaction. (Each of the

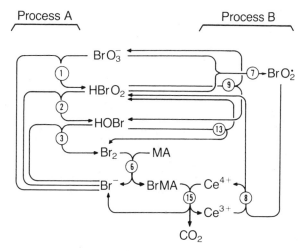

Figure 8. Schema of the mechanism for a BZ reaction.

steps is labelled with the number of the corresponding equation from the discussion of the mechanism which follows.) On the left side of the figure are the reactions of Process A, which consume bromate ions and bromide ions. The reactions of Process B are represented on the right side of the figure. These reactions consume bromate ions and oxidize Ce(III) to Ce(IV). The Ce(IV) ions participate in reactions that produce bromide ions. Process A occurs when the bromide ion concentration is above a certain critical level, while Process B dominates when the bromide ion concentration is low. Oscillations occur because Process A consumes bromide ions and, thus, leads to the conditions which favor Process B. Process B indirectly liberates bromide ions, which returns the reaction to control by Process A.

The direct reaction between malonic acid and bromate ions is very slow. However, bromate ions can oxidize bromide ions relatively rapidly. This is the first step in Process A. In acidic solution, bromate ions oxidize bromide ions to bromous acid and hypobromous acid, as indicated by equation 1.

PROCESS A

$$BrO_3^- + Br^- + 2\,H^+ \longrightarrow HBrO_2 + HOBr \qquad (1)$$

$$HBrO_2 + Br^- + H^+ \longrightarrow 2\,HOBr \qquad (2)$$

$$HOBr + Br^- + H^+ \longrightarrow Br_2 + H_2O \qquad (3)$$

The bromous acid produced by this reaction can react further with bromide ions, forming more hypobromous acid, as equation 2 indicates. Hypobromous acid itself reacts with bromide ions to produce elemental bromine, as depicted in equation 3. The net transformation occurring in Process A may be obtained by the stoichiometric addition of (equation 1) + (equation 2) + 3(equation 3). This net reaction is represented by equation 4.

$$BrO_3^- + 5\,Br^- + 6\,H^+ \longrightarrow 3\,Br_2 + 3\,H_2O \qquad (4)$$

The net reaction represented by equation 4 is the reduction of bromate ions by bromide ions in a series of oxygen transfers (two-electron reductions).

The elemental bromine in aqueous solution is consumed in a reaction with malonic acid, producing bromomalonic acid and hydrobromic acid. Malonic acid exists in solution mainly in the diacid form and is converted only relatively slowly to the reactive enol form, as indicated in equation 5.

$$(5)$$

diacid   enol

Bromine does not react with the diacid form of malonic acid, but it reacts very quickly with the enol form, brominating it at the central carbon atom and releasing bromide ions, as represented by equation 6.

$$Br_2 + H\!-\!C\!=\!C(OH)_2 \rightarrow Br\!-\!C\!=\!C(OH)_2 + Br^- + H^+ \qquad (6)$$
$$\qquad\qquad |\qquad\qquad\qquad\qquad |$$
$$\qquad\quad CO_2H \qquad\qquad\quad CO_2H$$

As the bromine reacts with the enol, the diacid slowly forms more enol. Therefore, the rate at which the bromine is consumed is determined by the rate of the enolization reaction.

The net effect of Process A combined with the reaction of bromine with malonic acid is a reduction in the concentration of bromide ions in the solution. As equation 4 indicates, Process A consumes five bromide ions for each three bromine molecules it produces. For each molecule of bromine consumed through the reaction with malonic acid, one bromide ion is returned to the solution. When the concentration of bromide ions is relatively high, the second step in Process A (equation 2), consumes the $HBrO_2$ virtually as fast as it is produced. However, when the Br^- concentration becomes low, the $HBrO_2$ is consumed through equation 2 only very slowly. Thus, the rate of Process A is negligible, and Process B takes over.

From a thermodynamic point of view, it appears that cerium(III) ions should reduce bromate ions to either Br^- or Br_2. However, Ce(III) is a one-electron reducing agent, and it is unable to transfer one electron to a bromate ion. Therefore, cerium(III) ions do not react directly with bromate ions. However, Ce(III) can react directly with $BrO_2^\cdot$ radicals. These radicals are produced in the first step of Process B.

PROCESS B

$$BrO_3^- + HBrO_2 + H^+ \longrightarrow 2\, BrO_2^\cdot + H_2O \qquad (7)$$

$$BrO_2^\cdot + Ce^{3+} + H^+ \longrightarrow HBrO_2 + Ce^{4+} \qquad (8)$$

$$2\, HBrO_2 \longrightarrow HOBr + BrO_3^- + H^+ \qquad (9)$$

$$2\, HOBr \longrightarrow HBrO_2 + Br^- + H^+ \qquad (10)$$

$$HOBr + Br^- + H^+ \longrightarrow Br_2 + H_2O \qquad (11)$$

The $HBrO_2$ produced in equation 1 reacts with BrO_3^-, producing $BrO_2^\cdot$ radicals, as indicated in equation 7. The $BrO_2^\cdot$ radicals react with Ce(III), as indicated by equation 8. The overall reaction effected by these two reactions is given by the sum of (equation 7) + 2(equation 8).

$$2\, Ce^{3+} + BrO_3^- + HBrO_2 + 3\, H^+ \longrightarrow 2\, Ce^{4+} + H_2O + 2\, HBrO_2 \quad (12)$$

The overall reaction indicates that this sequence generates $HBrO_2$ autocatalytically. This autocatalytic sequence provides the feedback which was described earlier as being essential to chemical oscillations. The autocatalysis does not continue until the reactants are depleted, because there is a second-order destruction of the autocatalytic species. This reaction is the disproportionation of bromous acid, represented by equation 9. The last two equations in Process B represent the disproportionation of hypobromous acid to bromous acid and elemental bromine; their sum is given in equation 13.

$$3\, HOBr \longrightarrow HBrO_2 + Br_2 + H_2O \qquad (13)$$

This equation indicates how elemental bromine can be formed by Process B. For the sake of simplicity, equations 10 and 11 have been condensed to equation 13 in Figure 8. The hypobromous acid also reacts with organic radicals in the solution (see below), which may limit the amount of elemental bromine formed through disproportionation.

The net transformation taking place in Process B may be obtained by the stoichiometric addition of 5(equation 7) + 10(equation 8) + 3(equation 9) + (equation 10) + (equation 11). This net reaction is represented by equation 14.

$$2\, BrO_3^- + 12\, H^+ + 10\, Ce^{3+} \longrightarrow Br_2 + 6\, H_2O + 10\, Ce^{4+} \qquad (14)$$

The Ce^{4+} from equation 14 is reduced by organic species. One of the reductants is the bromomalonic acid produced in reaction 6.

$$BrCH(CO_2H)_2 + 4 Ce^{4+} + 2 H_2O \longrightarrow$$
$$HCO_2H + 2 CO_2 + Br^- + 4 Ce^{3+} + 5 H^+ \qquad (15)$$

The organic radical intermediates of the process represented by equation 15 also reduce HOBr to Br^-, so that it cannot react via equation 3. When enough HOBr and Ce^{4+} have been produced, Br^- is produced so rapidly that the rate of equation 2 surpasses that of equation 7, and the reaction again follows Process A, as represented by equation 4.

Process A and Process B result in a competition between bromide ions and bromate ions for bromous acid. When the concentration of bromide ions is high, nearly all of the bromous acid reacts with it, following Process A. During this process, bromide ion concentration decreases, and the bromide ions become less and less successful at competing for the bromous acid. Eventually, Process B takes over. This process, however, produces bromide ions, and eventually the concentration of bromide ions becomes high enough to cause a shift back to Process A. This switching back and forth between these processes is the origin of the oscillations observed in this reaction.

As the reaction oscillates between Process A and Process B, triggered by changes in the bromide ion concentration, other concentrations oscillate as well. While Process A occurs, the cerium ions are in their reduced state, Ce(III), as a result of their reaction with bromomalonic acid. During Process B, the cerium ions are oxidized to Ce(IV) through the reaction with BrO_2^- radicals. Thus, the ratio of the concentration of Ce(III) to the concentration of Ce(IV) oscillates as well. As this ratio oscillates, so does the electrical potential of the Ce(III)/Ce(IV) couple. The potential of the Ce(III)/Ce(IV) couple, at 25°C, is given by the Nernst equation.

$$E = E° - 0.059 \log \frac{[Ce(III)]}{[Ce(IV)]}$$

The electrical potential changes logarithmically with the ratio of the concentrations. The oscillations in the electrical potential can be observed by measuring the potential of a platinum electrode versus a reference electrode (see below). The oscillations in the potential can also be observed by using an oxidation-reduction indicator, such as ferroin (tris(1,10-phenanthroline)iron(II) sulfate). As the ratio of [Ce(III)] to [Ce(IV)] decreases (i.e., [Ce(IV)] increases), the Ce(IV) can oxidize the iron in ferroin from iron(II) to iron(III). Because the iron(II) complex is red and the iron(III) complex is blue, the color of the solution changes as the iron is oxidized. When the [Ce(III)] to [Ce(IV)] ratio rises, the iron(III) is reduced to iron(II), and the solution returns to its original color. This description of the functioning of the ferroin indicator is perhaps oversimplified, because the ferroin is believed to participate in the reaction as a catalyst as well, and is known to serve as a catalyst in the absence of other metal ions [12].

A Kinetic Model

In order to investigate how well the above mechanism represents the actual reaction, it is necessary to estimate rate constants for all of the reactions and to calculate the values of the measurable properties of the reaction mixture that this mechanism predicts. The mechanism as described above is extremely complex, and simulating such a mechanism requires extensive calculations. The calculations are more easily performed if the mechanism can be reduced to a simpler model. The Oregonator model, described earlier, is a suitable model for this reaction. The model can be applied to the BZ reaction by making the following identifications: $A = BrO_3^-$, $Y = Br^-$, $X = HBrO_2$, $P = HBrO$, and $Z = 2\,Ce^{4+}$. Then, the steps become

step 1:	$BrO_3^- + Br^- \longrightarrow HBrO_2 + HBrO$
step 2:	$HBrO_2 + Br^- \longrightarrow 2\,HBrO$
step 3:	$BrO_3^- + HBrO_2 \longrightarrow 2\,HBrO_2 + 2\,Ce^{4+}$
step 4:	$2\,HBrO_2 \longrightarrow HBrO + BrO_3^-$
step 5:	$2\,Ce^{4+} \longrightarrow Br^-$

It is clear that these individual steps are not balanced chemical equations, but they do summarize what is happening in the reaction. Step 1 corresponds to the first step of Process A (equation 1, above), while step 2 corresponds to equation 2. Step 3 is a representation of equation 12, in which bromate ions are consumed in the autocatalytic production of bromous acid with the simultaneous production of cerium(IV). Step 4 is the check, which prevents the autocatalytic step from running away and consuming all of the reactants; this step corresponds to equation 9 above. Step 5 indicates that cerium(IV) is consumed in the production of bromide ions, and represents equation 15 and other processes producing Br^- from Ce^{4+} and HOBr.

The Oregonator model indicates how a mechanism is simplified, stripped to its bare bones, as it were, in order to facilitate mathematical modelling and computer calculations. The mathematical model is described in a series of papers by Noyes and several of his collaborators [8, 10, 13, 14]. In his analysis of this model, Noyes found that when [Y] is greater than a certain critical value, [X] attains a steady state value, and the majority of the chemical transformations occur through steps 1 and 2, represented by equation 16.

$$A + 2\,Y \longrightarrow 3\,P \tag{16}$$

When [Y] is less than the critical value, the steady state of [X] changes, and most of the chemical change takes place by steps 3 and 4, as summarized in equation 17.

$$X \longrightarrow P + Z \tag{17}$$

Step 5 is a relatively slow process, which introduces a delay before enough Y is generated to cause a return to the process of equation 16. Field and Noyes calculated the concentrations of the three intermediates and found sustained oscillations in all three. The variations in [X], [Y], and [Z] with time are depicted in Figure 5.

A more detailed description of the mechanism of the BZ reaction comprising over 20 steps has been presented by Edelson, Field, and Noyes [15, 16]. Complete numerical calculations of the concentration changes of the intermediates in this mechanism have been performed. The results of these calculations show many striking similarities to the experimentally measured results. These include the long induction time before oscillations begin and the sustained oscillations observed in this reaction. Although the calculated results are not in perfect agreement with the measured results, the agreement is close enough to instill confidence that the proposed mechanism is not too different from that actually followed by the reaction.

The discussion of the mechanism of the Belousov-Zhabotinsky reaction presented here oversimplifies what occurs in the reaction mixture. Much more is known about this system than is described here, and because this is an active field of research, much more is likely to be discovered soon. Those who are interested in studying this field can begin with any of a number of excellent reviews [17–27].

MEASURING THE OSCILLATIONS IN THE ELECTRICAL POTENTIAL

The oscillations in all of the demonstrations in this chapter produce some visible periodic changes, generally in the color of the solution. However, there are also oscillations in other properties of the solutions. One property which is easily measured is the electrical potential of the solution, as sensed by a platinum electrode. Figure 9 is a recording of the oscillations in the electrical potential of a Briggs-Rauscher reaction, as described in Demonstration 7.1. The equipment needed to observe these potential oscillations includes an inert sampling electrode, a reference electrode, a potential display device, and several wires with a clip on each end for interconnecting these components. The inert sampling electrode is best made of platinum—it can be a piece of wire or foil, or it can be a commercial platinum-disk electrode. The reference electrode must be one that does not leak chloride ions into the solution, because chloride ions interfere with many of the oscillating reactions and prevent the oscillations. A suitable reference electrode is any commercially available double-junction standard calomel electrode or double-junction silver–silver chloride electrode. The potential display device can be a voltmeter, a strip-chart recorder, or both. The voltages to be displayed oscillate over a range as narrow as 30 mV or as wide as 250 mV, and are centered anywhere from 0 to 2 volts, depending on the reaction and the reference electrode employed.

To display the oscillations in the electrical potential, use a piece of wire to connect the platinum electrode to the positive terminal of the voltmeter or recorder, and connect

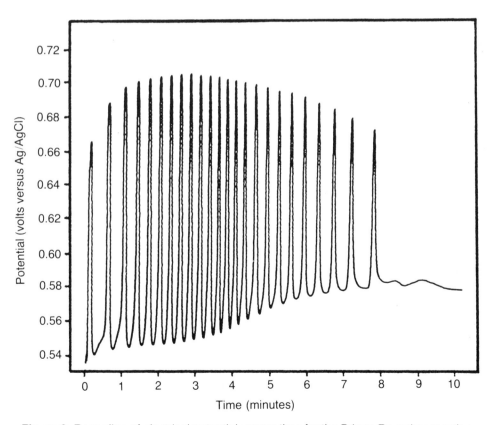

Figure 9. Recording of electrical potential versus time for the Briggs-Rauscher reaction.

the reference electrode to the negative terminal. Insert the electrodes into the solution containing the oscillating reaction. Adjust the range of the voltmeter or recorder so that the largest voltage produced by the electrodes is on scale. Then observe how the voltage oscillates in phase with the color changes. It may be desirable to increase the sensitivity of the chart recorder to amplify the oscillations, and this may necessitate adjusting the zero offset to keep the oscillations on scale. Adjust the chart speed of the recorder to provide a clear record of these oscillations.

ORGANIZATION OF THIS CHAPTER

The demonstrations in this chapter are largely variations of the Belousov-Zhabotinsky reaction. The first, however, is the Briggs-Rauscher reaction, which is perhaps the most visually impressive of the currently known oscillating reactions [28]. This reaction is a derivative of the oscillating reaction first reported by Bray [3]. Demonstrations 7.2 through 7.9 are variations of the BZ reaction, the last using a photofluorescent indicator that periodically glows under a black light. Demonstrations 7.10 and 7.11 are bromate oxidations of pyrogallol and tannic acid, respectively. Demonstration 7.12 uses an oscillating system to produce a pattern of moving bands in a thin layer of solution, while in Demonstration 7.13 a reaction exhibits oscillations in the evolution of a gas. Demonstration 7.14 is also a heterogeneous reaction, the formation of Liesegang rings of silver chromate crystals as silver ion diffuses into a gel containing chromate ion [29]. This is too slow a process to make an effective lecture demonstration; however, it does make an excellent corridor demonstration.

The demonstrations in this chapter do not include examples of all known oscillating chemical reactions. For example, oscillations have been reported in the concentration of sodium dithionite ($Na_2S_2O_4$) during its autocatalytic decomposition to sodium bisulfite ($NaHSO_3$) and sodium thiosulfate ($Na_2S_2O_3$) [30]. Oscillations have also been observed in the intensity of the blue color of methylene blue when it catalyzes the oxidation of sulfide ion to sulfur [31]. The oxidant chlorite ion (ClO_2^-) can drive several oscillating reactions [32, 33], and oscillations have been reported during the air oxidation of benzaldehyde catalyzed by cobalt and bromide ions [34]. None of these reactions is well understood, nor has any as yet been developed into a successful demonstration. The Morgan reaction, the dehydration of formic acid by concentrated sulfuric acid, requires carefully controlled conditions, which we have found make it unreliable as a lecture demonstration [35].

REFERENCES

1. A. J. Lotka, *J. Chem. Phys.* 14:271 (1910).
2. A. J. Lotka, *J. Am. Chem. Soc.* 42:1595 (1920).
3. W. C. Bray, *J. Am. Chem. Soc.* 43:1262 (1921).
4. B. P. Belousov, *Ref. Radiats. Med.* 1958:145 (1959).
5. A. M. Zhabotinsky, *Dokl. Akad. Nauk SSSR* 157:392 (1964).
6. A. M. Zhabotinsky, *Biofizika* 9:306 (1964).
7. A. N. Zaikin and A. M. Zhabotinsky, *Nature* 225:535 (1970).
8. R. J. Field and R. M. Noyes, *J. Chem. Phys.* 60:1877 (1974).
9. A. C. Vidal, J. C. Roux, and A. Rossi, *J. Am. Chem. Soc.* 102:1341 (1980).
10. R. J. Field, E. Körös, and R. M. Noyes, *J. Am. Chem. Soc.* 94:8649 (1972).

11. J. J. Tyson, "A Quantitative Account of Oscillations, Bistability, and Traveling Waves in the Belousov-Zhabotinskii Reaction," in *Oscillations and Traveling Waves in Chemical Systems*, R. J. Field and M. Burger, Eds., Interscience Publishers, John Wiley and Sons: New York (1985).

12. E. Körös, M. Burger, V. Friedrich, L. Ladányi, Z. Nagy, and M. Orbán, *Faraday Symp. Chem. Soc.* 9:28 (1974).

13. R. J. Field and R. M. Noyes, *Accts. Chem. Res.* 10:214 (1974).

14. R. M. Noyes, *J. Chem. Phys.* 80:6071 (1984).

15. D. Edelson, R. J. Field, and R. M. Noyes, *Int. J. Chem. Kinet.* 7:417 (1975).

16. D. Edelson, R. M. Noyes, and R. J. Field, *Int. J. Chem. Kinet.* 11:155 (1979).

17. H. Degn, *J. Chem. Educ.* 49:302 (1972).

18. G. Nicolis and J. Portnow, *Chem. Rev.* 73:365 (1973).

19. R. M. Noyes and R. J. Field, *Ann. Rev. Phys. Chem.* 25:95 (1974).

20. R. J. Field and R. M. Noyes, *Accts. Chem. Res.* 10:214 (1977).

21. R. M. Noyes and R. J. Field, *Accts. Chem. Res.* 10:273 (1977).

22. U. F. Franck, *Angew. Chem. Intl. Ed. Engl.* 17:1 (1978).

23. R. M. Noyes, *Ber. Bunsenges. phys. Chem.* 84:295 (1980).

24. A. M. Zhabotinsky, *Ber. Bunsenges. phys. Chem.* 84:303 (1980).

25. I. R. Epstein, K. Kustin, P. De Kepper, and M. Orbán, *Sci. Am.* 248:112 (1983).

26. I. R. Epstein, *J. Phys. Chem.* 88:187 (1984).

27. R. J. Field and M. Burger, Eds., *Oscillations and Traveling Waves in Chemical Systems*, Interscience Publishers, John Wiley and Sons: New York (1985).

28. T. S. Briggs and W. C. Rauscher, *J. Chem. Educ.* 50:496 (1973).

29. R. E. Liesegang, *Naturwiss. Wochenschr.* 11:353 (1896).

30. A. B. Corbet and D. M. Mason, *Ber. Bunsenges. phys. Chem.* 84:408 (1980).

31. M. Burger and R. J. Field, *Nature* 307:720 (1984).

32. M. Orbán, C. Dateo, P. De Kepper, and I. R. Epstein, *J. Am. Chem. Soc.* 104:5911 (1982).

33. M. Orbán, P. De Kepper, and I. R. Epstein, *J. Phys. Chem.* 86:431 (1982).

34. J. H. Jensen, *J. Am. Chem. Soc.* 105:2639 (1983).

35. J. S. Morgan, *J. Chem. Soc.* (London) 109:274 (1916).

7.1

Briggs-Rauscher Reaction

Three colorless solutions are combined in a large beaker and stirred on a magnetic stirrer. The solution becomes amber, then blue-black, and then colorless again. This sequence of color changes repeats with a period of approximately 15 seconds at 25°C. The period of the oscillation gradually increases, and after several minutes, the blue-black color persists. The electrical potential of the solution oscillates along with its color, and the range of these oscillations is about 60 mV.

MATERIALS FOR PROCEDURE A

3 liters distilled water

410 mL 30% hydrogen peroxide, H_2O_2

43 g potassium iodate, KIO_3

4.3 mL concentrated (18M) sulfuric acid, H_2SO_4

16 g malonic acid, $CH_2(CO_2H)_2$

3.4 g manganese(II) sulfate monohydrate, $MnSO_4 \cdot H_2O$

0.3 g soluble starch

20 g sodium thiosulfate, $Na_2S_2O_3$ (See Disposal section for use.)

4 2-liter beakers

gloves, plastic or rubber

hot plate

2 glass stirring rods

100-mL beaker

50-mL beaker

magnetic stirrer, with 2-inch stirring bar

platinum electrode (optional)

double-junction reference electrode † (optional)

strip-chart recorder (optional)

2-liter glass cylinder

---

† The reference electrode must not leak chloride ions into the solution, because chloride ions interfere with the mechanism of the oscillating reaction and inhibit the oscillations.

MATERIALS FOR PROCEDURE B

1.0 liter 3% hydrogen peroxide, H_2O_2

29 g potassium iodate, KIO_3

1 liter distilled water

8.6 mL 6.0M sulfuric acid, H_2SO_4 (To prepare 1.0 liter of stock solution, slowly and carefully pour 330 mL of concentrated [18M] H_2SO_4 into 500 mL of distilled water. After the mixture has cooled, dilute it to 1.0 liter with distilled water.)

10.4 g malonic acid, $CH_2(CO_2H)_2$

2.2 g manganese(II) sulfate monohydrate, $MnSO_4 \cdot H_2O$

0.2 g soluble starch

20 g sodium thiosulfate, $Na_2S_2O_3$ (See Disposal section for use.)

2 1-liter beakers

hot plate

2 glass stirring rods

100-mL beaker

50-mL beaker

3-liter beaker

magnetic stirrer, with 2-inch stirring bar

platinum electrode (optional)

double-junction reference electrode † (optional)

strip-chart recorder (optional)

PROCEDURE A

Preparation

Solution A-1. Pour 400 mL of distilled water into a 2-liter beaker. Wearing gloves, pour 410 mL of 30% hydrogen peroxide into the beaker of water. Dilute the solution to 1.0 liter with distilled water. This solution is 4.0M in H_2O_2.

Solution A-2. Place 43 g of potassium iodate and approximately 800 mL of distilled water in the second 2-liter beaker. Add 4.3 mL concentrated H_2SO_4 to this mixture. Warm and stir the mixture until the potassium iodate dissolves. Dilute the solution to 1.0 liter with distilled water. This solution is 0.20M in KIO_3 and 0.077M in H_2SO_4.

Solution A-3. Dissolve 16 g of malonic acid and 3.4 g of manganese(II) sulfate monohydrate in approximately 500 mL of distilled water in the third 2-liter beaker. In the 100-mL beaker, heat 50 mL of distilled water to a boil. In the 50-mL beaker, mix 0.3 g of soluble starch with about 5 mL of distilled water and stir the mixture to form a

---

† The reference electrode must not leak chloride ions into the solution, because chloride ions interfere with the mechanism of the oscillating reaction and inhibit the oscillations.

slurry. Pour the slurry into the boiling water and continue heating and stirring the mixture until the starch has dissolved (1–2 minutes). The solution may be slightly turbid. Pour this starch solution into the solution of malonic acid and manganese(II) sulfate. Dilute the mixture to 1.0 liter with distilled water. This solution is 0.15M in malonic acid and 0.020M in $MnSO_4$.

Set the remaining 2-liter beaker on the magnetic stirrer and place the stirring bar in the beaker.

To record the potential of the solution as the oscillating reaction proceeds, connect the platinum electrode to the positive terminal of the strip-chart recorder and the reference electrode to the negative terminal. Mount the electrodes on the 2-liter beaker so that they will be immersed when the beaker contains 1500 mL of liquid.

Presentation

Pour 500 mL of solution A-1 and 500 mL of solution A-2 into the beaker on the magnetic stirrer. Adjust the stirring rate to produce a large vortex in the mixture. Pour 500 mL of solution A-3 into the beaker. (This will produce a mixture whose nominal contents are 0.050M malonic acid, 0.0067M Mn^{2+}, 0.067M IO_3^{-}, 1.3M H_2O_2, and 0.038M H_2SO_4.) The initially colorless solution will become amber almost immediately. Then, it will suddenly turn blue-black. The blue-black will fade to colorless, and the cycle will repeat several times with a period which initially lasts about 15 seconds but gradually lengthens. After a few minutes, the solution will remain blue-black.

If the electrical potential of the solution is being recorded, adjust the chart recorder to produce the optimum display of the oscillations in the potential. The range of the potential oscillations is about 60 mV.

Pour 500 mL of solution A-1 into the 2-liter glass cylinder. Carefully pour 500 mL of solution A-2 down the inside of the cylinder to minimize the mixing of the two solutions. Carefully pour 500 mL of solution A-3 into the cylinder, again minimizing the mixing of the solutions. Regions of amber will develop in the solution, and these regions will turn blue-black. The blue-black layers will appear to drift through the solution in the cylinder.

PROCEDURE B

Preparation

Solution B-1. This is 1.0 liter of 3% hydrogen peroxide.

Solution B-2. Place 29 g of potassium iodate and approximately 400 mL of distilled water in a 1-liter beaker. Add 8.6 mL of 6.0M H_2SO_4 to this mixture. Warm and stir the mixture until the potassium iodate dissolves. Dilute the solution to 500 mL with distilled water. This solution is 0.27M in KIO_3 and 0.10M in H_2SO_4.

Solution B-3. Dissolve 10.4 g of malonic acid and 2.2 g of manganese(II) sulfate monohydrate in approximately 400 mL of distilled water in the other 1-liter beaker. In the 100-mL beaker, heat 50 mL of distilled water to a boil. In the 50-mL beaker, mix 0.2 g of soluble starch with about 5 mL of distilled water and stir the mixture to form a slurry. Pour the slurry into the boiling water and continue heating and stirring the mixture until the starch has dissolved (1–2 minutes). The solution may be slightly turbid.

Pour the starch solution into the solution of malonic acid and manganese(II) sulfate. Dilute the mixture to 500 mL with distilled water. This solution is 0.20M in malonic acid and 0.026M in $MnSO_4$.

Set the 3-liter beaker on the magnetic stirrer and place the stirring bar in the beaker.

To record the potential of the solution as the oscillating reaction proceeds, connect the platinum electrode to the positive terminal of the strip-chart recorder and the reference electrode to the negative terminal. Mount the electrodes on the 3-liter beaker so that they will be immersed when the beaker contains 2 liters of liquid.

Presentation

Pour solution B-1 and solution B-2 into the beaker on the magnetic stirrer. Adjust the stirring rate to produce a large vortex in the mixture. Pour solution B-3 into the beaker. (The nominal composition of this mixture is 0.050M malonic acid, 0.0065M Mn^{2+}, 0.067M IO_3^-, 0.44M H_2O_2, and 0.025M H_2SO_4.) The mixture will become amber almost immediately, and this color will gradually deepen. After 45–60 seconds, the mixture will turn deep blue. The blue will fade to colorless, and the cycle will repeat several times with a period which initially lasts about 15 seconds but gradually lengthens. After several minutes the solution will remain deep blue.

If the electrical potential of the solution is being recorded, adjust the chart recorder to produce the optimum display of the oscillations in the potential. The range of the potential oscillations is about 60 mV.

HAZARDS

The reaction produces iodine in solution, in suspension, and also as a vapor above the reaction mixture. The vapor or the solid is very irritating to the eyes, skin, and mucous membranes.

Because 30% hydrogen peroxide is a strong oxidizing agent, contact with skin and eyes must be avoided. In case of contact, immediately flush the affected area with water for at least 15 minutes; get immediate medical attention if the eyes are affected. Avoid contact between 30% hydrogen peroxide and combustible materials. Avoid contamination from any source, because any contaminant, including dust, will cause rapid decomposition and the generation of large quantities of oxygen gas. Store 30% hydrogen peroxide in its original, closed container, making sure that the container vent works properly.

Because sulfuric acid is a strong acid and a powerful dehydrating agent, it can cause burns. Spills should be neutralized with an appropriate agent, such as sodium bicarbonate ($NaHCO_3$), and then rinsed clean.

Malonic acid is a strong irritant to skin, eyes, and mucous membranes.

DISPOSAL

The reaction produces copious amounts of elemental iodine (I_2), which should be reduced to iodide ions before disposal. To do so, carefully add 10 g of sodium thiosulfate to each mixture and stir it until the mixture becomes colorless. **Caution! The reaction between iodine and thiosulfate is exothermic, and the mixture may become hot.** The cold solution should then be flushed down the drain with water.

DISCUSSION

The two procedures of this demonstration differ in that the first uses 30% hydrogen peroxide, while the second uses 3% solution. Procedure A has two notable advantages: the time that elapses before the oscillations begin is much shorter, and the color changes are much sharper with Procedure A than with Procedure B. However, for those who do not have access to 30% H_2O_2 or who do not wish to handle this hazardous material, Procedure B provides a satisfactory alternative.

The Briggs-Rauscher reaction was developed by Thomas S. Briggs and Warren C. Rauscher of Galileo High School in San Francisco [1].† It is perhaps the most visually impressive of the chemical oscillators. A stirred batch of solution goes through 15 or more cycles of colorless, to amber, to blue-black, before ending as a blue-black mixture with the odor of iodine.

The Briggs-Rauscher (BR) reaction is a hybrid of two other oscillating chemical reactions, the Bray-Liebhafsky (BL) reaction and the Belousov-Zhabotinsky (BZ) reaction. Bray was investigating the dual role of H_2O_2 as an oxidizing agent and a reducing agent when he discovered oscillations in the evolution of oxygen gas from the reaction mixture [2]. He mixed H_2O_2, KIO_3, and H_2SO_4, and in this mixture, hydrogen peroxide reduced iodate to iodine and was oxidized to oxygen gas in the process.

$$5\ H_2O_2(aq) + 2\ IO_3{}^-(aq) + 2\ H^+(aq) \longrightarrow I_2(aq) + 5\ O_2(g) + 6\ H_2O(l) \quad (1)$$

The hydrogen peroxide also oxidizes iodine to iodate.

$$5\ H_2O_2(aq) + I_2(aq) \longrightarrow 2\ IO_3{}^-(aq) + 2\ H^+(aq) + 4\ H_2O(l) \quad (2)$$

The net result of these reactions is the iodate catalysis of the disproportionation of hydrogen peroxide.

$$2\ H_2O_2(aq) \longrightarrow O_2(g) + 2\ H_2O(l) \quad (3)$$

The discovery of oscillations prompted Liebhafsky to study this reaction before 1933 and after 1969 [3], when he retired from a career as an industrial chemist, but few others paid much attention to this reaction until after the discovery of the BZ reaction. Belousov reported oscillations during the reaction of citric acid with acidic bromate ions and cerium(IV) ions [4]. This oscillatory behavior was exploited by Zhabotinsky, who discovered that oscillations still occurred if certain other organic compounds, such as malonic acid, were substituted for citric acid, and if other one-electron transfer agents, such as Mn(II) ions, were substituted for cerium ions [5–7].

Briggs and Rauscher combined the hydrogen peroxide and iodate of the BL reaction with the malonic acid and manganese ions of the BZ reaction, and discovered the oscillating reaction that bears their name. In the BR oscillating reaction, the evolution of oxygen and carbon dioxide gases and the concentrations of iodine and iodide ions oscillate. These oscillations are represented in the figure. Iodine is produced rapidly

---

†According to Henry A. Bent (253d annual meeting of the AAAS, May 1984, New York), Briggs and Rauscher discovered the reaction that bears their name while investigating the Bray reaction in the laboratories of Professor William Jolly of the University of California at Berkeley.

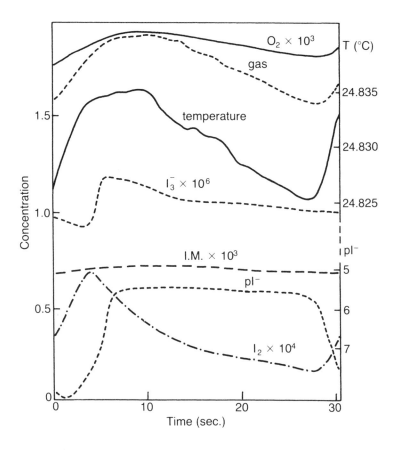

Variations with time of O_2, I_3^-, iodomalonic acid, and I_2 concentrations (in moles per liter), of pI^-, of gas evolution (in arbitrary units), and of temperature (in Celsius) during one oscillation of a BR reaction (Source: Figure 4 [8]).

when the concentration of iodide ions is low. As the concentration of iodine in the solution increases, the amber color of the solution intensifies. The production of I^- increases as $[I_2]$ increases, and these ions react with iodine molecules and starch to form a blue-black complex containing the pentaiodide ion (I_5^-) [9]. Most of the O_2 and CO_2 is produced during the formation of I_2. The $[I_2]$ reaches a maximum and begins to fall, although $[I^-]$ rises further and remains high as $[I_2]$ continues to decline until the solution clears. Then the $[I^-]$ suddenly falls and the cycle begins again. This cycle repeats a number of times until the solution ends as a deep blue mixture that liberates iodine vapors.

The mechanism of this reaction has been studied by Cooke [10–12], by Furrow and Noyes [13–15], and by De Kepper and Epstein [16]. The results of their investigations led them to similar conclusions about the nature of the mechanism. Much of the proposed mechanism is a direct transfer of the mechanism of the BZ reaction, as described in the introduction to this chapter, to the iodate–hydrogen peroxide system. The proposed mechanism is only a skeleton of what happens during the reaction. It does not account for the production of CO_2, nor does it identify the ultimate organic products of the reaction; these factors are still under investigation. However, it does explain the origin of the oscillations in the concentrations of I_2 and I^-.

The transformation that accounts for the oscillations in the BR reaction is represented in equation 4.

$$IO_3^- + 2\,H_2O_2 + CH_2(CO_2H)_2 + H^+ \longrightarrow$$
$$ICH(CO_2H)_2 + 2\,O_2 + 3\,H_2O \qquad (4)$$

This transformation is accomplished through two component reactions (equations 5 and 6).

$$IO_3^- + 2\,H_2O_2 + H^+ \longrightarrow HOI + 2\,O_2 + 2\,H_2O \qquad (5)$$

$$HOI + CH_2(CO_2H)_2 \longrightarrow ICH(CO_2H)_2 + H_2O \qquad (6)$$

The first of these two reactions can occur via two different processes, a radical process and a nonradical process. Which of these two processes dominates is determined by the concentration of iodide ions in the solution. When $[I^-]$ is low, the radical process dominates; when $[I^-]$ is high, the nonradical process is the dominant one. The second reaction (equation 6) couples the two processes. This reaction consumes HOI more slowly than that species is produced by the radical process when that process is dominant, but it consumes HOI more rapidly than it is produced by the nonradical process. Any HOI which does not react by equation 6 is reduced to I^- by hydrogen peroxide as one of the component steps of the nonradical process for reaction 5. When HOI is produced rapidly by the radical process, the excess forms the iodide ions, which shut off that radical process and start the slower nonradical process. Equation 6 then consumes the HOI so rapidly that not enough is available to produce the iodide ion necessary to keep the nonradical process going, and the radical process starts again. Each of the processes of equation 5 produces conditions favorable to the other process, and, therefore, the reaction oscillates between these two processes.

The detailed explanation requires attention to the individual steps of the two processes. If iodide ions are present in sufficient concentration, the reaction follows the nonradical process of equation 5. The iodide ions react rather slowly with iodate ions, as represented in equation 7.

$$IO_3^- + I^- + 2\,H^+ \longrightarrow HIO_2 + HOI \qquad (7)$$

The iodous acid (HIO_2) is further reduced to hypoiodous acid (HOI) by the reaction of equation 8.

$$HIO_2 + I^- + H^+ \longrightarrow 2\,HOI \qquad (8)$$

(These two equations are analogous to those proposed for the BZ reaction with bromate.) The hypoiodous acid is then reduced by hydrogen peroxide, as indicated by equation 9.

$$HOI + H_2O_2 \longrightarrow I^- + O_2 + H^+ + H_2O \qquad (9)$$

The net transformation as represented in equation 5 is obtained by the stoichiometric addition of (equation 7) + (equation 8) + 2(equation 9).

Because reaction 5 is slower than reaction 6 under these conditions, so much HOI is used up by reaction 6 that reaction 9 cannot replenish the I^- consumed in reactions 7 and 8; the $[I^-]$ keeps diminishing.

Once the iodide ions have been sufficiently depleted, the nonradical process becomes very slow, and the radical process for reaction 5 can take over. This process involves the five steps represented by equations 10 through 14 [16].

$$IO_3^- + HIO_2 + H^+ \longrightarrow 2\,IO_2^\cdot + H_2O \qquad (10)$$

$$IO_2^\cdot + Mn^{2+} + H_2O \longrightarrow HIO_2 + Mn(OH)^{2+} \qquad (11)$$

$$Mn(OH)^{2+} + H_2O_2 \longrightarrow Mn^{2+} + H_2O + HOO^{\cdot} \qquad (12)$$

$$2\,HOO^{\cdot} \longrightarrow H_2O_2 + O_2 \qquad (13)$$

$$2\,HIO_2 \longrightarrow IO_3^- + HOI + H^+ \qquad (14)$$

These steps, when combined in the stoichiometry of 2(equation 10) + 4(equation 11) + 4(equation 12) + 2(equation 13) + (equation 14), have the overall result given by equation 5. A significant feature of this process is that, taken together, the first two steps (equations 10 and 11) are autocatalytic—they produce 2 HIO_2 for each one consumed. Therefore, the rate of these steps increases as they occur. Because this radical process is autocatalytic, it causes a rapid increase in the concentration of HOI, which is produced by the disproportionation of HIO_2 (equation 14). This process does not rapidly consume all of the iodate in the solution, because the last step is second order in the catalytic species. Thus, as its concentration increases because of the autocatalytic nature of the early steps, HIO_2 is ever more rapidly consumed in this last step, and the sequence of reactions quickly reaches a steady state. (Equations 10, 11, and 14 are analogous to those for the BZ reaction involving bromate.)

Equations 11 and 12 indicate the function of the manganese catalyst. The manganese is oxidized in reaction 11 and reduced in reaction 12. Its catalytic effect in the reaction is accounted for through its providing the means for reducing $IO_2^{\cdot}$ radicals to HIO_2, thereby completing the autocatalytic cycle of equations 10 and 11.

The hypoiodous acid produced by the radical process reacts with malonic acid by reaction 6. However, the radical process is faster than reaction 6, and the excess HOI reacts with hydrogen peroxide by reaction 9 to create I^-, which shuts off the radical process and returns the system to the slow nonradical process initiated by reaction 7.

The dramatic color effects arise because reaction 6 does not take place in a single step, but by the sequence of reactions 15 and 16.

$$I^- + HOI + H^+ \longrightarrow I_2 + H_2O \qquad (15)$$

$$I_2 + CH_2(CO_2H)_2 \longrightarrow ICH(CO_2H)_2 + H^+ + I^- \qquad (16)$$

The solution turns amber from the I_2 produced through reaction 15, when the radical process maintains [HOI] greater than [I^-]. The excess HOI is converted to I^- through the reaction with H_2O_2 (equation 9). The solution suddenly turns dark blue when [I^-] becomes greater than [HOI], and the I^- can combine with I_2 to form a complex with the starch. With [I^-] high, reaction 5 switches to the slow nonradical process. The color then fades as reaction 6 consumes iodine faster than it is produced. When the system switches back to the rapid radical process, the cycle is repeated.

The above reaction steps constitute a skeleton mechanism for the BR oscillating reaction. Upon initial mixing of the solutions, IO_3^- reacts with H_2O_2 to produce a little HIO_2. The HIO_2 reacts with IO_3^- in the first step of the radical process (equation 10). The autocatalytic radical process follows, rapidly increasing the concentration of HOI. The HOI is reduced to I^- in a reaction with H_2O_2 (equation 9). The large amount of HOI reacts with I^-, producing I_2 (equation 15). The I_2 reacts slowly with malonic acid, but the concentrations of HOI, I_2, and I^- all increase, because reaction 5 is faster than reaction 6. As [I^-] increases, the rate of its reaction with HIO_2 (equation 8) surpasses that of the autocatalytic sequence of reactions 10 and 11. The radical process is then shut off, and the accumulation of reduced iodine is consumed by reaction 6 operating through the sequence of equations 15 and 16. Eventually [I^-] is reduced to such a low value that reactions 10 and 11 become faster than reaction 8, and the radical process takes over again. This oscillating sequence repeats until the malonic acid or IO_3^- is depleted.

In the procedures for this demonstration, the order of mixing the solutions is specified. However, this order is not crucial. The solutions can be mixed in any order, as long as they are combined quickly.

Warning About Chloride Ion Contamination

Chloride ion concentrations in excess of 0.07M suppress the oscillations [17]. Therefore, the vessels used for the preparation of the solutions must be clean. If the oscillations in the electrical potential of the solution are to be observed, a reference electrode that does not leak chloride ions must be used. An ordinary standard calomel electrode or silver–silver chloride electrode is *not* suitable. A double-junction version of one of these electrodes is adequate.

REFERENCES

1. T. S. Briggs and W. C. Rauscher, *J. Chem. Educ.* 50:496 (1973).
2. W. C. Bray, *J. Am. Chem. Soc.* 43:1262 (1921).
3. H. A. Liebhafsky, W. C. McGavock, R. J. Reyes, G. M. Roe, and L. S. Wu, *J. Am. Chem. Soc.* 100:87 (1978), and references cited therein.
4. B. P. Belousov, *Ref. Radiats. Med.* 1958:145 (1959).
5. A. M. Zhabotinsky, *Dokl. Akad. Nauk SSSR* 157:392 (1964).
6. A. M. Zhabotinsky, *Biofizika* 9:306 (1964).
7. A. N. Zaikin and A. M. Zhabotinsky, *Nature* (London) 225:535 (1970).
8. J. C. Roux and C. Vidal, *Nouv. j. chim.* 3:247 (1979).
9. R. C. Teitelbaum, S. L. Ruby, and T. J. Marks, *J. Am. Chem. Soc.* 102:3322 (1980).
10. D. O. Cooke, *Inorg. Chim. Acta* 37:259 (1979).
11. D. O. Cooke, *Int. J. Chem. Kinet.* 12:671 (1980).
12. D. O. Cooke, *Int. J. Chem. Kinet.* 12:683 (1980).
13. S. D. Furrow and R. M. Noyes, *J. Am. Chem. Soc.* 104:38 (1982).
14. S. D. Furrow and R. M. Noyes, *J. Am. Chem. Soc.* 104:43 (1982).
15. R. M. Noyes and S. D. Furrow, *J. Am. Chem. Soc.* 104:45 (1982).
16. P. De Kepper and I. R. Epstein, *J. Am. Chem. Soc.* 104:49 (1982).
17. D. O. Cooke, *React. Kinet. Catal. Lett.* 3:377 (1975).

7.2

Cerium-catalyzed Bromate–Malonic Acid Reaction

(The Classic Belousov-Zhabotinsky Reaction)

A clear colorless solution and a pale yellow solution are mixed, producing an amber solution, which becomes colorless after about 1 minute. Then a yellow solution is added, followed by a small amount of red solution, producing a green solution. The color of the solution gradually changes over a period of about 1 minute from green to blue, then violet, and finally red. The color then suddenly returns to green, and the cycle repeats more than 20 times. The electrical potential of the solution oscillates along with the color of the solution, and the range of these oscillations is about 200 mV.

MATERIALS

19 g potassium bromate, $KBrO_3$, or 17 g sodium bromate, $NaBrO_3$

1.0 liter distilled water

16 g malonic acid, $CH_2(CO_2H)_2$

3.5 g potassium bromide, KBr, or 3.0 g sodium bromide, NaBr

5.3 g cerium(IV) ammonium nitrate, $Ce(NH_4)_2(NO_3)_6$

500 mL 2.7M sulfuric acid, H_2SO_4 (To prepare 1.0 liter of solution, carefully pour 150 mL of concentrated [18M] H_2SO_4 into 500 mL of distilled water, and dilute the resulting solution to 1.0 liter with distilled water.)

30 mL 0.50% ferroin solution (To prepare 100 mL of stock solution, dissolve 0.23 g of iron(II) sulfate heptahydrate, $FeSO_4 \cdot 7H_2O$, in 100 mL of distilled water. In the resulting solution, dissolve 0.46 g of 1,10-phenanthroline, $C_{12}H_8N_2$.)

3 500-mL Erlenmeyer flasks

2-liter beaker (tall form, if possible)

magnetic stirrer, with 2-inch stirring bar

platinum electrode (optional)

double-junction reference electrode † (optional)

strip-chart recorder (optional)

---

† The reference electrode must not leak chloride ions into the solution, because chloride ions interfere with the mechanism of the oscillating reaction and inhibit the oscillations.

PROCEDURE

Preparation

Solution A. In a 500-mL flask, dissolve 19 g of potassium bromate in 500 mL of distilled water. This solution is 0.23M in $KBrO_3$.

Solution B. In the second 500-mL flask, dissolve 16 g of malonic acid and 3.5 g of potassium bromide in 500 mL of distilled water. This solution is 0.31M in malonic acid and 0.059M in KBr.

Solution C. In the third 500-mL flask, dissolve 5.3 g of cerium(IV) ammonium nitrate in 500 mL of 2.7M sulfuric acid. This solution is 0.019M in $Ce(NH_4)_2(NO_3)_6$ and 2.7M in H_2SO_4.

Set the 2-liter beaker on the magnetic stirrer and place the stirring bar in the beaker.

To record the potential of the solution as the oscillating reaction proceeds, connect the platinum electrode to the positive terminal of the strip-chart recorder and the reference electrode to the negative terminal. Mount the electrodes on the 2-liter beaker so that they will be immersed when the beaker contains 1500 mL of liquid.

Presentation

Pour solution A and solution B into the beaker, and adjust the stirrer to produce a vortex in the solution. The solution will become amber, and after about 1 minute it will become colorless. Once it has become colorless, add solution C and 30 mL of ferroin solution. (The nominal composition of this solution is 0.077M $BrO_3{}^-$, 0.10M malonic acid, 0.020M Br^-, 0.0063M Ce(IV), 0.90M H_2SO_4, and 0.17mM ferroin.) The solution will become green. Over a period of about a minute, the color of the solution will change from green to blue, then violet, and finally red. The color will suddenly return to green, and the cycle will repeat itself more than 20 times.

If the electrical potential of the solution is being recorded, adjust the chart recorder to produce the optimum display of the oscillations in the potential. The range of the potential oscillations in this solution is about 200 mV.

HAZARDS

Because sulfuric acid is a strong acid and a powerful dehydrating agent, it can cause burns. Spills should be neutralized with an appropriate agent, such as sodium bicarbonate ($NaHCO_3$), and then rinsed clean.

Malonic acid is a strong irritant to skin, eyes, and mucous membranes.

Bromates are strong oxidizing agents. Mixtures of bromates with finely divided organic materials, metals, carbon, or other combustible materials are easily ignited, sometimes explosively. Ingestion of potassium bromate can cause vomiting, diarrhea, and renal injury.

DISPOSAL

The reaction mixture should be neutralized with sodium bicarbonate ($NaHCO_3$) and flushed down the drain with water.

DISCUSSION

The oscillating reaction observed in this demonstration is one in a family of reactions, the first of which was discovered by Belousov in 1958 [1]. He mixed potassium bromate, cerium(IV) sulfate, and citric acid in dilute sulfuric acid and found that the ratio of concentration of the cerium(IV) and cerium(III) ions oscillated. Zhabotinsky studied this reaction extensively [2–4]. He found that the oscillations still occurred when citric acid was replaced by any of a number of carboxylic acids with the common structural feature

$$R-\overset{\overset{\displaystyle O}{\|}}{C}-CH_2-\overset{\overset{\displaystyle O}{\|}}{C}-OH$$

The cerium ions could also be replaced with manganese ions, and oscillations still occurred. Together, these reactions form a family called the Belousov-Zhabotinsky (BZ) reactions. BZ reactions are by far the most studied and the best understood of any chemical oscillators.

The overall reaction occurring in this demonstration is the cerium-catalyzed oxidation of malonic acid by bromate ions in dilute sulfuric acid. The bromate ions are reduced to bromide ions, while the malonic acid is oxidized to carbon dioxide and water. The overall reaction can be represented by the equation

$$3\ CH_2(CO_2H)_2 + 4\ BrO_3^- \longrightarrow 4\ Br^- + 9\ CO_2 + 6\ H_2O \tag{1}$$

This equation represents the ultimate chemical transformations that occur during this demonstration, and the free-energy difference between the products and the reactants is what drives the reaction. However, this equation certainly does not account for the most striking observations to be made during this demonstration—the periodic changes in the color of the solution as the reaction proceeds. Neither does it account for the catalytic effect of the cerium, nor for the role played by the bromide ions which are added at the start of the demonstration. In order to appreciate how this reaction can produce repetitive color changes, it is necessary to examine *how* the reactants are transformed into products, that is, to examine the reaction mechanism.

Experimental methods used to investigate this and other chemical oscillators and the mechanism of the BZ reaction are described in some detail in the introduction to this chapter. The basic elements of this mechanism have stood the test of time and are successful in explaining the major features of this reaction. The mechanism involves two different processes. One (Process A) involves ions, and the steps are two-electron transfers. The other (Process B) involves radicals and one-electron transfers. Which process is dominant at a particular time is determined by the bromide ion concentration. Process A occurs when the bromide ion concentration rises above a certain critical level, while Process B dominates when the bromide ion concentration falls below a certain level. Oscillations occur because Process A consumes bromide ions and, thus, leads to the conditions which favor Process B. Process B indirectly liberates bromide ions, which returns the reaction to control by Process A.

The net transformation taking place in Process A is represented by equation 2.

$$BrO_3^- + 5\ Br^- + 6\ H^+ \longrightarrow 3\ Br_2 + 3\ H_2O \tag{2}$$

The net reaction represented by equation 2 is the reduction of bromate ions by bromide ions through a series of oxygen transfers (two-electron reductions). This reaction is what occurs after solutions A and B are mixed. The amber color which develops is due to the elemental bromine. This color disappears as the bromine reacts with malonic acid, as represented by equation 3.

$$Br_2 + CH_2(CO_2H)_2 \longrightarrow BrCH(CO_2H)_2 + Br^- + H^+ \tag{3}$$

Together, reactions 2 and 3 result in a decline in the bromide ion concentration. Once Process A has generated the necessary intermediates and consumed most of the Br^-, its rate falls to a negligible value, and Process B dominates.

The overall reaction effected by Process B is given by equation 4.

$$2\ BrO_3^- + 12\ H^+ + 10\ Ce^{3+} \longrightarrow Br_2 + 6\ H_2O + 10\ Ce^{4+} \tag{4}$$

This process is initiated by reactions 5 and 6, which cause [HOBr] to increase autocatalytically.

$$BrO_3^- + HBrO_2 + H^+ \longrightarrow 2\ BrO_2 + H_2O \tag{5}$$

$$BrO_2 + Ce^{3+} + H^+ \longrightarrow HBrO_2 + Ce^{4+} \tag{6}$$

Autocatalysis seems to be an essential feature of oscillating chemical reactions (see the introduction to this chapter). The autocatalysis does not continue until the reactants are depleted, because there is a second-order destruction of the autocatalytic species. This reaction is represented by equation 7.

$$2\ HBrO_2 \longrightarrow HOBr + BrO_3^- + H^+ \tag{7}$$

Process B produces Ce(IV) and Br_2, both of which react at least in part to oxidize organic material with the formation of bromide ion. As the concentration of Br^- produced by this reaction increases, the rate of equation 2 increases, surpassing the rate of equations 5 and 6, and the reaction again follows the pathway of Process A.

Process A and Process B result in a competition between bromide ions and bromate ions for bromous acid. When the concentration of bromide ions is high, nearly all of the bromous acid reacts with it, following Process A. During this process, bromide ion concentration decreases, and the bromide ion becomes less and less successful at competing for the bromous acid. Eventually, Process B takes over. This process, however, produces bromide ion indirectly, and eventually the concentration of bromide ion becomes high enough to cause a shift back to Process A. This switching back and forth between these processes is the origin of the oscillations observed in this reaction.

As the reaction oscillates between Process A and Process B, triggered by changes in the bromide ion concentration, other concentrations oscillate as well. Figure 7 of the introduction to this chapter illustrates how the concentrations of various substances oscillate. While Process A occurs, the cerium ions are in their reduced state, Ce(III), as a result of their reaction with bromomalonic acid. During Process B, some of the cerium ions are oxidized to Ce(IV). Thus, the ratio of [Ce(III)] to [Ce(IV)] oscillates as well. As this ratio oscillates, so does the electrical potential of the Ce(III)–Ce(IV) couple. The potential of the Ce(III)–Ce(IV) couple, at 25°C, is given by the Nernst equation.

$$E = E° - 0.059 \log \frac{[Ce(III)]}{[Ce(IV)]}$$

The electrical potential of the solution, as registered by a platinum electrode, changes logarithmically with the ratio of the concentrations. The oscillations in the electrical potential can be observed by measuring the potential of a platinum electrode versus a reference electrode, as described in the procedure. The oscillations in the potential can also be observed by using an oxidation-reduction indicator. The indicator used in this demonstration is ferroin (tris(1,10-phenanthroline)iron(II) sulfate). As the ratio of

[Ce(III)] to [Ce(IV)] decreases (i.e., as [Ce(IV)] increases), the Ce(IV) can oxidize the iron in ferroin from iron(II) to iron(III). Because the iron(II) complex is red and the iron(III) complex is blue, the color of the solution changes as the iron is oxidized. When the [Ce(III)] to [Ce(IV)] ratio rises, the iron(III) is reduced to iron(II), and the solution returns to its original color. The color changes that occur in this demonstration are more complex than a simple red-blue oscillation, because in addition to the changes in the color of the iron complex, there are changes in color due to the cerium ions themselves; cerium(III) is colorless, and cerium(IV) is yellow. This description of the functioning of the ferroin indicator is perhaps oversimplified, because the ferroin is known to catalyze the oscillating reaction in the absence of Ce ions, and it appears to enter the catalytic cycle involving Ce ions [5].

The discussion of the mechanism of the Belousov-Zhabotinsky reaction presented here oversimplifies what occurs in the reaction mixture. Much more is known about this system than is described here, and because this is an active field of research, much more is likely to be discovered soon.

Warning About Chloride Ion Contamination

Even small amounts of chloride ions in the solution can interfere with the mechanism of the BZ reaction, inhibiting the oscillations [6]. Therefore, the vessels used for the preparation of the solutions must be clean. In the preparation of the ferroin solution, 1,10-phenanthroline must be used in its free form, not in the hydrochloride salt, in order to prevent the introduction of chloride ions into the solution. If the oscillations in the electrical potential of the solution are to be observed, a reference electrode that does not leak chloride ions must be used. An ordinary standard calomel electrode or silver–silver chloride electrode is *not* suitable. A double-junction version of one of these electrodes is adequate.

REFERENCES

1. B. P. Belousov, *Ref. Radiats. Med.* 1958:145 (1959).
2. A. M. Zhabotinsky, *Dokl. Akad. Nauk SSSR* 157:392 (1964).
3. A. M. Zhabotinsky, *Biofizika* 9:306 (1964).
4. A. N. Zaikin and A. M. Zhabotinsky, *Nature* 225:535 (1970).
5. E. Körös, M. Burger, V. Friedrich, L. Ladányi, Z. Nagy, and M. Orbán, *Faraday Symp. Chem. Soc.* 9:28 (1974).
6. S. S. Jacobs and I. R. Epstein, *J. Am. Chem. Soc.* 98:1721 (1976).

7.3

Cerium-catalyzed Bromate–Methylmalonic Acid Reaction

(A Modified Belousov-Zhabotinsky Reaction)

A yellow solution is added to a clear solution being stirred in a large beaker. Then a red solution is added. The mixture turns green, and over the next 3–5 minutes the color gradually changes to blue, then violet, and finally red. Then the color quickly returns to green. This cycle will repeat for 8 hours, with an initial period of about 2 minutes, increasing to 10–15 minutes at the end of the reaction. The electrical potential of the solution oscillates along with the color of the solution, and the range of these oscillations is about 240 mV.

MATERIALS

5.0 g potassium bromate, $KBrO_3$, or 4.5 g sodium bromate, $NaBrO_3$

500 mL 0.90M sulfuric acid, H_2SO_4 (To prepare 1.0 liter of stock solution, slowly and carefully add 33 mL of concentrated [18M] H_2SO_4 to 500 mL of distilled water, and dilute the mixture to 1.0 liter with distilled water.)

6.0 g methylmalonic acid, $CH_3CH(CO_2H)_2$

1.2 g potassium bromide, KBr, or 1.1 g sodium bromide, NaBr

250 mL distilled water

1.7 g cerium(IV) ammonium nitrate, $Ce(NH_4)_2(NO_3)_6$

15 mL 0.50% ferroin solution (To prepare 100 mL of stock solution, dissolve 0.23 g of iron(II) sulfate heptahydrate, $FeSO_4 \cdot 7H_2O$, in 100 mL of distilled water. In this solution, dissolve 0.46 g of 1,10-phenanthroline, $C_{12}H_8N_2$.)

3 500-mL Erlenmeyer flasks

1-liter beaker

magnetic stirrer, with 2-inch stirring bar

platinum electrode (optional)

double-junction reference electrode † (optional)

strip-chart recorder (optional)

---

† The reference electrode must not leak chloride ions into the solution, because chloride ions interfere with the mechanism of the oscillating reaction and inhibit the oscillations.

PROCEDURE

Preparation

Solution A. In a 500-mL flask, dissolve 5.0 g of potassium bromate in 250 mL of 0.90M sulfuric acid. This solution is 0.12M in BrO_3^- and 0.90M in H_2SO_4.

Solution B. In the second 500-mL flask, dissolve 6.0 g of methylmalonic acid and 1.2 g of potassium bromide in 250 mL of distilled water. This solution is 0.20M in methylmalonic acid and 0.040M in Br^-.

Solution C. In the third 500-mL flask, dissolve 1.7 g of cerium(IV) ammonium nitrate in 250 mL of 0.90M sulfuric acid. This solution is 0.012M in Ce^{4+} and 0.90M in H_2SO_4.

In a hood, set the 1-liter beaker on the magnetic stirrer, and place the stirring bar in the beaker. One-half hour before presenting the demonstration, pour solution A and solution B into the beaker, and adjust the stirring rate to create a deep vortex in the mixture. The mixture will become amber and, after about 20 minutes, will return to colorless. Once the solution is colorless, place the beaker and magnetic stirrer at the site of the demonstration.

Connect the platinum electrode to the positive terminal of the recorder and the reference electrode to the negative terminal.

Presentation

Add solution C and 15 mL of the ferroin solution to the colorless solution being stirred in the beaker. (The nominal composition of this solution is 0.040M BrO_3^-, 0.067M methylmalonic acid, 0.013M Br^-, 0.0040M Ce^{4+}, and 0.60M H_2SO_4.) The mixture will become green and then, over a span of 3–5 minutes, gradually turn to blue, then violet, and finally red. Then the color will quickly return to green, and the cycle will repeat for about 8 hours, its period lengthening gradually to 10–15 minutes.

Insert the electrodes into the solution soon after adding solution C, and adjust the recorder to produce the optimum display of the oscillations in the electrical potential of the reaction mixture. The range of the potential oscillations is about 240 mV.

HAZARDS

Bromates are strong oxidizing agents. Mixtures of bromates with finely divided organic materials, metals, carbon, or other combustible materials are easily ignited, sometimes explosively. Ingestion of potassium bromate can cause vomiting, diarrhea, and renal injury.

Because sulfuric acid is a strong acid and a powerful dehydrating agent, it can cause burns. Spills should be neutralized with an appropriate agent, such as sodium bicarbonate ($NaHCO_3$), and then rinsed clean.

Methylmalonic acid is a strong irritant to skin, eyes, and mucous membranes.

DISPOSAL

The reaction mixture should be neutralized with sodium bicarbonate ($NaHCO_3$) and flushed down the drain with water.

DISCUSSION

The reaction in this demonstration is a derivative of the reaction used in Demonstration 7.2. This demonstration uses methylmalonic acid in place of malonic acid (structures shown below). One of the hydrogen atoms of the central carbon atom of malonic acid is replaced by a methyl group in methylmalonic acid. When malonic acid is oxidized in the Belousov-Zhabotinsky (BZ) reaction in Demonstration 7.2, the products include carbon dioxide and formic acid. The oxidation of methylmalonic acid in this demonstration produces acetic acid instead of formic acid.

$$
\underset{\text{methylmalonic acid}}{HO-\overset{\overset{\displaystyle O}{\|}}{C}-\underset{\underset{\displaystyle CH_3}{|}}{CH}-\overset{\overset{\displaystyle O}{\|}}{C}-OH}
\qquad\qquad
\underset{\text{malonic acid}}{HO-\overset{\overset{\displaystyle O}{\|}}{C}-\underset{\underset{\displaystyle H}{|}}{CH}-\overset{\overset{\displaystyle O}{\|}}{C}-OH}
$$

In the mechanism of the BZ reaction, as described in the introduction to this chapter, the malonic acid reacts with elemental bromine to produce bromomalonic acid, as represented in equation 1.

$$CH_2(CO_2H)_2(aq) + Br_2(aq) \longrightarrow$$
$$BrCH(CO_2H)_2(aq) + H^+(aq) + Br^-(aq) \qquad (1)$$

Malonic acid exists in aqueous solution as an equilibrium between the diacid form and the enol form, as indicated by equation 2.

$$
\underset{\text{diacid}}{HO-\overset{\overset{\displaystyle O}{\|}}{C}\underset{\underset{\underset{H\quad H}{|}}{C}}{\qquad}\overset{\overset{\displaystyle O}{\|}}{C}-OH}
\quad\rightleftharpoons\quad
\underset{\text{enol}}{HO-\overset{\overset{\displaystyle O-H}{|}}{C}\underset{\underset{\underset{H}{|}}{C}}{\qquad}\overset{\overset{\displaystyle O}{\|}}{C}-OH}
\qquad (2)
$$

The bromine reacts, not with the diacid form of malonic acid, but with its enol form, through an electrophilic attack on the carbon-carbon double bond. Thus, the rate at which bromine reacts with malonic acid depends not only on the rate of this electrophilic attack but also on the rate of the conversion from the diacid to the enol forms of malonic acid. This conversion is much slower than the electrophilic attack by bromine, and, therefore, the rate of the reaction is determined by the rate of the conversion.

Methylmalonic acid also exists in solution in an equilibrium between the diacid and the enol form (equation 3).

$$
\underset{\text{diacid}}{HO-\overset{\overset{\displaystyle O}{\|}}{C}\underset{\underset{\underset{H\quad CH_3}{|}}{C}}{\qquad}\overset{\overset{\displaystyle O}{\|}}{C}-OH}
\quad\rightleftharpoons\quad
\underset{\text{enol}}{HO-\overset{\overset{\displaystyle O-H}{|}}{C}\underset{\underset{\underset{CH_3}{|}}{C}}{\qquad}\overset{\overset{\displaystyle O}{\|}}{C}-OH}
\qquad (3)
$$

For methylmalonic acid, the electrophilic attack by bromine on the carbon-carbon double bond of the enol form is also much more rapid than the enolization reaction. In fact, the enolization reaction for methylmalonic acid is much slower than the enolization of malonic acid. The slowness of the enolization is apparent in the length of time required for the amber color of bromine to disappear after solutions A and B are mixed in preparation for the demonstration. The color of bromine disappears as the bromine reacts with methylmalonic acid. The slowness of the enolization also lengthens the time required for the completion of one part of an oscillation (Process A in the introduction to this chapter) over the time required when malonic acid is used. Therefore, the period of the oscillations is much longer with methylmalonic acid than with malonic acid. Because the period is longer, the oscillations continue for a longer time, up to several hours. Therefore, this demonstration is suitable for use as a protracted display.

Warning About Chloride Ion Contamination

Even small amounts of chloride ions in the solution can interfere with the mechanism of the BZ reaction, inhibiting the oscillations [1]. Therefore, the vessels used for the preparation of the solutions must be clean. In the preparation of the ferroin solution, 1,10-phenanthroline must be used in its free form, not in the hydrochloride salt, in order to prevent the introduction of chloride ions into the solution. If the oscillations in the electrical potential of the solution are to be observed, a reference electrode that does not leak chloride ions must be used. An ordinary standard calomel electrode or silver–silver chloride electrode is *not* suitable. A double-junction version of one of these electrodes is adequate.

REFERENCE

1. S. S. Jacobs and I. R. Epstein, *J. Am. Chem. Soc.* 98:1721 (1976).

7.4

Cerium-catalyzed Bromate-Ethylacetoacetate Reaction

(A Modified Belousov-Zhabotinsky Reaction)

A clear colorless solution and a light yellow solution are mixed. Then a small amount of a red-orange solution is added, producing a blue mixture. A yellow solution is added to this blue mixture, resulting in a green solution which returns to blue in about 20 seconds. After several oscillations between blue and green, the color of the solution will start to oscillate from green, to blue, to violet, to red, and back to green, with an initial period of about 20 seconds. These oscillations will continue for about 20 minutes, by which time the period will have lengthened to about 60 seconds. The electrical potential of the solution oscillates along with the color, and these oscillations cover a range of about 180 mV.

MATERIALS

19 g potassium bromate, $KBrO_3$, or 17 g sodium bromate, $NaBrO_3$

1.0 liter 1.5M sulfuric acid, H_2SO_4 (To prepare 1.0 liter of solution, slowly and carefully pour 83 mL of concentrated [18M] H_2SO_4 into 500 mL of distilled water and dilute the mixture to 1.0 liter with distilled water.)

11 mL ethylacetoacetate, $CH_3COCH_2CO_2CH_2CH_3$

500 mL distilled water

4.5 g cerium(IV) ammonium nitrate, $Ce(NH_4)_2(NO_3)_6$

30 mL 0.50% ferroin solution (To prepare 100 mL of stock solution, dissolve 0.23 g of iron(II) sulfate heptahydrate, $FeSO_4 \cdot 7H_2O$, in 100 mL of distilled water. In the resulting solution, dissolve 0.46 g of 1,10-phenanthroline, $C_{12}H_8N_2$.)

100 mL 2M potassium hydroxide, KOH, in ethanol (See Disposal section for use. To prepare this solution, dissolve 11 g KOH in 100 mL of 95% ethanol.)

3 500-mL Erlenmeyer flasks

2-liter beaker

magnetic stirrer, with 2-inch stirring bar

platinum electrode (optional)

double-junction reference electrode † (optional)

strip-chart recorder (optional)

---

† The reference electrode must not leak chloride ions into the solution, because chloride ions interfere with the mechanism of the oscillating reaction and inhibit the oscillations.

PROCEDURE

Preparation

Solution A. In a 500-mL flask, dissolve 19 g of potassium bromate in 500 mL of 1.5M sulfuric acid. This solution is 0.23M in BrO_3^- and 1.5M in H_2SO_4.

Solution B. This solution should be prepared within 24 hours of use to minimize hydrolysis of the ethylacetoacetate. In the second 500-mL flask, dissolve 11 mL of ethyl-acetoacetate in 500 mL of distilled water. This solution is 0.17M in ethylacetoacetate.

Solution C. In the third 500-mL flask, dissolve 4.5 g of cerium(IV) ammonium nitrate in 500 mL of 1.5M sulfuric acid. This solution is 0.16M in Ce^{4+} and 1.5M in H_2SO_4.

Set the 2-liter beaker on the magnetic stirrer, and place the stirring bar in the beaker.

Connect the platinum electrode to the positive terminal of the chart recorder and the reference electrode to the negative terminal.

Presentation

Pour solution A and solution B into the 2-liter beaker. Adjust the magnetic stirrer to produce a vortex in the solution. Add 30 mL of ferroin solution; the mixture will turn blue. Then, add solution C. (The nominal composition of this mixture is 0.077M BrO_3^-, 0.057M ethylacetoacetate, 0.053M Ce^{4+}, 1.0M H_2SO_4, and 1.7×10^{-4}M ferroin.) The mixture will turn green and, within 1 minute, return to blue. After several oscillations between blue and green, the color of the solution will start to oscillate from green, to blue, to violet, to red, and back to green, with an initial period of about 20 seconds. During this time, the solution will become turbid. The oscillations will con-tinue for about 20 minutes, by which time the period will have lengthened to about 60 seconds.

Insert the electrodes into the solution soon after solution C has been added. Adjust the recorder to produce the optimum display of the oscillations in the electrical poten-tial of the reaction mixture. The range of these potential oscillations is about 180 mV.

HAZARDS

Bromates are strong oxidizing agents. Mixtures of bromates with finely divided organic materials, metals, carbon, or other combustible materials are easily ignited, sometimes explosively. Ingestion of potassium bromate can cause vomiting, diarrhea, and renal injury.

Ethylacetoacetate is moderately irritating to skin and mucous membranes.

Because sulfuric acid is a strong acid and a powerful dehydrating agent, it can cause burns. Spills should be neutralized with an appropriate agent, such as sodium bicarbonate ($NaHCO_3$), and then rinsed clean.

DISPOSAL

The reaction mixture should be neutralized with sodium bicarbonate ($NaHCO_3$) and flushed down the drain with water. The organic residue on the beaker and stirring bar should be dissolved in a 2M solution of potassium hydroxide in ethanol, and then flushed down the drain with water.

DISCUSSION

The reaction in this demonstration is a derivative of the reaction used in Demonstration 7.2. This demonstration uses ethylacetoacetate in place of malonic acid (structures shown below). The structures of the molecules are similar, in that they both

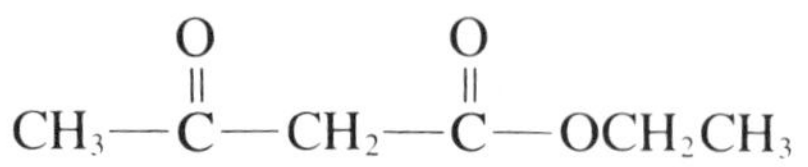

ethylacetoacetate

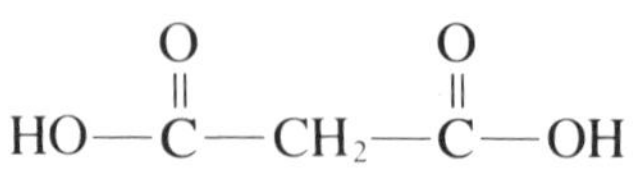

malonic acid

contain a methylene (—CH_2—) group between two carbonyl (—C(=O)—) groups. When malonic acid is oxidized in the Belousov-Zhabotinsky (BZ) reaction of Demonstration 7.2, the products include carbon dioxide and formic acid. The reaction of ethylacetoacetate in this demonstration produces $CH_3COCBr_2CO_2CH_2CH_3$ and nonvolatile oxidation products [*1*]. Because these reaction products are not oxidized further to gaseous carbon dioxide, the ethylacetoacetate substrate is particularly useful for studies in a closed-flow reactor, where the production of gas is undesirable because it changes the volume of liquid in the reactor.

In view of the similarity in the structures of malonic acid and ethylacetoacetate, the mechanism of the reaction in this demonstration is likely to be similar to the mechanism of the reaction with malonic acid, as described in the introduction to this chapter. Like malonic acid, ethylacetoacetate undergoes enolization in solution, as represented in equation 1.

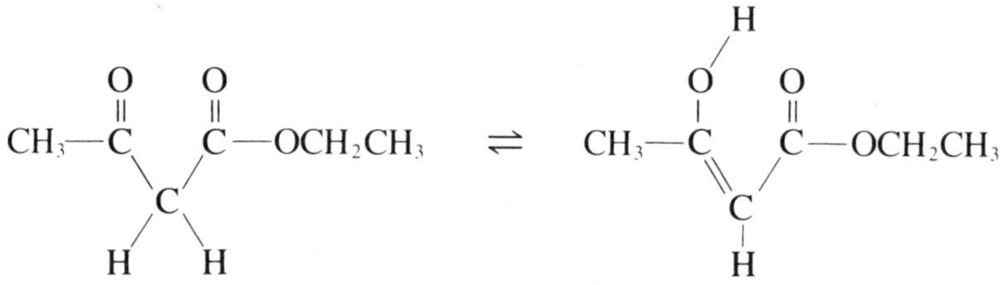

(1)

dicarbonyl enol

The enol form of ethylacetoacetate undergoes electrophilic attack by elemental bromine at the carbon-carbon double bond, as does malonic acid in the reaction of Demonstration 7.2. The enolization reaction is an equilibrium reaction, with most of the ethylacetoacetate in the dicarbonyl form in aqueous solution. The rate of the reaction with bromine is determined by the rate of enolization, which is much slower than the electrophilic attack by bromine.

Ethylacetoacetate as a substrate in BZ reactions has been investigated by a number of researchers [*1–3*]. For more information, these works and the references cited therein can be consulted.

Warning About Chloride Ion Contamination

Even small amounts of chloride ions in the solution can interfere with the mechanism of the BZ reaction, inhibiting the oscillations [4]. Therefore, the vessels used for the preparation of the solutions must be clean. In the preparation of the ferroin solution, 1,10-phenanthroline must be used in its free form, not in the hydrochloride salt, in order to prevent the introduction of chloride ions into the solution. If the oscillations in the electrical potential of the solution are to be observed, a reference electrode that does not leak chloride ions must be used. An ordinary standard calomel electrode or silver–silver chloride electrode is *not* suitable. A double-junction version of one of these electrodes is adequate.

REFERENCES

1. E. J. Heilweil, M. J. Henchman, and I. R. Epstein, *J. Am. Chem. Soc.* 101:3698 (1979).
2. L. Treindl and P. Kaplan, *Chem. zvesti* 35:145 (1981).
3. L. F. Slater and J. G. Sheppard, *Int. J. Chem. Kinet.* 14:815 (1982).
4. S. S. Jacobs and I. R. Epstein, *J. Am. Chem. Soc.* 98:1721 (1976).

7.5

Manganese-catalyzed Bromate-Ethylacetoacetate Reaction

(A Modified Belousov-Zhabotinsky Reaction)

A colorless solution and a pale yellow solution are mixed, and soon after, a small amount of red solution is added. The resulting mixture is blue. To this another colorless solution is added. The color of the mixture begins to oscillate from blue to violet, then to red, and back to blue. The oscillations continue for about 15 minutes, with an initial period of about 30 seconds and a final period of about 40 seconds. The range of the oscillations in the electrical potential of the solution is about 180 mV.

MATERIALS

19 g potassium bromate, $KBrO_3$, or 17 g sodium bromate, $NaBrO_3$

1.0 liter of 1.5M sulfuric acid, H_2SO_4 (To prepare 1.0 liter of solution, slowly and carefully pour 83 mL of concentrated [18M] H_2SO_4 into 500 mL of distilled water, and dilute the solution to 1.0 liter with distilled water.)

11 mL ethylacetoacetate, $CH_3COCH_2CO_2CH_2CH_3$

500 mL distilled water

6.0 g manganese(II) sulfate monohydrate, $MnSO_4 \cdot H_2O$

30 mL 0.5% ferroin solution (To prepare 100 mL of stock solution, dissolve 0.23 g of iron(II) sulfate heptahydrate, $FeSO_4 \cdot 7H_2O$, in 100 mL of distilled water. In this solution, dissolve 0.46 g of 1,10-phenanthroline, $C_{12}H_8N_2$.)

100 mL 2M potassium hydroxide, KOH, in ethanol (See Disposal section for use. To prepare this solution, dissolve 11 g KOH in 100 mL of 95% ethanol.)

platinum electrode (optional)

double-junction reference electrode† (optional)

strip-chart recorder (optional)

3 500-mL Erlenmeyer flasks

2-liter beaker

magnetic stirrer, with 2-inch stirring bar

---

† The reference electrode must not leak chloride ions into the solution, because chloride ions interfere with the mechanism of the oscillation reaction and inhibit the oscillations.

PROCEDURE

Preparation

Solution A. In a 500-mL flask, dissolve 19 g of potassium bromate in 500 mL of 1.5M sulfuric acid. This solution is 0.23M in BrO_3^- and 1.5M in H_2SO_4.

Solution B. This solution should be prepared within 24 hours of use to minimize hydrolysis of the ethylacetoacetate. In the second 500-mL flask, dissolve 11 mL of ethylacetoacetate in 500 mL of distilled water. This solution is 0.17M in ethylacetoacetate.

Solution C. In the third 500-mL flask, dissolve 6.0 g of manganese(II) sulfate monohydrate in 500 mL of 1.5M sulfuric acid. This solution is 0.071M in Mn^{2+} and 1.5M in H_2SO_4.

Connect the platinum electrode to the positive terminal of the recorder and the reference electrode to the negative terminal.

Presentation

Place the 2-liter beaker on the magnetic stirrer, and put the stirring bar in the beaker. Pour solution A and solution B into the beaker, and adjust the stirrer to produce a vortex in the solution. Add 30 mL of ferroin solution to the beaker. This will result in a blue solution in the beaker. Then pour solution C into the beaker. (The nominal composition of this mixture is 0.077M BrO_3^-, 0.057M ethylacetoacetate, 0.024M Mn^{2+}, 1.0M H_2SO_4, and 1.7×10^{-4}M ferroin.) After about 2 minutes, the color of the mixture will begin to oscillate from blue, to violet, to red, and back to blue. The oscillations will continue for about 15 minutes, with an initial period of about 30 seconds and a final period of about 40 seconds.

Insert the electrodes into the solution soon after adding solution C, and adjust the recorder to produce the optimum display of the oscillations in the electrical potential of the reaction mixture. The range of the potential oscillations is about 180 mV.

HAZARDS

Bromates are strong oxidizing agents. Mixtures of bromates with finely divided organic materials, metals, carbon, or other combustible materials are easily ignited, sometimes explosively. Ingestion of potassium bromate can cause vomiting, diarrhea, and renal injury.

Ethylacetoacetate is moderately irritating to skin and mucous membranes.

Because sulfuric acid is a strong acid and a powerful dehydrating agent, it can cause burns. Spills should be neutralized with an appropriate agent, such as sodium bicarbonate ($NaHCO_3$), and then rinsed clean.

DISPOSAL

The reaction mixture should be neutralized with sodium bicarbonate ($NaHCO_3$) and flushed down the drain with water. The organic residue on the beaker and stirring bar should be dissolved in a 2M solution of potassium hydroxide in ethanol, and then flushed down the drain with water.

DISCUSSION

The reaction in this demonstration is very similar to that in Demonstration 7.4, but here manganese ions are used as a catalyst in place of cerium ions. Presumably, the manganese ions function in a fashion similar to that of the cerium ions. As described in the introduction to this chapter, the cerium ions take part in a reaction involving $BrO_2^.$ radicals, as indicated in equation 1.

$$Ce^{3+}(aq) + BrO_2^.(aq) + H^+(aq) \longrightarrow Ce^{4+}(aq) + HBrO_2(aq) \qquad (1)$$

This is a one-electron transfer in which the cerium ions are oxidized. By analogy, the manganese ions can be expected to fill the same role, as illustrated in equation 2.

$$Mn(II)(aq) + BrO_2^.(aq) + H^+(aq) \longrightarrow Mn(III)(aq) + HBrO_2(aq) \qquad (2)$$

However, no Mn(III) has yet been detected in the reaction mixture. The standard reduction potentials of the two couples are similar [1].

$$Ce^{4+} + e^- \longrightarrow Ce^{3+} \qquad E° = 1.44 \text{ volts}$$
$$Mn^{3+} + e^- \longrightarrow Mn^{2+} \qquad E° = 1.51 \text{ volts}$$

Therefore, the two metals can fulfill thermodynamically similar roles. The slight difference in these standard potentials indicates that, at a given potential, the ratio of [Ce(IV)] to [Ce(III)] is greater than that of [Mn(III)] to [Mn(II)]. This suggests that the concentration of Mn(III) may not reach a detectable level in the oscillating reaction mixture, accounting for the failure to observe it in the mixture.

Warning About Chloride Ion Contamination

Even small amounts of chloride ions in the solution can interfere with the mechanism of the BZ reaction, inhibiting the oscillations [2]. Therefore, the vessels used for the preparation of the solutions must be clean. In the preparation of the ferroin solution, 1,10-phenanthroline must be used in its free form, not in the hydrochloride salt, in order to prevent the introduction of chloride ions into the solution. If the oscillations in the electrical potential of the solution are to be observed, a reference electrode that does not leak chloride ions must be used. An ordinary standard calomel electrode or silver–silver chloride electrode is *not* suitable. A double-junction version of one of these electrodes is adequate.

REFERENCES

1. R. C. Weast, Ed., *CRC Handbook of Chemistry and Physics*, 59th ed., CRC Press: Boca Raton, Florida (1978).
2. S. S. Jacobs and I. R. Epstein, *J. Am. Chem. Soc.* 98:1721 (1976).

7.6

Manganese-catalyzed Bromate–Malonic Acid Reaction[†]

(A Modified Belousov-Zhabotinsky Reaction)

Three white solids are added sequentially to a colorless solution being stirred in a beaker. After the third solid is added, the solution turns orange, and after about 75 seconds, it becomes colorless. The color then oscillates between colorless and orange, with an initial period of about 20 seconds. The oscillations will continue for about 10 minutes [1].

MATERIALS

750 mL distilled water

75 mL concentrated (18M) sulfuric acid, H_2SO_4

9 g malonic acid, $CH_2(CO_2H)_2$

8 g potassium bromate, $KBrO_3$

1.8 g manganese(II) sulfate monohydrate, $MnSO_4 \cdot H_2O$

1-liter beaker

magnetic stirrer, with 1-inch stirring bar

PROCEDURE

Preparation

Pour 750 mL of distilled water into the 1-liter beaker. Slowly and carefully pour 75 mL of concentrated (18M) sulfuric acid into the water. The solution will become warm as the acid is added. Allow the solution to cool to room temperature. Place the stirring bar in the beaker and set the beaker on the magnetic stirrer.

Presentation

Turn on the magnetic stirrer and add 9 g of malonic acid to the sulfuric acid solution in the beaker. After the malonic acid has dissolved, add 8 g of potassium bromate. When the potassium bromate has dissolved, add 1.8 g of manganese(II) sulfate mono-

[†] We wish to thank Professor David A. Katz of the Community College of Philadelphia for calling this demonstration to our attention while he was an instructor at the Institute for Chemical Education at the University of Wisconsin–Madison in the summer of 1984.

hydrate. (The nominal composition of the mixture is 0.1M malonic acid, 0.06M BrO_3^-, 0.013M Mn^{2+}, and 1.6M H_2SO_4.) The solution will become orange as the manganese sulfate is added, and after about 75 seconds, it will become colorless. The color of the solution will then oscillate between orange and colorless, with an initial period of about 20 second, lengthening to about 80 seconds after 20 minutes. The oscillations will continue for about another 10 minutes.

HAZARDS

Bromates are strong oxidizing agents. Mixtures of bromates with finely divided organic materials, metals, carbon, or other combustible materials are easily ignited, sometimes explosively. Ingestion of potassium bromate can cause vomiting, diarrhea, and renal injury.

Because sulfuric acid is a strong acid and a powerful dehydrating agent, it can cause burns. Spills should be neutralized with an appropriate agent, such as sodium bicarbonate ($NaHCO_3$), and then rinsed clean.

Malonic acid is a strong irritant to skin, eyes, and mucous membranes.

DISPOSAL

The reaction mixture should be neutralized with sodium bicarbonate ($NaHCO_3$) and flushed down the drain with water.

DISCUSSION

The reaction in this demonstration is very similar to the reaction employed in Demonstration 7.2. However, in this demonstration, manganese ions are used as the catalyst instead of cerium ions, no bromide ions are used in the preparation of the solutions, and the redox indicator, ferroin, is omitted. These modifications change the appearance of the demonstration but do not greatly alter the chemical reactions.

Presumably, the manganese ions function in a fashion similar to that of the cerium ions. As described in the introduction to this chapter, the cerium ions take part in a reaction involving $BrO_2^{\cdot}$ radicals, as indicated in equation 1.

$$Ce^{3+}(aq) + BrO_2^{\cdot} + H^+(aq) \longrightarrow Ce^{4+}(aq) + HBrO_2(aq) \qquad (1)$$

This is a one-electron transfer in which the cerium ions are oxidized. By analogy, the manganese ions can be expected to fill the same role, as illustrated in equation 2.

$$Mn(II)(aq) + BrO_2^{\cdot}(aq) + H^+(aq) \longrightarrow Mn(III)(aq) + HBrO_2(aq) \qquad (2)$$

However, no Mn(III) has yet been detected in the reaction mixture. The standard reduction potentials of the two couples are similar [2].

$$Ce^{4+} + e^- \longrightarrow Ce^{3+} \qquad E° = 1.44 \text{ volts}$$

$$Mn^{3+} + e^- \longrightarrow Mn^{2+} \qquad E° = 1.51 \text{ volts}$$

Therefore, the two metals can fulfill thermodynamically similar roles. The slight difference in these standard potentials indicates that, at a given potential, the ratio of [Ce(IV)] to [Ce(III)] is greater than that of [Mn(III)] to [Mn(II)]. This suggests that the concentration of Mn(III) may not reach a detectable level in the oscillating reaction mixture, accounting for the failure to observe it in the mixture.

No bromide salt is added in the preparation of the solutions used in this demonstration, unlike Demonstration 7.2. However, bromide ions, which are necessary for oscillations, are produced in a series of reactions that occur in the mixture. The BrO_3^- ions react in a series of steps with Mn(II) ions, producing HOBr, as indicated in equation 3 [3].

$$BrO_3^-(aq) + 5 H^+(aq) + 4 Mn(II)(aq) \longrightarrow$$
$$HOBr(aq) + 4 Mn(III)(aq) + 2 H_2O(l) \qquad (3)$$

The HOBr reacts with malonic acid to form bromomalonic acid, as illustrated by equation 4 [4].

$$CH_2(CO_2H)_2(aq) + HOBr(aq) \longrightarrow BrCH(CO_2H)_2(aq) + H_2O(l) \qquad (4)$$

This bromomalonic acid is oxidized by the Mn(III) which was produced in the previous reaction, releasing bromide ions.

$$BrCH(CO_2H)_2(aq) + 4 Mn(III)(aq) + 2 H_2O(l) \longrightarrow$$
$$Br^-(aq) + HCO_2H(aq) + 2 CO_2(g) + 4 Mn(II)(aq) + 5 H^+(aq) \qquad (5)$$

(This reaction is analogous to equation 15 in the introduction to this chapter, in which Ce(IV) is the oxidant.) Thus, bromide ions are introduced into the solution by a series of reactions, and the pathway of the oscillating reaction is similar to that described in the introduction to this chapter.

This demonstration uses no ferroin indicator, so the colors produced by that indicator are not observed. Instead, the oscillatory appearance and disappearance of the orange color of elemental bromine are observed. The bromine is produced by a series of reactions similar to those in Demonstration 7.2.

This demonstration is perhaps one of the easiest oscillating reactions to present. Except for the sulfuric acid solution, all of the ingredients are used as solids; no preparation of solutions is required. Thus, all preparations may be done long in advance and easily transported to the site of the demonstration.

Warning About Chloride Ion Contamination

Even small amounts of chloride ions in the solution can interfere with the mechanism of the BZ reaction, inhibiting the oscillations [5]. Therefore, the vessels used for the preparation of the solutions must be clean.

REFERENCES

1. *My Favorite Lecture Demonstrations*, A Symposium at the Science Teachers Short Course, W. Hutton, Chairman; Iowa State University, Ames, Iowa, March 6–7, 1977.
2. R. C. Weast, Ed., *CRC Handbook of Chemistry and Physics*, 59th ed., CRC Press: Boca Raton, Florida (1978).
3. C. M. Singh, H. C. Mishra, and R. N. Upadhyay, *J. Indian. Chem. Soc.* 57:835 (1979).
4. H. C. Mishra and C. M. Singh, *J. Indian. Chem. Soc.* 55:857 (1978).
5. S. S. Jacobs and I. R. Epstein, *J. Am. Chem. Soc.* 98:1721 (1976).

7.7

Manganese-catalyzed Bromate–2,4-Pentanedione Reaction

(A Modified Belousov-Zhabotinsky Reaction)

A yellow solution and a colorless solution are mixed in a large beaker. Then a red solution is added, resulting in a blue mixture, which fades to violet within 15 seconds. A colorless solution is added to the mixture, and the mixture turns orange. The color of the mixture oscillates between orange and olive for 15–20 minutes. The initial period of the oscillation lasts about 2 minutes, and it decreases to about 30 seconds by the end of the reaction. The range of the oscillations in the electrical potential of the solution is initially 60 mV, decreasing to 30 mV by the end of the reaction.

MATERIALS

18 g potassium bromate, $KBrO_3$, or 16 g sodium bromate, $NaBrO_3$

1.0 liter 1.5M sulfuric acid, H_2SO_4 (To prepare 1.0 liter of stock solution, slowly and carefully add 83 mL of concentrated [18M] H_2SO_4 to 500 mL of distilled water, and dilute the mixture to 1.0 liter with distilled water.)

7.5 g 2,4-pentanedione, $CH_3COCH_2COCH_3$

500 mL distilled water

4.3 g manganese(II) sulfate monohydrate, $MnSO_4 \cdot H_2O$

15 mL 0.5% ferroin solution (To prepare 100 mL of stock solution, dissolve 0.23 g of iron(II) sulfate heptahydrate, $FeSO_4 \cdot 7H_2O$, in 100 mL of distilled water. In this solution, dissolve 0.46 g of 1,10-phenanthroline, $C_{12}H_8N_2$.)

100 mL 2M potassium hydroxide, KOH, in ethanol (See Disposal section for use. To prepare this solution, dissolve 11 g KOH in 100 mL of 95% ethanol.)

3 500-mL Erlenmeyer flasks

platinum electrode (optional)

double-junction reference electrode† (optional)

strip-chart recorder (optional)

2-liter beaker

magnetic stirrer, with 2-inch stirring bar

---

276 † The reference electrode must not leak chloride ions into the solution, because chloride ions interfere with the mechanism of the oscillating reaction and inhibit the oscillations.

PROCEDURE

Preparation

Solution A. In a 500-mL flask, dissolve 18 g of potassium bromate in 500 mL of 1.5M sulfuric acid. This solution is 0.22M in BrO_3^- and 1.5M in H_2SO_4.

Solution B. In the second 500-mL flask, dissolve 7.5 g of 2,4-pentanedione in 500 mL of distilled water. This solution is 0.15M in 2,4-pentanedione.

Solution C. In the third 500-mL flask, dissolve 4.3 g of manganese(II) sulfate monohydrate in 500 mL of 1.5M sulfuric acid. This solution is 0.051M in Mn(II) and 1.5M in H_2SO_4.

Connect the platinum electrode to the positive terminal of the recorder and the reference electrode to the negative terminal.

Presentation

Set the 2-liter beaker on the magnetic stirrer, and place the stirring bar in the beaker. Pour solution A and solution B into the beaker, and adjust the stirrer to produce a vortex in the solution. Add 15 mL of ferroin solution to the colorless solution in the beaker. The mixture will become blue and then, over a span of about 15 seconds, gradually turn red-orange. Then add solution C. (The nominal composition of the mixture is 0.073M BrO_3^-, 0.05M 2,4-pentanedione, 0.017M Mn(II), 1.0M H_2SO_4, and 8.5×10^{-5}M ferroin.) The color of the mixture will become orange and then begin to oscillate from orange, through blue and gray, to olive, and back to orange, with a period of about 1 minute.

Insert the electrodes into the solution soon after adding solution C, and adjust the recorder to produce the optimum display of the oscillations in the electrical potential of the reaction mixture. The initial range of the potential oscillations is about 60 mV, diminishing to 30 mV by the end of the reaction in 15–20 minutes. About 3 minutes after the oscillations in potential of the solution have ceased, they may reappear, although this will not occur in every presentation of this demonstration.

HAZARDS

Bromates are strong oxidizing agents. Mixtures of bromates with finely divided organic materials, metals, carbon, or other combustible materials are easily ignited, sometimes explosively. Ingestion of potassium bromate can cause vomiting, diarrhea, and renal injury.

Liquid 2,4-pentanedione is an irritant to skin and eyes. The vapor can irritate the respiratory system.

Because sulfuric acid is a strong acid and a powerful dehydrating agent, it can cause burns. Spills should be neutralized with an appropriate agent, such as sodium bicarbonate ($NaHCO_3$), and then rinsed clean.

DISPOSAL

Because the products of the reaction in this demonstration are lacrymatory, the disposal procedure should be performed in a hood.

The reaction mixture should be neutralized with sodium bicarbonate ($NaHCO_3$) and flushed down the drain with water. The organic residue on the beaker and stirring bar should be dissolved in a 2M solution of potassium hydroxide in ethanol, and then flushed down the drain with water.

DISCUSSION

The reaction in this demonstration is similar to the reactions employed in Demonstrations 7.2 through 7.6, all of which are in the family of Belousov-Zhabotinsky (BZ) reactions. In this demonstration, manganese ions are used as the catalyst, and 2,4-pentanedione is used as the substrate for oxidation by bromate ions.

The use of 2,4-pentanedione as a substrate in a BZ reaction was first reported by Bowers, et al. [1]. They observed that no gas was evolved in the reaction, unlike the situation in the previous demonstrations, where carbon dioxide is evolved. The 2,4-pentanedione used here is similar to the substrates used in the previous demonstrations, in that in solution it exists in equilibrium with its enol form.

$$CH_3-C(=O)\overset{\underset{|}{C}}{\underset{H\ \ H}{}}C(=O)-CH_3 \rightleftharpoons CH_3-C\overset{\underset{|}{C}}{\underset{H}{}}C(=O)-CH_3 \qquad (1)$$

dicarbonyl   enol

The elemental bromine reacts with the enol form through an electrophilic attack on its carbon-carbon double bond, producing 3-bromo-2,4-pentanedione and bromide ions [2], as illustrated by equation 2.

$$CH_3COCH_2COCH_3(aq) + Br_2(aq) \longrightarrow$$
$$CH_3COCHBrCOCH_3(l) + Br^-(aq) + H^+(aq) \qquad (2)$$

(This equation is analogous to equation 6 in the introduction to this chapter.) The 3-bromo-2,4-pentanedione is not oxidized to carbon dioxide by Mn(III), and the brominated product builds up in the mixture. The oily residue which forms during this reaction is most likely a mixture of 3-bromo-2,4-pentanedione and 3,3-dibromo-2,4-pentanedione [3]. (These brominated waste products are hydrolyzed to acetate in the disposal procedure by digesting the residue in alcoholic potassium hydroxide.) The exact nature of the reaction which returns bromide ions to the solution is still under investigation.

Warning About Chloride Ion Contamination

Even small amounts of chloride ions in the solution can interfere with the mechanism of the BZ reaction, inhibiting the oscillations [4]. Therefore, the vessels used for the preparation of the solutions must be clean. In the preparation of the ferroin solution, 1,10-phenanthroline must be used in its free form, not in the hydrochloride salt, in order to prevent the introduction of chloride ions into the solution. If the oscillations in the electrical potential of the solution are to be observed, a reference electrode that does not leak chloride ions must be used. An ordinary standard calomel electrode or silver–silver chloride electrode is *not* suitable. A double-junction version of one of these electrodes is adequate.

REFERENCES

1. P. G. Bowers, K. E. Caldwell, and D. F. Predergast, *J. Phys. Chem.* 76:2185 (1972).
2. I. Tkac and L. Treindl, *Coll. Czech. Chem. Commun.* 48:13 (1983).
3. E. J. Heilweil, M. J. Henchman, and I. R. Epstein, *J. Am. Chem. Soc.* 101:3698 (1979).
4. S. S. Jacobs and I. R. Epstein, *J. Am. Chem. Soc.* 98:1721 (1976).

7.8

Manganese-catalyzed Bromate–Citric Acid Reaction

(A Modified Belousov-Zhabotinsky Reaction)

A clear solution is combined with a pale yellow solution, producing a deeper yellow solution. To this another colorless solution is added, resulting in a solution which becomes turbid and brown within 15 seconds and then begins to oscillate between brown and tan. The oscillations last for about 5 minutes, with an initial period of 20 seconds, lengthening to 40 seconds near the end of the reaction. After the addition of a white powder, the oscillations reappear. The range of the oscillations in the electrical potential of the solution is approximately 180 mV.

MATERIALS

14.3 g potassium bromate, $KBrO_3$

1.0 liter 4.5M sulfuric acid, H_2SO_4 (To prepare 1.0 liter of solution, slowly and carefully pour 250 mL of concentrated [18M] H_2SO_4 into 500 mL of distilled water, and dilute the resulting solution to 1.0 liter with distilled water.)

29 g citric acid, $HO_2CCH_2C(OH)(CO_2H)CH_2CO_2H$

0.34 g potassium bromide, KBr

500 mL distilled water

2.1 g manganese(II) sulfate monohydrate, $MnSO_4 \cdot H_2O$

100 mL 2M potassium hydroxide, KOH, in ethanol (See Disposal section for use. To prepare this solution, dissolve 11 g KOH in 100 mL of 95% ethanol.)

3 500-mL Erlenmeyer flasks

platinum electrode (optional)

double-junction reference electrode† (optional)

strip-chart recorder (optional)

2-liter beaker (tall form, if possible)

magnetic stirrer, with 2-inch stirring bar

---

† The reference electrode must not leak chloride ions into the solution, because chloride ions interfere with the mechanism of the oscillating reaction and inhibit the oscillations.

PROCEDURE

Preparation

Solution A. In a 500-mL flask, dissolve 11 g of potassium bromate in 500 mL of 4.5M sulfuric acid. This solution is 0.13M in BrO_3^- and 4.5M in H_2SO_4.

Solution B. In another 500-mL flask, dissolve 29 g of citric acid and 0.34 g of potassium bromide in 500 mL of distilled water. This solution is 0.30M in citric acid and 5.7×10^{-3} in Br^-.

Solution C. In the remaining 500-mL flask, dissolve 2.1 g of manganese(II) sulfate monohydrate in 500 mL of 4.5M sulfuric acid. This solution is 0.025M in Mn^{2+} and 4.5M in H_2SO_4.

Connect the platinum electrode to the positive terminal of the strip-chart recorder and the reference electrode to the negative terminal.

Presentation

Set the 2-liter beaker on the magnetic stirrer and place the stirring bar in the beaker. Pour solution A and solution B into the beaker, and adjust the stirrer to produce a vortex in the solution. The solution will become yellow. Add solution C to the mixture in the beaker. (The nominal composition of the mixture is 0.043M BrO_3^-, 0.10M citric acid, 1.9×10^{-3}M Br^-, 0.0083M Mn^{2+}, and 3.0M H_2SO_4.) The solution will become brown, and its color will begin to oscillate between brown and tan. By this time, the mixture will be turbid. The oscillations last for about 5 minutes, with an initial period of 20 seconds which lengthens to about 40 seconds at the end of the reaction.

Insert the two electrodes into the solution soon after solution C is added. Adjust the chart recorder to produce the optimum display of the oscillations in the electrical potential of the solution. The range of the potential oscillations in this solution is about 180 mV.

After the oscillations have stopped, add 3.3 g of solid potassium bromate to the stirred solution. The oscillations in color and electrical potential will reappear for about 4 minutes.

HAZARDS

Bromates are strong oxidizing agents. Mixtures of bromates with finely divided organic materials, metals, carbon, or other combustible materials are easily ignited, sometimes explosively. Ingestion of potassium bromate can cause vomiting, diarrhea, and renal injury.

Because sulfuric acid is a strong acid and a powerful dehydrating agent, it can cause burns. Spills should be neutralized with an appropriate agent, such as sodium bicarbonate ($NaHCO_3$), and then rinsed clean.

DISPOSAL

The reaction mixture should be neutralized with sodium bicarbonate ($NaHCO_3$) and flushed down the drain with water. The organic residue on the beaker and stirring bar should be dissolved in a 2M solution of potassium hydroxide in ethanol, and then flushed down the drain with water.

DISCUSSION

The reaction in this demonstration is similar to that in Demonstration 7.6. However, in this demonstration, citric acid has been substituted for malonic acid. The structure of citric acid follows:

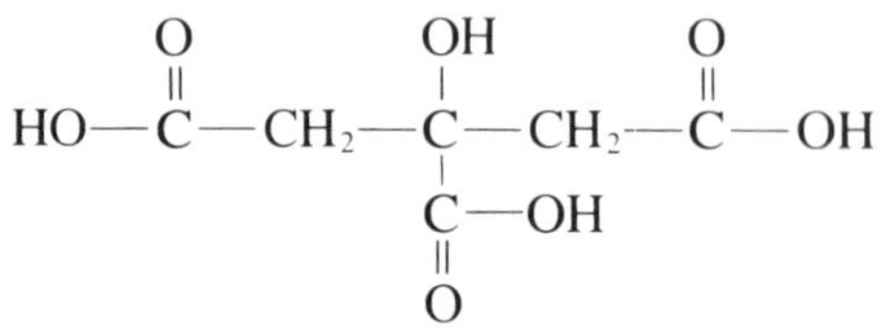

$$HO-\underset{\underset{O}{\|}}{C}-CH_2-\underset{\underset{\underset{\|}{O}}{C-OH}}{\overset{\overset{OH}{|}}{C}}-CH_2-\underset{\underset{O}{\|}}{C}-OH$$

citric acid

In the presence of Ce(IV) ions, citric acid can be oxidized by bromate ions to acetone-dicarboxylic acid [1].

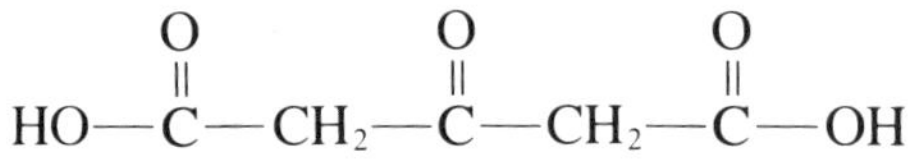

$$HO-\underset{\underset{O}{\|}}{C}-CH_2-\underset{\underset{O}{\|}}{C}-CH_2-\underset{\underset{O}{\|}}{C}-OH$$

acetonedicarboxylic acid

The structure of acetonedicarboxylic acid is similar to that of malonic acid, in that both have —CH$_2$— groups between carbonyl groups.

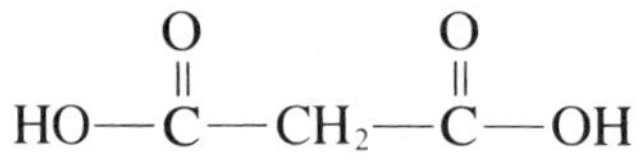

$$HO-\underset{\underset{O}{\|}}{C}-CH_2-\underset{\underset{O}{\|}}{C}-OH$$

malonic acid

Therefore, it is reasonable to assume that the reaction in this demonstration occurs by a mechanism similar to that described in the introduction to this chapter. However, the nature of the final products of the reaction with citric acid has not been resolved, although they may include acetone, pentabromoacetone, carbon dioxide, and formic acid [2–4].

Warning About Chloride Ion Contamination

Even small amounts of chloride ions in the solution can interfere with the mechanism of the BZ reaction, inhibiting the oscillations [5]. Therefore, the vessels used for the preparation of the solutions must be clean. If the oscillations in the electrical potential of the solution are to be observed, a reference electrode that does not leak chloride ions must be used. An ordinary standard calomel electrode or silver–silver chloride electrode is *not* suitable. A double-junction version of one of these electrodes is adequate.

REFERENCES

1. B. P. Belousov, "A Periodic Reaction and Its Mechanism," in *Oscillations and Traveling Waves in Chemical Systems*, ed. R. J. Field and M. Burger, Interscience Publishers, John Wiley and Sons: New York (1985).
2. R. P. Rastogi, P. Rastogi, and R. B. Rai, *Indian J. Chem., Sect. A* 16A: 374 (1978).
3. R. Ramaswarmy and N. Ganapathisubramanian, *Trans. SAEST* 13:70 (1978).
4. H. C. Mishra and C. M. Singh, *J. Indian Chem. Soc.* 55:857 (1978).
5. S. S. Jacobs and I. R. Epstein, *J. Am. Chem. Soc.* 98:1721 (1976).

7.9

Photofluorescent Cerium-catalyzed Bromate–Malonic Acid Reaction

(A Modified Belousov-Zhabotinsky Reaction)

Two clear, colorless solutions, followed by a green solution, are combined in a beaker. The color of the mixture begins to oscillate immediately between green and orange with a period of about 10 seconds. When the room lights are dimmed and a black light is immersed in the solution and turned on, the solution shows oscillations between a bright orange emission and a dark state. The range of the oscillations in the electrical potential of the solution is about 200 mV.

MATERIALS

16 g malonic acid, $CH_2(CO_2H)_2$

0.30 g potassium bromide, KBr, or 0.26 g sodium bromide, NaBr

600 mL distilled water

13 g potassium bromate, $KBrO_3$, or 12 g sodium bromate, $NaBrO_3$

0.045 g tris(2,2′-bipyridyl)ruthenium(II) chloride hexahydrate, $Ru(C_{10}H_8N_2)_3Cl_2 \cdot 6H_2O$

0.38 g cerium(IV) ammonium nitrate, $Ce(NH_4)_2(NO_3)_6$

150 mL 6M sulfuric acid, H_2SO_4 (To prepare 1.0 liter of stock solution, slowly and carefully add 330 mL of concentrated [18M] H_2SO_4 to 500 mL of distilled water, and dilute the mixture to 1.0 liter with distilled water.)

6-watt fluorescent lamp fixture

    or

    transformer for fluorescent lamp, 8 watts, 118 volts, 0.17 ampere

    starter for fluorescent lamp, 6 watts

    socket for starter

    2 sockets for lamp

    4 60-cm lengths of wire, 18 gauge

    18-gauge lamp cord, with plug

    switch, 115 volt, 15 ampere, single pole, single throw

fluorescent black light lamp, 6 watts, 1.5 cm in diameter, 21 cm in length (e.g., Sylvania F6T5/BLB)

magnetic stirrer, with 1-inch stirring bar

ring stand

1-liter, tall-form beaker

2 clamps with holders

test tube, 38 mm × 200 mm

3 500-mL Erlenmeyer flasks

platinum electrode (optional)

standard calomel reference electrode (optional)

strip-chart recorder (optional)

PROCEDURE

Preparation

A specially arranged lamp is used in this demonstration to irradiate the reaction mixture. The components of this lamp may be obtained from a small fluorescent lamp fixture that uses a 6-watt lamp, 21 cm long and 1.5 cm across. Alternatively, the components can be obtained separately. They are an 8-watt, 118-volt, 0.17-amp transformer, such as Robertson L8; a 6-watt starter, such as Sylvania FS-5; a starter socket; two lamp sockets; four 60-cm lengths of 18-gauge wire; an 18-gauge lamp cord and plug; and a 115-volt, 15-amp switch. These components are connected according to the wiring diagram shown in Figure 1.

Assemble the following apparatus as illustrated in Figure 2. Place the magnetic stirrer on the base of the ring stand, and set the 1-liter, tall-form beaker on the stirrer. Put the stirring bar in the beaker. Clamp the test tube onto the stand so that it is concentric with the beaker and the bottom of the tube is just above the stirring bar. Connect

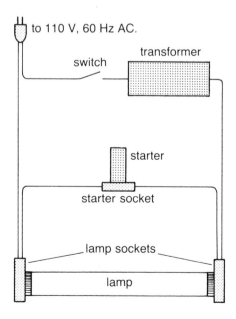

Figure 1. Wiring diagram for a fluorescent black light.

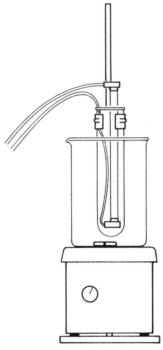

Figure 2.

one of the lamp sockets to one end of the black light, and insert the socket into the test tube so it is at the bottom of the tube. Connect the other socket to the end of the black light extending from the test tube. Using a second clamp, secure the lamp and sockets in place in the test tube. Test the lamp by making sure it lights when the switch is turned on. Leave the lamp turned off.

Solution A. In a 500-mL flask, dissolve 16 g of malonic acid and 0.30 g of potassium bromide in 300 mL of distilled water. This solution is 0.51M in malonic acid and 0.0084M in Br^-.

Solution B. In the second 500-mL flask, dissolve 13 g of potassium bromate in 300 mL of distilled water. This solution is 0.26M in BrO_3^-.

Solution C. In the third 500-mL flask, dissolve 0.045 g of tris(2,2'-bipyridyl)-ruthenium(II) chloride hexahydrate and 0.38 g of cerium(IV) ammonium nitrate in 150 mL of 6M sulfuric acid. This solution is 4.0×10^{-4}M in $Ru(bipy)_3^{2+}$, 4.6×10^{-3}M in Ce(IV), and 6M in H_2SO_4.

Insert the electrodes into the beaker, and connect the platinum electrode to the positive terminal of the recorder and the reference electrode to the negative terminal.

Presentation

Pour solution A into the beaker and start the magnetic stirrer. Add solution B and then solution C to the beaker. (The nominal composition of the mixture is 0.20M malonic acid, 3.4×10^{-3}M Br^-, 0.10M BrO_3^-, 8.0×10^{-5}M $Ru(bipy)_3^{2+}$, 9.2×10^{-4}M Ce(IV), and 1.2M H_2SO_4.) The color of the solution will begin to oscillate almost immediately between orange and green, with a period of about 10 seconds. If a chart recorder is being used, adjust it to give the optimum display of the oscillations in the electrical potential of the solution. The range of the potential oscillations is about 200 mV.

Darken the room lights and turn on the black light. The solution will oscillate between bright orange fluorescence and a dark state. The oscillations will continue for 2–3 hours.

HAZARDS

Bromates are strong oxidizing agents. Mixtures of bromates with finely divided organic materials, metals, carbon, or other combustible materials are easily ignited, sometimes explosively. Ingestion of potassium bromate can cause vomiting, diarrhea, and renal injury.

Because sulfuric acid is a strong acid and a powerful dehydrating agent, it can cause burns. Spills should be neutralized with an appropriate agent, such as sodium bicarbonate ($NaHCO_3$), and then rinsed clean.

Malonic acid is a strong irritant to skin, eyes, and mucous membranes.

The toxicity and carcinogenicity of tris(2,2'-bipyridyl)ruthenium(II) chloride hexahydrate are not known.

DISPOSAL

The reaction mixture should be neutralized with sodium bicarbonate ($NaHCO_3$) and flushed down the drain with water.

DISCUSSION

The reaction used in this demonstration is nearly identical to that used in Demonstration 7.2. The only difference, other than slight differences in the starting concentrations of the reagents, is that tris(2,2'-bipyridyl)ruthenium(II) is used as an indicator in place of ferroin [1]. In Demonstration 7.2, the ferroin indicates the electrical potential of the solution by changing color. When the solution has a high [Ce(IV)] to [Ce(III)] ratio, the cerium(IV) oxidizes the iron(II) in ferroin to iron(III). The oxidized iron complex is blue. When the solution contains a low [Ce(IV)] to [Ce(III)] ratio, then the iron in the complex is reduced to iron(II), and the complex is red. In this demonstration, the ruthenium complex serves a similar indicating function. When the [Ce(IV)] to [Ce(III)] ratio is high, the ruthenium in the complex is oxidized to ruthenium(III). The complex of Ru(III) is green. When the [Ce(IV)] to [Ce(III)] ratio is low, the ruthenium in the complex is reduced to Ru(II). The color of the Ru(II) complex is orange. These complexes of ruthenium are responsible for the colors, green and orange, which oscillate in this demonstration.

The photofluorescence, which oscillates along with the color of the solution, is also due to the ruthenium complex. The complex of Ru(III) is not fluorescent, while the complex of Ru(II) emits orange light when it is irradiated by ultraviolet light, such as that from a black light. The light energy absorbed by the Ru(II) complex is reemitted as orange light. The mechanism of this luminescence is discussed in Demonstration 2.4 of Volume 1 in this series.

REFERENCE

1. E. Körös, M. Burger, V. Friedrich, L. Ladányi, Zs. Nagy, and M. Orbán, *Faraday Symp. Chem. Soc.* 9:28 (1974).

7.10

Oxidation of Pyrogallol by Bromate

A yellow solution, a colorless solution, and a red solution are mixed, producing a deep orange solution. After about 1 minute, the solution suddenly becomes yellow. The color of the solution then oscillates between yellow and orange for about 10 minutes. The period of the oscillations is about 20 seconds initially, lengthening to about 1 minute near the end. The range of the oscillations in the electrical potential of the solution is about 250 mV initially, decreasing to about 75 mV at the end.

MATERIALS

25 g potassium bromate, $KBrO_3$, or 23 g sodium bromate, $NaBrO_3$

1.5 liter 1.8M sulfuric acid, H_2SO_4 (To prepare, slowly and carefully pour 150 mL of concentrated [18M] H_2SO_4 into 750 mL of distilled water, and dilute the mixture to 1.5 liter with distilled water.)

9.8 g pyrogallol (1,2,3-trihydroxybenzene), $C_6H_6O_3$

7.5 mL 0.5% ferroin solution (To prepare 100 mL of stock solution, dissolve 0.23 g of iron(II) sulfate heptahydrate, $FeSO_4 \cdot 7H_2O$, in 100 mL of distilled water. In the resulting solution, dissolve 0.45 g of 1,10-phenanthroline, $C_{12}H_8N_2$.)

100 mL 2M potassium hydroxide, KOH, in ethanol (See Disposal section for use. To prepare this solution, dissolve 11 g KOH in 100 mL of 95% ethanol.)

2 1-liter Erlenmeyer flasks

gloves, plastic or rubber

platinum electrode (optional)

double-junction reference electrode† (optional)

strip-chart recorder (optional)

2-liter beaker

magnetic stirrer, with 2-inch stirring bar

PROCEDURE

Preparation

Solution A. In a 1-liter flask, dissolve 25 g of potassium bromate in 750 mL of 1.8M sulfuric acid. The solution is 0.20M in BrO_3^- and 1.8M in H_2SO_4.

---

† The reference electrode must not leak chloride ions into the solution, because chloride ions interfere with the mechanism of the oscillating reaction and inhibit the oscillations.

Solution B. Wear gloves while preparing and handling this solution, which should be prepared immediately before use. In the other 1-liter flask, dissolve 9.8 g of pyrogallol in 750 mL of 1.8M sulfuric acid. This solution is 0.10M in pyrogallol and 1.8M in H_2SO_4.

Connect the platinum electrode to the positive terminal of the recorder, and the reference electrode to the negative terminal.

Presentation

Wear gloves while presenting this demonstration. Set the 2-liter beaker on the magnetic stirrer, and place the stirring bar in the beaker. Pour solution A into the beaker, and adjust the stirrer so that a vortex forms in the solution. Add solution B and 7.5 mL of ferroin solution to the beaker. (The composition of the mixture is 0.10M BrO_3^-, 0.050M pyrogallol, 1.8M H_2SO_4, and 4.3×10^{-5}M ferroin.) This will produce a deep red, nearly black, solution. In 1–2 minutes, the solution will suddenly turn yellow and then return to deep red. The color of the solution will oscillate between orange and red-orange for about 10 minutes, and the period of the oscillations will increase from about 20 seconds initially to about 1 minute at the end.

Insert the electrodes into the solution in the beaker shortly after the addition of the ferroin solution. Adjust the recorder to produce the optimum display of the oscillations in the electrical potential of the solution. The range of the potential oscillations is about 250 mV initially, decreasing to about 75 mV near the end of the reaction.

HAZARDS

Bromates are strong oxidizing agents. Mixtures of bromates with finely divided organic materials, metals, carbon, or other combustible materials are easily ignited, sometimes explosively. Ingestion of potassium bromate can cause vomiting, diarrhea, and renal injury.

Pyrogallol (pyrogallic acid; 1,2,3-trihydroxybenzene) is highly toxic and can be absorbed through the skin. Skin contact can cause severe irritation. Inhalation of the powder can cause acute poisoning with symptoms of cyanosis, anemia, convulsions, vomiting, and liver and kidney damage. Ingestion can be fatal.

Because sulfuric acid is a strong acid and a powerful dehydrating agent, it can cause burns. Spills should be neutralized with an appropriate agent, such as sodium bicarbonate ($NaHCO_3$), and then rinsed clean.

DISPOSAL

The reaction mixture should be neutralized with sodium bicarbonate ($NaHCO_3$) and flushed down the drain with water. The organic residue on the beaker and stirring bar should be dissolved in a 2M solution of potassium hydroxide in ethanol, and then flushed down the drain with water.

DISCUSSION

The reaction in this demonstration, the oxidation of pyrogallol by bromate ions, is an example of an uncatalyzed oscillator controlled by the concentration of bromide ions [1, 2]. The organic substrate in this demonstration is pyrogallol (1,2,3-trihydroxybenzene).

pyrogallol

The oxidation of pyrogallol by bromate ions can proceed by either of two processes which are similar to the processes involved in the bromate oxidation of malonic acid in the standard Belousov-Zhabotinsky (BZ) reaction, as described in the introduction to this chapter. During the initial induction period of this reaction, the bromate ions react with pyrogallol, liberating bromide ions. When the bromide ion concentration reaches a sufficiently high level, Process A occurs.

PROCESS A

$$BrO_3^- + Br^- + 2 H^+ \longrightarrow HBrO_2 + HBrO \qquad (1)$$

$$HBrO_2 + Br^- + H^+ \longrightarrow 2 HBrO \qquad (2)$$

$$HBrO + Br^- + H^+ \longrightarrow Br_2 + H_2O \qquad (3)$$

These steps are identical to those of Process A of the BZ reaction described in the introduction to this chapter. The bromine produced in the third step reacts with pyrogallol, liberating bromide ions, as indicated in equation 4.

$$Br_2 + C_6H_3(OH)_3 \longrightarrow BrC_6H_2(OH)_3 + H^+ + Br^- \qquad (4)$$

The three adjacent hydroxy groups on the benzene ring in pyrogallol activate the ring toward electrophilic attack by bromine, whereby hydrogen on the benzene ring is replaced with bromine. (Actually, two of the hydrogen atoms—the two on either side of the three hydroxy groups—can be readily replaced by bromine atoms [3]. One of the effects of Process A is a reduction in the concentration of bromide ion. As the concentration of bromide ion decreases, the rate of Process A diminishes, and Process B takes over.

PROCESS B

$$BrO_3^- + HBrO_2 + H^+ \longrightarrow 2 BrO_2^\bullet + H_2O \qquad (5)$$

$$BrO_2^\bullet + C_6H_3(OH)_3 \longrightarrow HBrO_2 + C_6H_3(OH)_2O^\bullet \qquad (6)$$

$$2 HBrO_2 \longrightarrow HOBr + BrO_3^- + H^+ \qquad (7)$$

These steps are similar to steps in Process B, as described in the introduction to this chapter. Here, in equation 6, rather than reacting with a metal-ion catalyst, the $BrO_2^\bullet$ radicals oxidize the pyrogallol to a radical such as

These pyrogallol radicals can react with another BrO$_2^-$ radical to form a quinone such as

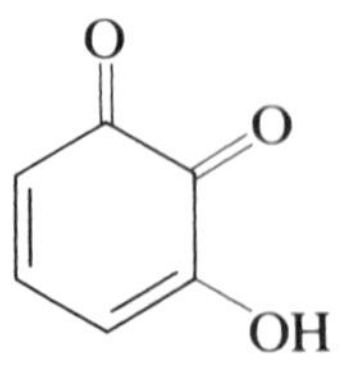

They can also react with HOBr to produce a bromine atom, which is rapidly reduced to a bromide ion, as represented in the following equations:

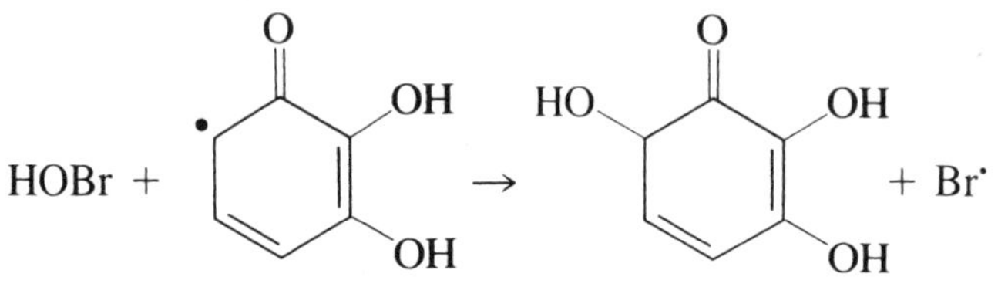

Therefore, Process B indirectly increases the concentration of bromide ions, so that eventually the rate of Process A will increase to the point where it dominates again. This switching back and forth between Processes A and B is what leads to the oscillations in the color of the solution.

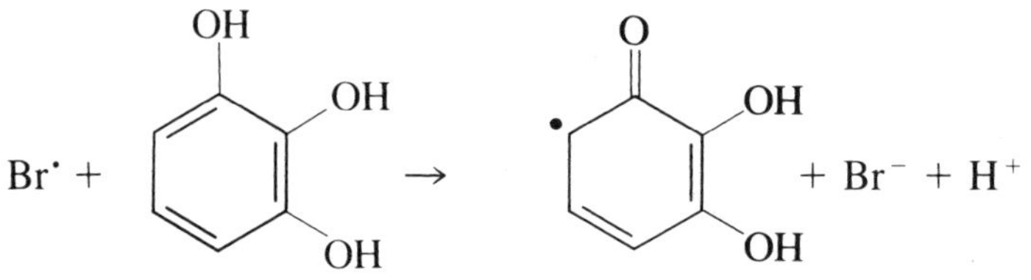

The ferroin (tris(1,10-phenanthroline)iron(II) sulfate) in this demonstration is used as a redox indicator, rendering the oscillations between Process A and Process B visible through oscillations in the color of the solution. Because ferroin is a one-electron reductant, it can enter into the mechanism of the reaction, and it appears to do so, because the period of the oscillations is different when ferroin is present from when it is absent [4].

Warning About Chloride Ion Contamination

Even small amounts of chloride ions in the solution can interfere with the mechanism of the BZ reaction, inhibiting the oscillations [5]. Therefore, the vessels used for the preparation of the solutions must be clean. In the preparation of the ferroin solution, 1,10-phenanthroline must be used in its free form, not in the hydrochloride salt, in order to prevent the introduction of chloride ions into the solution. If the oscillations in the electrical potential of the solution are to be observed, a reference electrode that does not leak chloride ions must be used. An ordinary standard calomel electrode or silver–silver chloride electrode is *not* suitable. A double-junction version of one of these electrodes is adequate.

REFERENCES

1. J. S. Babu and K. Srinivasulu, *Z. phys. Chem.* (Leipzig) 259:1191 (1978).
2. E. Körös and M. Orbán, *Nature* 273:371 (1978).
3. K. Friedrich and H. Mirbach, *Chem. Ber.* 92:2574 (1959).
4. E. Körös, M. Orbán, and I. Habon, *J. Phys. Chem.* 84:559 (1980).
5. S. S. Jacobs and I. R. Epstein, *J. Am. Chem. Soc.* 98:1721 (1976).

7.11

Oxidation of Tannic Acid by Bromate

A white solid is added to water being stirred in a beaker. Once this solid has dissolved, a small amount of a colorless solution is added, followed by a larger amount of another colorless solution. The mixture turns purple and cloudy. Shortly thereafter, the mixture becomes yellow. At this time a small amount of red solution is poured into the beaker, and the mixture turns green. Within a minute, the solution turns orange. The color of the solution then oscillates between green and orange for about 10 minutes, with a period of about 6 seconds initially, increasing slightly during the course of the reaction. The range of the oscillations in the electrical potential of the mixture is about 200 mV initially, and it decreases throughout the reaction.

MATERIALS

21 g potassium bromate, $KBrO_3$, or 19 g sodium bromate, $NaBrO_3$

750 mL 4.0M sulfuric acid, H_2SO_4 (To prepare, slowly and carefully pour 165 mL of concentrated [18M] H_2SO_4 into 500 mL of distilled water, and dilute the mixture to 750 mL with distilled water.)

0.69 g potassium bromide, KBr, or 0.60 g sodium bromide, NaBr

760 mL distilled water

9.9 g tannic acid, average molecular formula: $C_{76}H_{52}O_{46}$

20 mL 0.5% ferroin solution (To prepare 100 mL of stock solution, dissolve 0.23 g of iron(II) sulfate heptahydrate, $FeSO_4 \cdot 7H_2O$, in 100 mL of distilled water. In the resulting solution, dissolve 0.46 g of 1,10-phenanthroline, $C_{12}H_8N_2$.)

1-liter Erlenmeyer flask

50-mL beaker

platinum electrode (optional)

strip-chart recorder (optional)

double-junction reference electrode † (optional)

2-liter beaker

magnetic stirrer, with 2-inch stirring bar

---

† The reference electrode must not leak chloride ions into the solution, because chloride ions interfere with the mechanism of the oscillating reaction and inhibit the oscillations.

PROCEDURE

Preparation

In the 1-liter flask, dissolve 21 g of potassium bromate in 750 mL of 4.0M sulfuric acid. This solution is 0.17M in BrO_3^- and 4.0M in H_2SO_4.

In the 50-mL beaker, dissolve 0.69 g of potassium bromide in 10 mL of distilled water. This solution is 0.60M in Br^-.

Connect the platinum electrode to the positive terminal of the recorder and the reference electrode to the negative terminal.

Presentation

Set the 2-liter beaker on the magnetic stirrer and place the stirring bar in the beaker. Pour 750 mL of distilled water into the beaker. Adjust the stirrer to produce a vortex in the water. Add 9.9 g of tannic acid to the water. After the tannic acid has dissolved, pour the potassium bromide solution into the tannic acid solution. Then, pour the solution of potassium bromate into the beaker. (The nominal composition of the solution is 3.9×10^{-3}M tannic acid, 0.085M BrO_3^-, 4.0×10^{-3}M Br^-, 2.0M H_2SO_4, and 1.1×10^{-4}M ferroin.) The solution will become purple, and a precipitate will form. Within 1 minute, the mixture will become yellow but remain turbid. Once the mixture turns yellow, add 20 mL of ferroin solution. The mixture will turn green, and after about 1 minute, it will become orange. The color of the solution will oscillate between green and orange for about 10 minutes, with a period of about 6 seconds initially, increasing slightly during the course of the reaction.

Insert the electrodes into the solution in the beaker shortly after the addition of the ferroin solution. Adjust the recorder to produce the optimum display of the oscillations in the electrical potential of the solution. The range of the potential oscillations is about 200 mV initially, and it decreases throughout the course of the reaction.

HAZARDS

Bromates are strong oxidizing agents. Mixtures of bromates with finely divided organic materials, metals, carbon, or other combustible materials are easily ignited, sometimes explosively. Ingestion of potassium bromate can cause vomiting, diarrhea, and renal injury.

Because sulfuric acid is a strong acid and a powerful dehydrating agent, it can cause burns. Spills should be neutralized with an appropriate agent, such as sodium bicarbonate ($NaHCO_3$), and then rinsed clean.

DISPOSAL

The reaction mixture should be neutralized with sodium bicarbonate ($NaHCO_3$) and flushed down the drain with water.

DISCUSSION

The reaction used in this demonstration is very similar to that used in Demonstration 7.10. That demonstration uses pyrogallol (1,2,3-trihydroxybenzene) as the oxidizable organic substrate, while this demonstration uses tannic acid instead. Tannic acid is not a single pure substance but a mixture of substances of vegetable origin [1]. The elemental composition of tannic acid is consistent with a formula of $C_{76}H_{52}O_{46}$. Tannic acid contains esters of gallic acid (3,4,5-trihydroxybenzoic acid) and glucose, and the structure of a typical ester follows:

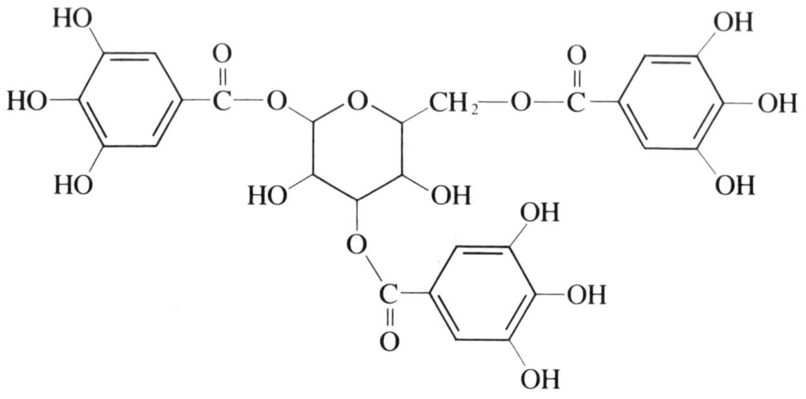

The gallic acid portions of this ester have a structure similar to that of pyrogallol, specifically, the three adjacent hydroxy groups bonded to a benzene ring. These three hydroxy groups activate the benzene ring toward electrophilic substitution at the two adjacent positions. Thus, the tannic acid used in this demonstration can function in a fashion quite similar to the pyrogallol used in Demonstration 7.10.

Warning About Chloride Ion Contamination

Even small amounts of chloride ions in the solution can interfere with the mechanism of this reaction, inhibiting the oscillations [2]. Therefore, the vessels used for the preparation of the solutions must be clean. In the preparation of the ferroin solution, 1,10-phenanthroline must be used in its free form, not in the hydrochloride salt, in order to prevent the introduction of chloride ions into the solution. If the oscillations in the electrical potential of the solution are to be observed, a reference electrode that does not leak chloride ions must be used. An ordinary standard calomel electrode or silver–silver chloride electrode is *not* suitable. A double-junction version of one of these electrodes is adequate.

REFERENCES

1. M. Windholz, Ed., *The Merck Index*, 9th ed., Merck and Co., Rahway, New Jersey (1976).
2. S. S. Jacobs and I. R. Epstein, *J. Am. Chem. Soc.* 98:1721 (1976).

7.12

Travelling Waves of Color

A blue dot spontaneously appears in a red-orange solution thinly distributed over the bottom of a Petri dish. As the dot grows larger, a red-orange dot appears at the center of the blue region, producing a blue ring in the solution. As the red-orange dot grows larger, a blue dot reappears at its center, producing a red-orange ring. In this manner, successive concentric rings appear. More than one set of concentric rings will develop in the solution, and where the rings of different sets meet, they annihilate each other. Bubbles of carbon dioxide gas appear in the solution and gradually grow and escape.

MATERIALS

6.0 mL 0.50M potassium bromate, $KBrO_3$ (To prepare 1.0 liter of stock solution, dissolve 84 g of $KBrO_3$ in 750 mL of distilled water and dilute the mixture to 1.0 liter with distilled water.)

0.60 mL 6.0M sulfuric acid, H_2SO_4 (To prepare 1.0 liter of stock solution, carefully pour 330 mL of concentrated [18M] H_2SO_4 into 500 mL of distilled water and dilute the solution to 1.0 liter with distilled water.)

1.0 mL 0.50M potassium bromide, KBr (To prepare 1.0 liter of stock solution, dissolve 59 g of KBr in 750 mL of distilled water and dilute the solution to 1.0 liter with distilled water.)

2.5 mL 0.50M malonic acid, $CH_2(CO_2H)_2$ (To prepare 1.0 liter of stock solution, dissolve 52 g of $CH_2(CO_2H)_2$ in 750 mL of distilled water and dilute the solution to 1.0 liter with distilled water.)

1.0 mL 0.025M ferroin solution (To prepare 1.0 liter of stock solution, dissolve 6.9 g of iron(II) sulfate heptahydrate, $FeSO_4 \cdot 7H_2O$, in 750 mL of distilled water. In the resulting solution, dissolve 13.5 g of 1,10-phenanthroline and dilute the solution to 1.0 liter with distilled water.)

10-mL graduated cylinder

50-mL beaker

5-mL Mohr pipette

pipette bulb

stirring rod

Petri dish

white card, 5 cm × 5 cm, or overhead projector and screen

PROCEDURE

Preparation

These preparations should be done about 10 minutes before presenting the demonstration. Using the 10-mL graduated cylinder, measure 6.0 mL of 0.50M potassium bromate and pour it into the 50-mL beaker. Using a 5-mL graduated pipette, add 0.60 mL of 6.0M sulfuric acid to the solution in the beaker. Rinse the pipette and use it to add 1.0 mL of 0.50 M potassium bromide to the beaker. Rinse the pipette again, and use it to add 2.5 mL of 0.50M malonic acid. Stir the mixture until the amber color disappears. Rinse the pipette and add 1.0 mL of 0.025M ferroin. The solution will turn blue, then red-orange. Pour the solution into the Petri dish and swirl the dish to coat the bottom of it uniformly. (The nominal composition of this solution is 0.27M BrO_3^-, 0.045M Br^-, 0.018M malonic acid, 0.32M H_2SO_4, and 2.2×10^{-3}M ferroin.)

Presentation

For observation by a small group, place the Petri dish on the white card. For observation by a large group, place the Petri dish on an overhead projector and adjust the projector so that the solution is in focus on the screen. Within several minutes, one or more blue dots will appear in the solution. These dots will grow slowly, and after a time, the center of the blue dot will turn red-orange, forming a blue ring. The blue ring will expand, its red-orange interior will grow, and another blue dot will appear at the center. By alternately turning blue, then red-orange, the center will produce a series of concentric rings which gradually advance through the solution. If the blue rings from two different sets should meet in the solution, they will annihilate each other. When several sets of rings develop at different centers in the dish, they will eventually interact to form a complex pattern of blue and orange bands. The dish may be swirled to mix the solution and return it to a red-orange color throughout. The blue dots will reappear and generate a new pattern.

After some time, bubbles of carbon dioxide will appear in the solution, making it difficult to observe the patterns. Eventually the solution will become completely blue, and no more patterns will develop.

HAZARDS

Bromates are strong oxidizing agents. Mixtures of bromates with finely divided organic materials, metals, carbon, or other combustible materials are easily ignited, sometimes explosively. Ingestion of potassium bromate can cause vomiting, diarrhea, and renal injury.

Because sulfuric acid is a strong acid and a powerful dehydrating agent, it can cause burns. Spills should be neutralized with an appropriate agent, such as sodium bicarbonate ($NaHCO_3$), and then rinsed clean.

Malonic acid is a strong irritant to skin, eyes, and mucous membranes.

DISPOSAL

The reaction mixture should be neutralized with sodium bicarbonate ($NaHCO_3$) and flushed down the drain with water.

DISCUSSION

Oscillations in the concentrations of the components of a stirred solution of potassium bromate, cerium sulfate, and citric acid in aqueous sulfuric acid were first reported by Belousov. These were temporal oscillations that occurred throughout the solution and produced an oscillation in the color of the solution. Other organic compounds, such as malonic acid, could be substituted for citric acid, and oscillations still occurred. A redox indicator, such as ferroin, made the oscillations in color more dramatic. Cerium sulfate could be replaced by a number of other catalysts, such as ferroin itself. Concentrations could be varied somewhat, producing changes in the period of the oscillations, the lifetime of the oscillations, and in the time required before oscillations began. But when the solution was not stirred, the temporal oscillations did not occur. Instead, Zaikin and Zhabotinsky reported spatial oscillations—rings of high [Ce(IV)] moving through regions of high [Ce(III)] [1]. This is such a striking phenomenon that it was soon developed into an effective demonstration [2–5]. Showalter, Noyes, and Turner were able to initiate rings deliberately by pulsing the potential of an electrode in the solution, and they discovered a number of interesting phenomena associated with the initiation and propagation of successive rings [6].

The reactions that occur in the generation of the moving waves are similar to those that produce the temporal oscillations in a stirred solution. (These reactions are described in the introduction to this chapter.) There are two processes, one which occurs at high [Br$^-$] and consumes Br$^-$, and a second which occurs at low [Br$^-$] and produces Br$^-$. In a stirred solution, these two processes temporally alternate with each other. In a quiescent solution, these two processes can alternate spatially.

A description of how the spatial waves occur has been provided by Rovinsky and Zhabotinsky [7]. When the reactants are mixed initially, the composition of the solution is uniform, and the iron in the ferroin complex is oxidized by bromate to Fe(III), making the solution blue. The oxidation takes place in a sequence of steps whose net result is represented by equation 1.

$$2\ Fe^{2+}(aq) + BrO_3^-(aq) + HBrO_2(aq) + 3\ H^+(aq) \longrightarrow$$
$$2\ Fe^{3+}(aq) + 2\ HBrO_2(aq) + H_2O(l) \qquad (1)$$

The oxidized Fe(III) is reduced to Fe(II) by a reaction with malonic acid (equation 2), and the reduced ferroin turns the solution red.

$$6\ Fe^{3+}(aq) + CH_2(CO_2H)_2(aq) + 2\ H_2O(l) \longrightarrow$$
$$6\ Fe^{2+}(aq) + HCO_2H(aq) + 2\ CO_2(g) + 6\ H^+(aq) \qquad (2)$$

In a series of steps, the net result of which is represented in equation 3, hypobromous acid reacts with malonic acid, producing bromomalonic acid.

$$2\ HBrO_2(aq) + CH_2(CO_2H)_2(aq) \longrightarrow$$
$$BrCH(CO_2H)_2(aq) + BrO_3^-(aq) + H^+(aq) + H_2O(l) \qquad (3)$$

Because of surface imperfections in the container, or dust, or trapped gas bubbles, the rates of the reaction can vary from one spot to another in the container. This destroys the uniformity of the solution. Eventually, at one or more spots, the concentration of bromate will become high and oxidize the ferroin, turning the solution blue in these spots. In the red area around the blue spots, the concentration of bromide ion is high, and it diffuses into the blue area. Bromide ions react with bromous acid according to equation 4.

$$Br^-(aq) + HBrO_2(aq) + H^+(aq) \longrightarrow 2\ HBrO(aq) \qquad (4)$$

However, as equation 1 indicates, the $HBrO_2$ is produced autocatalytically; therefore, it moves out from the blue areas faster than the Br^- diffuses in. This causes the $[Br^-]$ to decrease around the blue spot, which allows more Fe(III) to be produced, because Br^- is no longer competing with Fe(II) for $HBrO_2$. As [Fe(III)] increases, it diffuses outward from the blue spot, causing the surrounding solution to turn blue, enlarging the blue spot.

Meanwhile, at the center of the blue spot, the Fe(III) reacts with brominated malonic acid, as represented in equation 5.

$$4\ Fe^{3+}(aq) + BrCH(CO_2H)_2(aq) + 2\ H_2O(l) \longrightarrow$$
$$Br^-(aq) + 4\ Fe^{2+}(aq) + HCO_2H(aq) + 2\ CO_2(g) + 5\ H^+(aq) \qquad (5)$$

This reaction produces Br^- at the center of the spot, and this Br^-, through the reaction of equation 4, destroys $HBrO_2$. As $HBrO_2$ is destroyed, the reaction producing Fe(III) (equation 1) slows, and the Fe(III) is destroyed, through equation 5, more rapidly than it is formed. Thus, most of the iron at the center of the blue area becomes reduced, and the center turns red. This produces in effect a blue ring in the solution.

The blue area will continue to expand into the red area, causing the blue ring to increase in size. At the center of the ring, the solution undergoes temporal oscillations, alternately turning blue, then red. Each blue ring moves outward from the center, and this process eventually produces a series of blue concentric rings in a red solution. In order to produce easily visible rings of color, the solution must be in a shallow layer. If the solution is deep, the added third dimension will obscure the bands.

Warning About Chloride Ion Contamination

Even small traces of chloride ions in the solution will interfere with the mechanism of the BZ reaction, inhibiting the formation of rings [8]. Therefore, the vessels used for the preparation of the solutions must be clean. Even the chloride ions in fingerprints in the Petri dish can have deleterious effects on the oscillations. In the preparation of the ferroin solution, 1,10-phenanthroline must be used in its free form, not in the hydrochloride salt, in order to prevent the introduction of chloride ions into the solution.

REFERENCES

1. A. N. Zaikin and A. M. Zhabotinsky, *Nature* (London) 225:535 (1935).
2. A. T. Winfree, *Science* 175:634 (1972).
3. A. T. Winfree, *Sci. Am.* 230(6):82 (1974).
4. J. Walker, *Sci. Am.* 239(1):152 (1978).
5. J. Walker, *Chemtech*: 320 (1980).
6. K. Showalter, R. M. Noyes, and H. Turner, *J. Am. Chem. Soc.* 101:7463 (1979).
7. A. B. Rovinsky and A. M. Zhabotinsky, *J. Phys. Chem.* 88:6081 (1984).
8. S. S. Jacobs and I. R. Epstein, *J. Am. Chem. Soc.* 98:1721 (1976).

7.13

Nitrogen Gas Evolution Oscillator[†]

Two colorless solutions are mixed in a small beaker. Effervescence occurs almost immediately and then disappears and reappears, with a period of about 8 seconds. The oscillations in effervescence continue for about 4 minutes [1].

MATERIALS

26 g ammonium sulfate, $(NH_4)_2SO_4$

100 mL 0.2M sulfuric acid, H_2SO_4 (To prepare 1.0 liter of stock solution, carefully pour 11 mL of concentrated [18M] H_2SO_4 into 500 mL of distilled water, and dilute the resulting solution to 1.0 liter with distilled water.)

28 g sodium nitrite, $NaNO_2$

100 mL distilled water

2 250-mL Erlenmeyer flasks

2 rubber stoppers to fit Erlenmeyer flasks

100-mL beaker

magnetic stirrer, with 1-inch stirring bar

overhead projector, with vertical-mounting adapter (optional)

special test tube holder for overhead projection (optional, see Procedure)

15-cm test tube (optional)

PROCEDURE

Preparation

The following solutions may be prepared as much as 2 weeks in advance:

Solution A. In one of the 250-mL Erlenmeyer flasks, dissolve 26 g of $(NH_4)_2SO_4$ in 100 mL of 0.2M H_2SO_4. Stopper the flask.

Solution B. In the other 250-mL Erlenmeyer flask, dissolve 28 g of $NaNO_2$ in 100 mL of distilled water. Stopper the flask.

Place the magnetic stirring bar in the 100 mL beaker, and place the beaker on the magnetic stirrer.

If the demonstration is to be displayed by overhead projection, construct a test

† We wish to thank Professor Richard E. Noyes and co-workers of the University of Oregon for providing us with a prepublication copy of their article describing this demonstration.

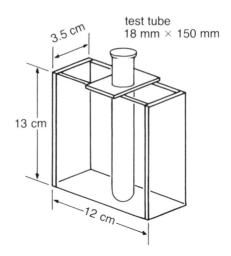

tube holder from an acrylic sheet according to the specifications in the figure. Fill the holder with room-temperature water and place it on the vertically mounted overhead projector.

Presentation

Pour 10 mL of solution A into the beaker. Add 10 mL of solution B and stir the mixture rapidly for about 5 seconds. Effervescence will appear in the solution almost immediately. Adjust the stirrer to a slow rotation rate. The effervescence will subside somewhat and then redevelop. This cycle will repeat, and the periods of effervescence will become more vigorous, while the intermissions will become clearer. After about 1 minute, quite clear oscillations in the effervescence will be apparent, having a period of about 8 seconds. The oscillations will continue for about 4 minutes.

To display the reaction with an overhead projector, pour 3 mL of solution A into the test tube, and add 3 mL of solution B. Stir the mixture thoroughly and place the test tube in the holder on the overhead projector. Effervescence will appear in the solution almost immediately. The effervescence will subside somewhat and then redevelop. After about 1 minute, quite clear oscillations in the effervescence will be apparent, having a period of about 8 seconds. The oscillations will continue for about 4 minutes.

HAZARDS

Because sulfuric acid is a strong acid and a powerful dehydrating agent, it can cause burns. Spills should be neutralized with an appropriate agent, such as sodium bicarbonate ($NaHCO_3$), and then rinsed clean.

Mixtures of sodium nitrite with combustible materials are easily ignited, sometimes explosively. In case of spills, clothing or other materials such as wood or paper which have come in contact with sodium nitrite solutions must be washed thoroughly, because they are likely to be dangerously combustible when dry. Mixtures of solid sodium nitrite with solid ammonium salts can explode upon heating.

DISPOSAL

The reaction mixture should be neutralized with sodium bicarbonate ($NaHCO_3$) and flushed down the drain with water.

DISCUSSION

Oscillatory gas evolution from a chemical reaction was first reported by Morgan in 1916 [2]. During the recently renewed interest in oscillating reactions, the Morgan reaction has been studied by a number of researchers [3–5]. The reaction involves the dehydration of formic acid by warm concentrated sulfuric acid, and the gas liberated is carbon monoxide. Because of the hazardous nature of the reactants and the products, the Morgan reaction is not as well suited to demonstration purposes as the reaction used in this demonstration.

The reaction used in this demonstration involves the decomposition of aqueous ammonium nitrite, as represented by equation 1.

$$NH_4^+(aq) + NO_2^-(aq) \longrightarrow N_2(aq) + 2\,H_2O(l) \tag{1}$$

Oscillatory behavior in this reaction was first reported by Degn [6] and was studied qualitatively by Smith [7].

Smith and Noyes have proposed an explanation for gas-evolution oscillators based on the release of gas in a supersaturated solution through homogeneous nucleation and subsequent bubble growth and escape [8]. The process, as applied to the reaction in this demonstration, is represented by the following equations:

$$N_2(solution) \rightleftharpoons N_2(nuclei) \tag{2}$$

$$N_2(nuclei) + N_2(solution) \rightleftharpoons 2\,N_2(bubbles) \tag{3}$$

$$N_2(bubbles) + N_2(solution) \rightleftharpoons 2\,N_2(larger\ bubbles) \tag{4}$$

$$N_2(large\ bubbles) \longrightarrow N_2(gas) \tag{5}$$

The preceding equations may be used to explain qualitatively what happens during this demonstration. Dissolved nitrogen, N_2(solution), is produced at a steady rate by the reaction of equation 1. When the solution becomes saturated in dissolved N_2, further production of N_2 results in a supersaturated solution, and this supersaturation is relieved by the spontaneous nucleation of bubbles, as represented by equation 2. Because the formation of these tiny bubble nuclei consumes very little of the dissolved nitrogen, nuclei form throughout the solution, and the solution remains saturated in dissolved N_2.

As N_2 is produced by the decomposition reaction, the bubble nuclei begin to grow into larger bubbles (equation 3). As the bubbles become larger, they have a larger surface area, and the rate of diffusion of N_2 from the solution into the bubbles increases (equation 4). Eventually the rate of diffusion of N_2 into the bubbles exceeds the rate of its formation by the decomposition reaction. Then the concentration of dissolved N_2 falls below the saturation point, and nucleation of new bubbles stops. The large bubbles rise to the surface and are released (equation 5), and because the solution is no longer saturated in N_2, the tiny bubbles redissolve (reverse of equations 3 and 2). This clears the solution of bubbles. As the decomposition reaction continues, the solution again becomes supersaturated in N_2, and bubble formation resumes, leading to another burst of bubbles.

The feedback which is necessary for oscillations is provided by the growing bubbles. As the bubbles grow, they reduce the concentration of dissolved N_2 to below the level required for nucleation, thereby stopping the formation of new bubbles and turning off the gas evolution.

Equation 1 is not sufficient to describe all of the chemical transformations taking place in this demonstration. Some brown fumes of NO_2 appear during the reaction, suggesting that NO also is evolved and then reacts with oxygen in the air. The production of NO may be due to the disproportionation of nitrous acid, as represented in equation 6 [7].

$$3 \, HNO_2 \longrightarrow 2 \, NO + NO_3^- + H^+ + H_2O \tag{6}$$

REFERENCES

1. S. M. Kaushik, Z. Yuan, and R. M. Noyes, *J. Chem. Educ.*, in press (1986).
2. J. S. Morgan, *J. Chem. Soc. Trans.* 109:274 (1916).
3. P. G. Bowers and G. J. Rawji, *J. Phys. Chem.* 81:1549 (1977).
4. K. Showalter and R. M. Noyes, *J. Am. Chem. Soc.* 100:1042 (1978).
5. K. W. Smith, R. M. Noyes, and P. G. Bowers, *J. Phys. Chem.* 87:1514 (1983).
6. H. Degn, informal report at European Molecular Biology Organization Workshop, Dortmund, Federal Republic of Germany, October 4–6, 1976.
7. K. W. Smith, Ph.D. dissertation, University of Oregon (1981).
8. K. W. Smith and R. M. Noyes, *J. Phys. Chem.* 87:1520 (1983).

7.14

Liesegang Rings

Over the span of several weeks, bands of crystals appear at regular intervals down a gel-filled test tube.

MATERIALS

9 mL 40% sodium silicate solution, $Na_2Si_3O_7$ (commercial "water glass")

21 mL distilled water

30 mL 0.6M acetic acid, $HC_2H_3O_2$ (To prepare 100 mL of stock solution, dissolve 3.5 mL of glacial [17M] acetic acid in 100 mL of distilled water.)

5 mL 1M potassium chromate, K_2CrO_4 (To prepare 100 mL of stock solution, dissolve 20 g K_2CrO_4 in 100 mL of distilled water.)

5 mL 1M silver nitrate, $AgNO_3$ (To prepare 100 mL of stock solution, dissolve 17 g $AgNO_3$ in 100 mL of distilled water.)

glass stirring rod

100-mL beaker

magnetic stirrer, with stirring bar

test tube, 25 mm × 200 mm

stand for test tube

solid rubber stopper for test tube

PROCEDURE

Preparation

Dilute the 40% sodium silicate solution by pouring 9 mL of it into 21 mL of distilled water and stirring the mixture thoroughly.

Place the stirring bar in the 100-mL beaker and set the beaker on the magnetic stirrer. Pour 30 mL of 0.6M acetic acid into the 100-mL beaker. Adjust the stirrer to produce a gentle stirring of the solution. Pour 5 mL of 1M K_2CrO_4 into the beaker. After the solutions have thoroughly mixed, gradually pour the diluted sodium silicate solution into the beaker, allowing the solutions to mix as they are combined. Transfer the mixture from the beaker into the test tube on a stand. Stopper the test tube and allow it to rest undisturbed until the mixture has set into a firm gel (about 15 minutes to 1 hour).

Presentation

Remove the stopper from the test tube and pour 5 mL of 1M silver nitrate solution onto the top of the gel. Replace the stopper and display the tube where it can be observed undisturbed for several weeks. During this time, fine, dark, needle-like crystals of silver chromate will form in distinct layers in the gel, starting at the top of the tube and progressing downward. As the crystals form, the gel will fade from yellow to colorless.

HAZARDS

Potassium chromate is irritating to the skin and eyes and is poisonous. Hexavalent chromium compounds are cancer-suspect agents.

Silver nitrate and its solutions are irritating to the skin and eyes and can cause burns. Also, they may stain skin or clothing an unsightly brown to black color. If a spill is recognized when it occurs, rinse the spot with some sodium thiosulfate solution followed by water. If a black spot develops later, try removing it as follows: Prepare a 10% solution of potassium ferricyanide in water and a 10% solution of sodium thiosulfate in 1% ammonia water. Mix equal volumes of these two solutions and scrub the black stain thoroughly with the resulting solution. Rinse well with water.

Sodium silicate solution is strongly alkaline and can cause burns to the skin and eyes.

Glacial acetic acid can cause burns to the skin, eyes, and mucous membranes.

DISPOSAL

The gel will last indefinitely in the sealed test tube, or it can be scraped from the tube and discarded in a solid waste receptacle.

DISCUSSION

In 1896 Liesegang reported the formation of bands in gels when precipitates formed by the interdiffusion of ions [1]. Liesegang produced rings in a flat gel in a shallow dish by placing several drops of one reagent on a gel impregnated with another reagent; the two together formed an insoluble precipitate. Bands of precipitate formed around the drops. These bands are a form of static spatial oscillation in precipitate formation, in contrast to the dynamic spatial oscillations displayed in Demonstration 7.12. In this demonstration, the gel is placed in a test tube and the precipitating agent added to the top of the tube. The precipitate forms in layered bands down the tube. It is presented this way to allow greater visibility than is possible with the original method. Band formation has been observed with a number of other sparingly soluble salts in addition to silver chromate [2].

The formation of bands is due to the diffusion of two different reacting species. In this demonstration, silver ions begin to diffuse into the gel from the top of the tube. When the concentration of silver ions becomes sufficiently high, crystals of silver chromate begin to form in the gel. This reduces the concentration of chromate ions in the gel around the crystals, and the chromate ions begin to diffuse up the tube, from a higher-concentration region to a lower-concentration region. This depletes the chromate ions from the gel just below the crystals, and the silver ions diffuse through this region without forming crystals. Eventually, the diffusing silver ions reach another region of high chromate ion concentration, and another band of crystals forms. This repeats until the supply of silver ions or chromate ions has been exhausted. A more detailed description of the processes occurring here has been provided by Flicker and Ross [3].

REFERENCES

1. R. E. Liesegang, *Naturwiss. Wochenschr.* 11:353 (1896).
2. K. H. Stern, *Chem. Rev.* 54:79 (1954).
3. M. Flicker and J. Ross, *J. Chem. Phys.* 60:3458 (1974).

Demonstrations in Volume 1

2   CHEMILUMINESCENCE

3   POLYMERS

4   COLOR AND EQUILIBRIA OF METAL ION PRECIPITATES AND COMPLEXES

Demonstrations in Future Volumes

The following demonstration topics will be included in future volumes:

acids, bases, and salts
atomic structure
chemical periodicity
chromatography
clock reactions
colloids and gels
corridor demonstrations and exhibits
cryogenics
electrochemistry
fluorescence and phosphorescence
kinetics and catalysis
lasers in chemistry
organic chemistry
overhead projector demonstrations
photochemistry
properties of liquids, solids, and solutions
radioactivity
spectroscopy and color